Dipole Moments and
Birefringence of Polymers

PRENTICE HALL
Polymer Science and Engineering Series
James E. Mark, Series Editor

MARK AND ERMAN, eds. *Elastomeric Polymer Networks*
MARK, ALLCOCK, AND WEST *Inorganic Polymers: An Introduction*
RIANDE AND SAIZ *Dipole Moments and Birefringence of Polymers*
ROE, ED. *Computer Simulation of Polymers*
VERGNAUD *Liquid Transport Processes in Polymeric Materials: Modeling and Industrial Applications*

FORTHCOMING BOOKS IN THIS SERIES (*tentative titles*)

BALLAUF *Stiff Chain Polymers*
CLARSON AND SEMLYEN, EDS. *Siloxane Polymers*
FRIED *Polymer Science and Technology*
GALIATSATOS, BOUE, AND SIESLER *Molecular Characterization of Networks*
VILGIS AND BOUE *Random Fractals*

Dipole Moments and Birefringence of Polymers

Evaristo Riande
Institute of Science and Technology of Polymers
National Council of Scientific Research (CSIC)
Madrid, Spain

Enrique Saiz
Department of Physical Chemistry
University of Alcalá de Henares
Madrid, Spain

PRENTICE HALL, Englewood Cliffs, New Jersey 07632

Library of Congress Cataloging-in-Publication Data

Riande, Evaristo.
  Dipole moments and birefringence of polymers / Evaristo Riande.
Enrique Saiz.
    p. cm. — (Prentice Hall polymer science and engineering)
  Includes bibliographical references and index.
  ISBN 0-13-218199-1
    1. Polymers—Dipole moments. 2. Refraction, Double. 3. Polymers–
Electric properties. 4. Polymers—Optical properties. I. Saiz,
Enrique. II. Title. III. Series.
QD381.9.E38R53 1992
620.1'9204295—dc20   91-22048
                                        CIP

Acquisitions Editor: *Michael Hays*
Production Editor: *Raeia Maes*
Copy Editor: *Peter Zurita*
Cover Designer: *Wanda Lubelska, Design*
Prepress Buyer: *Mary McCartney*
Manufacturing Buyer: *Susan Brunke*
Supplements Editor: *Alice Dworkin*

© 1992 by Prentice-Hall, Inc.
A Simon & Schuster Company
Englewood Cliffs, New Jersey 07632

Printed in the United States of America

10 9 8 7 6 5 4 3 2 1

ISBN   0-13-218199-1

Prentice-Hall International (UK) Limited, *London*
Prentice-Hall of Australia Pty. Limited, *Sydney*
Prentice-Hall Canada Inc., *Toronto*
Prentice-Hall Hispanoamericana, S.A., *Mexico*
Prentice-Hall of India Private Limited, *New Delhi*
Prentice-Hall of Japan, Inc., *Tokyo*
Simon & Schuster Asia Pte Ltd., *Singapore*
Editora Prentice-Hall do Brasil, Ltda., *Rio de Janeiro*

To Pilar and Pilar, Our Wives

Contents

Preface

High molecular-weight chains can present an almost unlimited number of conformations in space and this conformational wealth governs the physical properties of polymers. Among the most important conformation-dependent properties, the mean-square end-to-end distance, the mean-square dipole moment, the optical anisotropy, the molar Kerr constant, the stress-optical coefficient, etc., stand out. In the late 1950s and early 1960s statistical mechanical methods were developed that permitted predicting the equilibrium conformational properties of molecular chains provided that the energies associated with the rotational states of the skeletal bonds are known. Theoretical conformational analysis provides valuable information concerning the relationship between nonbonded molecular energetics and the spatial distribution of macromolecules.

Usually, skeletal bonds vary much more in polarity and polarizability than they do in length, and, consequently, the dielectric and optical conformational properties are more sensitive to structure than more traditional properties such as the molecular dimensions. A shortcoming in the theoretical evaluation of the former properties lies in that the contribution of each bond to the property is often affected by the environment. However, with some limitations, methods can be devised to estimate the environmental effects. This book addresses the experimental determination of dielectric and optical properties of polymers with special emphasis on dipole moments, the molar Kerr constant, and the optical configurational parameter. This information will be especially valuable to graduate students who during their university studies in chemistry, physics, or engineering did not become acquainted with the experimental details involved in the measurement of dielectric and optical properties. The application of the rotational isomeric state model to the study of these properties in flexible polymers is given considerable emphasis. A comprehensive discussion of equilibrium dielectric and optical properties is given that can be used for reference and review by experienced researchers, and also it provides a complete guide to those just entering the field.

Linear dielectric and optical relaxation phenomena are discussed within the framework of theoretical and phenomenological approaches, which can account for the response as a function of the probes. However, this subject cannot be treated to an extent proportional to its volume in the current literature.

The book is divided into two parts: Part One deals with the development of equations based on the cavity model that relate the static dielectric constant with the mean-square dipole moment. Experimental methods to determine the complex dielectric permittivity are discussed. Special consideration is given to the excluded volume effects in the dipole moments and to the evaluation of conformational energies. The relationship between structure and polarity for a considerable number of flexible polymers is described in some detail. Attention is also paid to the dielectric relaxations of polymers in solution and in bulk. Phenomenological and theoretical interpretations of relaxations are discussed for amorphous, crystalline, and liquid–crystal polymers.

Part Two of the book is concerned with the birefringence of polymers under: electric fields (the Kerr effect), magnetic fields (the Cotton–Mouton effect), mechanical forces (stress birefringence), and shear force fields (flow or streaming birefringence). Equations that relate the molar Kerr constant with the electric field, the Cotton–Mouton constant with the magnetic field, and the stress-optical coefficient with stress are discussed. Special emphasis is given to experimental details in the determination of the molar Kerr constant under static and alternating electric fields, and to the validity of the valence optical scheme for the theoretical determination of the conformational optical properties of polymers. The influence of structure on the optical properties, especially the molar Kerr constant and the optical configuration parameter, is studied for many polymers. Some attention is also paid to both experimental details and equations that relate birefringence of molecular chains in a shear field and molecular motions in that field.

An appendix describing the theoretical evaluation of the conformation-dependent properties is also included.

Evaristo Riande

Enrique Saiz

1
Introduction

A dipole can be defined as an entity in which a positive charge q is separated by a relatively short distance r from an equal negative charge. The dipole acts along the line connecting the two separating charges and, therefore, it is a vectorial quantity of magnitude [1]:†

$$\boldsymbol{\mu} = q\mathbf{r} \qquad (1.1)$$

and, by convention, its positive direction points from the negatively charged end toward the positively charged end. Traditionally, the Debye, which is 10^{-18} e.s.u., has been used as the unit of the dipole moment. In international units,

$$1 \text{ D} = 3.338 \times 10^{-30} \text{ C-m} \qquad (1.2)$$

Two unit electronic charges ($q = 1.6022 \times 10^{-19}$ C) separated by 0.1 nm (1 Å) yield a dipole moment of magnitude

$$\begin{aligned} \mu &= (1.6022 \times 10^{-19} \text{ C}) \times (1.0000 \times 10^{-10} \text{ m}) \\ &= 1.6022 \times 10^{-29} \text{ C-m} \approx 4.8 \text{ D} \end{aligned} \qquad (1.3)$$

†Numbers in brackets refer to References at the end of Part One.

Accordingly, the dipole moment of an ionic diatomic molecule seems to be relatively easy to predict once the internuclear separation is known from some independent measurement, such as the rotational spectrum. However, the predicted values are larger than the real ones because the cation has a distorting effect upon the electronic cloud of the anion and, as a result, the dipole moment is shortened. The percentage of ionic character X in the bond can be written as

$$X = (\mu_{obs.}/\mu_{ion}) \times 100 \tag{1.4}$$

The dipole moments of polyatomic molecules are dependent on the geometric structure and can be calculated in first approximation from the vectorial summation of the bond dipoles. However, since dipole moments measure the asymmetry of certain sections of a molecule, they are affected by the environment. In this case, group dipoles rather than bond dipoles give better results. The rotational spectrum of a molecule in the gas phase shifts its lines if the sample is under the effect of a strong electric field [1, 2]. The value of this Stark effect is dependent on the molecule's dipole moment and, as a consequence, the rotational spectra can be used to obtain accurate values of this quantity. When the molecule is too complex, nonvolatile, or unstable in the gas phase so that it cannot be examined by rotational spectroscopy, the dipole moment is usually obtained from dielectric or permittivity measurements carried out on solutions of the sample or in the bulk.

The term *dielectric* was first used by Faraday [3] in 1837 for the insulating material separating the two conducting plates of a condenser. Faraday realized that the capacity of a condenser was dependent on the nature of the material separating the conducting surfaces and he used the term *specific inductive capacity* to denote the ratio between the capacity of a condenser filled with a dielectric and the capacity of the same condenser when empty. The specific inductive capacity was later called the *permittivity*, or the dielectric constant, and it is currently denoted by ϵ.

In the midnineteenth century, attempts were made to correlate the dielectric permittivity with the microscopic structure of the matter. By considering the dielectric to be composed of conducting spheres imbedded in nonconducting medium, a relationship between the dielectric constant and the volume fraction occupied by the conducting particles in the dielectric was derived by Mosotti [4]. This expression remained relatively unknown and it was later derived independently by Clausius [5].

In the unified theory of electromagnetic phenomena developed by Maxwell [6] in 1860, the permittivity was conceived to be the ratio between the dielectric displacement and the electrical field intensity. Because light is a form of electromagnetic radiation, the dielectric permittivity should be related to the refractive index n of the dielectric. The Maxwell relationship $\epsilon = n^2$ was found to hold reasonably well for some solids as well as for some classes of gases and liquids, but for many other substances, which were called *associating*, the dielectric permittivity was found to be significantly higher than that predicted by the Maxwell relationship.

A corresponding expression to Mosotti's equation for the squares of the refractive indices was given by Lorentz [7] in 1880. By taking the density of particles containing electrically bound electrons proportional to the density of the compound, an equation accessible to the experimental verification was developed; this expression is known as the Lorentz–Lorentz [8] equation. If n^2 is replaced in this equation by the dielectric constant, the expression derived by Mosotti and Clausius is obtained. The failure of the Clausius–Mosotti equation for associating liquids was attributed to the presence of a permanent electric dipole moment in the molecules.

Berzelius [9] had much earlier assumed the molecules to be composed of oppositely charged regions. Although this concept was discussed by several scientists in the nineteenth century, the concept remained vague until Debye [10] published a quantitative theory in 1912. Although the Debye theory afforded the possibility of calculating the molecular dipole moments of polar compounds from dielectric measurements on the gaseous phase or on dilute solutions of these compounds in nonpolar solvents, the method failed for polar liquids. By introducing the concept of the reaction field, Onsager [11] succeeded in developing a fundamental modification of Debye's equation that gave a good account of the dipole moment of liquids whenever real association between molecules is not present. A theory that takes into account short-range interactions between neighboring molecules was lately given by Kirkwood [12]. On this basis, permittivity studies were used to obtain information on the specific interactions between molecules.

The dielectric dispersion observed in the dielectric measurements by some authors at the end of the past century was rationalized by Debye [13] by assuming that changes in the field require a definite time for the dipole to orientate in the direction of the field. It follows that after removal of an external applied field, the average dipole decays with time, and the characteristic time of the exponential decay is called the *relaxation time*. It was also Debye who deduced that for an alternating field, the lag between the average orientation of the dipole and the electric field becomes noticeable for frequencies of the field of the same order of magnitude as the reciprocal of the relaxation time.

In the mid-1920s, the concept of the *macromolecule* was established [14]. It became evident about 1930 that the special properties exhibited by polymers, such as their anomalous dielectric and viscoelastic behavior, was caused by the spatial character of polymeric chains. Kuhn [15] and Gut and Mark [16] made the first attempts at a mathematical description of the spatial configuration of random chains. The skeletal bonds were considered as steps in a random walk of three dimensions, the steps being uncorrelated one to another. A most realistic approach to the description of the conformation-dependent properties of polymer chains, which rests on the rotational states scheme, was developed in large measure by Volkenstein [17] and others [18] at the Leningrad school in the late 1950s and early 1960s. This model, which uses skeletal bond lengths and angles, rotational angles, and contributions of the bonds to the property to be measured, was rationalized by

Flory and co-workers [19] in the 1960s. The model was suitable to calculate any conformation-dependent property at equilibrium, such as the mean-square dipole moment, the mean-square end-to-end distance, etc., for chains of any length.

In addition to the mean-square end-to-end distance $\langle r^2 \rangle$, the mean-square dipole moment $\langle \mu^2 \rangle$ stands out to describe the spatial conformations of the chains [19]. If the dipoles of molecular chains are sufficiently separated one from another by nonpolar molecules that reduce the interactions among them, the system would resemble the dielectric behavior of a gaseous system. Equations were developed, based on the Debye theory [20–22], that permit evaluation of the mean-square dipole moments of polymer chains from dielectric measurements in solutions. By comparing theoretical and experimental results, information was obtained about the conformational parameters that condition the spatial conformations of macromolecules.

Whereas the theoretical basis of the conformation-dependent properties at equilibrium is well stated, a molecular model is still lacking to describe the dynamics of the chains at low and high frequencies. The response of a chain to external mechanical excitation involves motions of parts of the chain close to each other, probably dependent on the internal structure of the molecule, and those involving relative motions of parts widely separated, presumably dependent on the viscous resistance of the medium and therefore on molecular weight [23]. Rouse and others [24] formulated in the 1950s the spring-bead model, which describes the long-range motion of isolated chains. The effects of entanglements on the dynamics of molecular chains was later examined by De Gennes [25, 26] and Edwards [27–29] by formulating a model that conceives long-range motions in the entangled system as a reptation of the chain along a tube whose walls are formed by neighboring chains. The spring-bead model was used by Stockmayer [30] to study long-range dipole relaxations of molecular chains in solution in which one component of the dipole is parallel to the chain contour and, consequently, the time-dependent correlation function $\langle \mu(0) \cdot \mu(t) \rangle$ should show similar molecular-weight dependence as $\langle r(0) \cdot r(t) \rangle$, where r and μ are the end-to-end distance and the dipole moment of the chain, respectively. In general, the model gives a good account of the first mode of the dielectric relaxation process. In the same way, dielectric relaxation studies on entangled systems of this kind show that the dielectric behavior at low frequencies can be described in terms of the reptation model [31]. For chains in which the dipole moment is perpendicular to the chain contour, μ is not proportional to r and consequently neither the spring-bead model nor the reptation model can be used to describe the time-dependent autocorrelation function $\langle \mu(0) \cdot \mu(t) \rangle$ of these polymers in solution and in the bulk, respectively.

An important problem arises on the molecular interpretation of the subglass absorption [32], which is independent on molecular weight. Computer simulations suggest that cooperative motions of few skeletal bonds, in such a way that the molecular tails hardly move, presumably cause the relaxation. Great efforts are being made at present to understand the chain dynamics at the molecular level.

2
Dielectric Permittivity

GENERAL CONSIDERATIONS

The *dielectric permittivity* of a material can be defined as the ratio of the electric field E_0 in a vacuum to that in the material E for the same distribution of charge [33]. This quantity is measured in capacitors, the most simple of which consists of parallel plates of area A separated by a distance d, which is small in comparison with A (Fig. 2.1). If a difference of potential ϕ between the plates is established, the charges on the plates in a vacuum will be $+q$ and $-q$. The electric field E_0 inside the condenser will be perpendicular to the plates and its intensity in static units is

$$E_0 = 4\pi\sigma \qquad (2.1)$$

where $\sigma = q/A$ is the surface-charge density. Because

$$E = -\text{grad } \phi \qquad (2.2)$$

the capacity of the condenser in a vacuum, C_0, is

$$C_0 = q/\phi_0 = A/4\pi d \qquad (2.3)$$

If the charges on the plates remain unchanged and the space between them is filled with a dielectric material, the electric field strength decreases to

$$E = 4\pi\sigma/\epsilon \qquad (2.4)$$

5

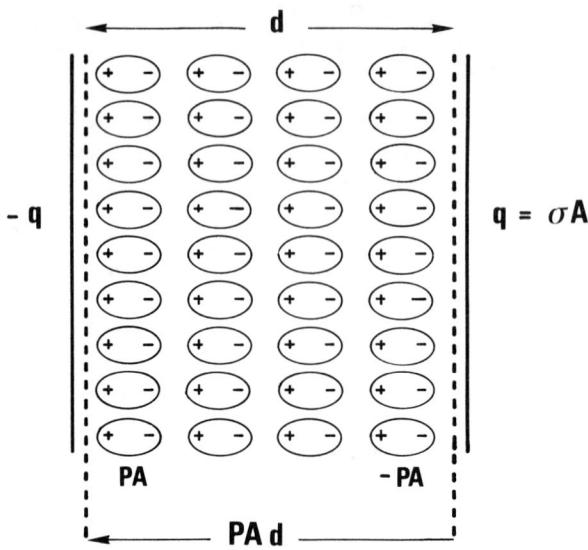

Figure 2.1   Polarization of the dielectric in a parallel-plate condenser.

where ϵ is the ratio of the dielectric constant of the medium between plates to that of free space. In this case, the capacitance C of the condenser is

$$C = A\epsilon/4\pi d \tag{2.5}$$

so that the dielectric permittivity of the material can be obtained from the relationship

$$\epsilon = C/C_0 \tag{2.6}$$

In the international system, the electric field and capacitance in vacuum would be given by

$$E_0 = \sigma/\epsilon_0 \quad \text{and} \quad C_0 = \epsilon_0 A/d \tag{2.7}$$

where ϵ_0 is the dielectric permittivity of free space, its value being 8.854×10^{-12} $C^2 Kg^{-1} m^{-3} sec^2$. In the same way E and C are expressed as

$$E = \sigma/\epsilon_0\epsilon \quad \text{and} \quad C = \epsilon_0\epsilon A/d \tag{2.8}$$

From Eqs. (2.1) and (2.4), the decrease in electric field strength by effect of the dielectric is found to be

$$E_0 - E = 4\pi\sigma(\epsilon - 1)/\epsilon \tag{2.9}$$

suggesting that the dielectric reduces the surface electric density by an amount

$$P = \sigma(\epsilon - 1)/\epsilon \tag{2.10}$$

In other words, the individual dipoles of the dielectric medium align with the electric field, giving rise to a change on the charge surface from σ to σ/ϵ. The parameter P can then be considered as the polarization, or the charge density on the surface of the dielectric, so that the dielectric can be considered as a dipole with total charge $+PA$ on one face and $-PA$ on the other, separated by a distance d. The moment μ of the dipole will be PAd and, consequently,

$$\mathbf{P} = PA\mathbf{d}/Ad \qquad (2.11)$$

that is, the polarization of the dielectric can also be regarded as an average dipole moment per unit volume. Therefore, $\mathbf{P}$ is straightforwardly related to the dipole moment of individual molecules. It follows from Eqs. (2.1), (2.4), (2.9), and (2.10) that the electric susceptibility χ of the material is

$$\chi = P/E = (\epsilon - 1)/4\pi \qquad (2.12)$$

Another useful quantity is the *electric displacement*, defined as $D = 4\pi\sigma$. From Eqs. (2.4) and (2.10), one obtains

$$\mathbf{D} = \mathbf{E} + 4\pi\mathbf{P} = \epsilon\mathbf{E}$$

POLARIZATION OF A DILUTE GAS: THE DEBYE EQUATION

If the dielectric is a gas with N molecules per unit volume, each of them having a permanent dipole moment $\boldsymbol{\mu}_p$ randomly orientated, the application of an external electric field $\mathbf{E}$ will produce a partial orientation of the permanent dipoles in the direction of the field together with the separation of the residual charges of the molecules. As can be seen in Figure 2.2 the dipole moment $\boldsymbol{\mu}$ is

$$\boldsymbol{\mu} = \boldsymbol{\mu}_p + \boldsymbol{\mu}_d \qquad (2.13)$$

and the total polarization can be expressed by

$$\mathbf{P} = N \langle \mu_z \rangle \mathbf{k} = N \langle \mu \cos \theta \rangle \mathbf{k} \qquad (2.14)$$

where θ is the angle that $\boldsymbol{\mu}$ forms with the Z axis, and $\mathbf{k}$ is the unit vector in the Z direction. It follows from Eqs. (2.13) and (2.14) that

$$\mathbf{P} = N\mu_p \langle \cos \theta \rangle \mathbf{k} + N\mu_d \mathbf{k} \qquad (2.15)$$

In the presence of a field, some orientations are energetically more favorable than others and $\langle \cos \theta \rangle$ can be evaluated by using the Boltzmann distribution for the dipoles at temperature T. Accordingly,

$$\langle \cos \theta \rangle = \frac{\int \cos \theta \exp(-\mu_p E \cos \theta / k_B T) \, d\Omega}{\int \exp(-\mu_p E \cos \theta / k_B T) \, d\Omega} \qquad (2.16)$$

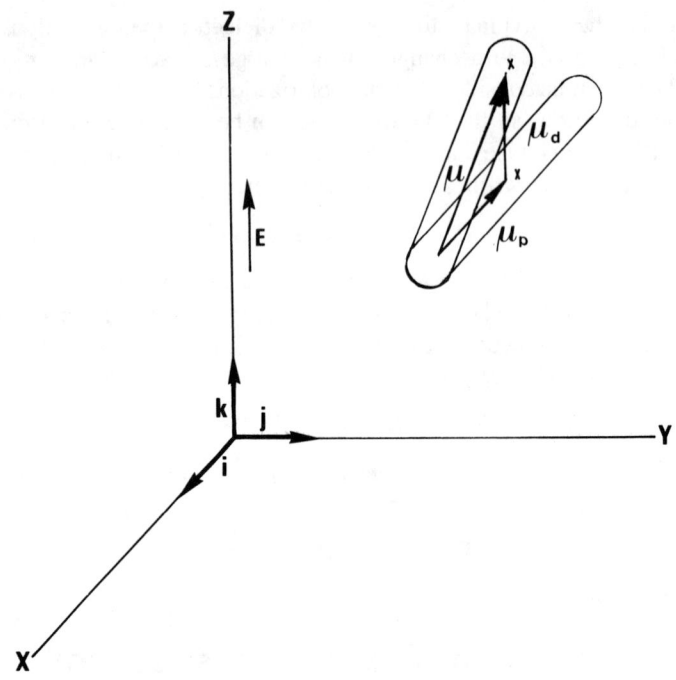

Figure 2.2   Components of the dipole of a molecule in an electric field E.

where Ω is the solid angle, and $-\mu_p E \cos \theta$ is the energy of a molecule that forms an angle θ with the electric field. The calculation of this equation gives

$$\langle \cos \theta \rangle = L(x) \tag{2.17}$$

where x is $\mu_p E/k_B T$, and $L(x)$ is the Langevin function [34]:

$$L(x) = [(e^x + e^{-x})/(e^x - e^{-x})] - 1/x \tag{2.18}$$

In most conventional measurements, $x \ll 1$, so that

$$L(x) \approx x/3 \tag{2.19}$$

and

$$\langle \cos \theta \rangle \approx \mu_p E/3k_B T \tag{2.20}$$

Therefore, the value obtained for **P** from Eqs. (2.15) and (2.20) is

$$\mathbf{P} = N[(E\mu_p^2/3k_B T) + \mu_d]\mathbf{k} \tag{2.21}$$

The induced dipole moment μ_d produced by distortion of the electronic distribution caused by **E** is related to the electric field by

$$\mu_d = \alpha_d E \tag{2.22}$$

where the coefficient of proportionality α_d is called the polarizability of the molecule, this parameter being often quoted in units of cm^3 or $Å^3$. Then Eq. (2.21) can be written as

$$\mathbf{P} = N[(\mu_p^2/3k_B T) + \alpha_d]\mathbf{E} = N[\alpha_0 + \alpha_d]\mathbf{E} \tag{2.23}$$

where

$$\alpha_0 = \mu_p^2/3k_B T \tag{2.24}$$

is called the *orientation polarizability*. Each molecule, located somewhere inside the dielectric, experiences the macroscopic field E together with a field E_{sph} that arises from the charges on the surface of a cavity that surrounds it, as indicated in Figure 2.3. Field E_{sph} is calculated by subdividing the boundaries in infinitesimally small rings in which the apparent surface charge density is $-P \cos \theta$. The total charge in the ring will be

$$dq = -2\pi r^2 \sin \theta \, P \cos \theta \, d\theta \tag{2.25}$$

A charge element dq on the boundary of the cavity contributes to the external field by an amount

$$dE = (dq/r^2) \cos \theta \tag{2.26}$$

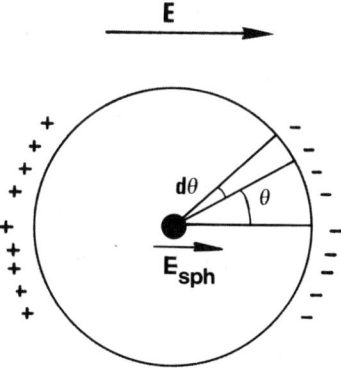

Figure 2.3 The electric field inside a spherical cavity.

The value of the electric field, in the direction of E, induced by the charges of the cavity, is obtained from Eqs. (2.25) and (2.26) as

$$E_{\text{sph}} = 2P\pi \int_0^\pi \sin\theta \cos^2\theta \, d\theta = (4/3)\pi P \tag{2.27}$$

Symmetry dictates that the other components of E_{sph} are zero and the local field in the molecule located in the center of the cavity is

$$E_1 = E + 4\pi P/3 \tag{2.28}$$

From Eqs. (2.23) and (2.28), one obtains

$$P = n(\alpha_d + \alpha_0)(E + 4\pi P/3) \tag{2.29}$$

which leads to

$$\frac{\epsilon - 1}{\epsilon + 2} = \frac{4\rho}{3M}\pi N_A(\alpha_0 + \alpha_d) \tag{2.30a}$$

if Eqs. (2.12) and (2.29) are considered. In this expression, called the Debye equation [10, 35], ρ and M represent the density and the molecular weight of the polar gas, respectively, N_A is Avogadro's number, and α_0 is given by Eq. (2.24). The first term in Eq. (2.30a) is called the molar polarization P, a parameter that evidently is different from the **P** used to define the dipole moment of unit volume. P has two contributions: the orientation polarization P_0 and the induced polarization P_d, which are given by

$$P_0 = (\tfrac{4}{3})\pi N_A\alpha_0 \quad \text{and} \quad P_d = (\tfrac{4}{3})\pi N_A\alpha_d \tag{2.30b}$$

Both the molar polarization and the polarizability have units of volume. The combination of Eqs. (2.24) and (2.30a) gives

$$P = \frac{\epsilon - 1}{\epsilon + 2}\frac{M}{\rho} = \frac{4}{3}\pi N_A\left(\frac{\mu_p^2}{3k_BT} + \alpha_d\right) \tag{2.31}$$

The inductive polarizability is governed by the strength with which the nuclear charges prevent the distortion of the electronic cloud by the applied field. The polarizability will increase with the atomic number, atomic size, and low ionization potential of the atoms of the molecules. Therefore, the induced polarizability can be considered as the sum of the electronic α_e and atomic α_a polarizabilities, the latter being originated by small displacements of atoms and groups of atoms in the molecule under the effects of the electric field. The magnitude of α_e can directly be obtained from the Debye equation by making $\mu_p = 0$ and considering that ϵ at the wavelength λ is related to the refractive index n at the same frequency by the equation

$$\epsilon(\lambda) = n^2(\lambda) \tag{2.32}$$

Because α_d in Eq. (2.30a) corresponds to a static electric field, n should be obtained at different wavelengths and its value extrapolated to $\lambda \rightarrow \infty$. Accordingly,

$$\frac{n^2 - 1}{n^2 + 2} = \frac{4\pi\rho N_A}{3M}\alpha_e \qquad (2.33)$$

This expression is called the Clausius–Mosotti equation [4, 5]. The atomic polarizability cannot be determined directly, but, fortunately, is small and often negligible. Combining Eqs. (2.30a) and (2.33) gives

$$\frac{\epsilon - 1}{\epsilon + 2} - \frac{n^2 - 1}{n^2 + 2} = \frac{4\pi\rho N_A \mu_p^2}{3k_B MT} \qquad (2.34)$$

which is another way of expressing the Debye equation. From Eqs. (2.24) and (2.31), the expression

$$PT = \frac{\epsilon - 1}{\epsilon + 2}\frac{M}{\rho} = P_0 + P_d T \qquad (2.35)$$

is obtained, which suggests that the plot PT against T should give a straight line of slope P_d (the induced polarization) and intercept P_0 (the orientation polarization). Since P_e can be obtained from Eq. (2.33), P_a can directly be determined from the difference between P_d and P_e.

The Debye equation was found to hold for a wide variety of gases and vapors at ordinary pressures. The equation has also been successfully used to calculate approximate values of the molecular dipole moment from the dielectric constants and densities of dilute solutions of the polar molecules in nonpolar solvents, but it fails for pure liquids and for gases and vapors in which association, dissociation, or changing intermolecular energy occur [33].

COMPLEX POLARIZATION

The static dielectric constant, which is observed when the dielectric is at equilibrium with the applied electric field, is used in Eq. (2.30a). However, when an alternating field is applied, an observable lag in the attainment of equilibrium, dependent on frequency, can be detected. This lag, usually referred to as *relaxation* [36], becomes apparent when the magnitude of the rate of response of the system is not far from that of the change of the applied force. As a consequence, both the polarizability and the permittivity will be complex numbers, which are expressed as

$$\alpha^* = \alpha' - i\alpha'' \qquad (2.36)$$

$$\epsilon^* = \epsilon' - i\epsilon'' \qquad (2.37)$$

At high frequencies, the permanent dipole moments may be unable to reorientate themselves with the electric field and their contribution to the polarization of the medium will decrease with increasing frequency. The frequency at which a drop in α_0^* is detected is evidently dependent on the nature of the medium, its value decreasing as the viscosity of the medium increases. The dependence of α_0^* on frequency was analyzed by Debye [35] for a sphere of radius a, immersed in a fluid of viscosity η, obtaining the following relationship:

$$\alpha_0^* = \frac{\mu_p^2}{3k_BT}\left(\frac{1}{1 + i\omega\tau}\right) = \frac{\mu_p^2}{3k_BT}\left(\frac{1}{1 + \omega^2\tau^2} - i\frac{\omega\tau}{1 + \omega^2\tau^2}\right) \qquad (2.38)$$

where τ is the relaxation time, which is given by

$$\tau = 4\pi\eta a^3/k_BT \qquad (2.39)$$

Since a simple molecule in the gas requires about 10^{-12} seconds to rotate, the complex orientation polarizability α_0^* for a gas will generally not be detected at frequencies above 10^{12} Hz. At higher frequencies, the contribution to the polarization of the medium will come from the induced polarization α_d. For small molecules in liquids of low viscosity, the time required for the orientation polarization process is 10^{-10} to 10^{-11} seconds, whereas for large molecules in dilute solutions and for viscous liquids, the time is of the order of 10^{-6} seconds. By considering electron and ions as classical oscillators with natural frequencies ω_{0e} and ω_{0a}, respectively, one can obtain the following expressions for the electronic and atomic complex polarizabilities, respectively:

$$\alpha_e^* = A\left[\frac{\omega_{0e}^2 - \omega^2}{(\omega_{0e}^2 - \omega^2)^2 + \omega^2\gamma_e} + i\frac{\omega\gamma_e}{(\omega_{0e}^2 - \omega)^2 + \omega^2\gamma_e^2}\right] \qquad (2.40)$$

$$\alpha_a^* = B\left[\frac{\omega_{0a}^2 - \omega^2}{(\omega_{0a}^2 - \omega^2)^2 + \omega^2\gamma_a} + i\frac{\omega\gamma_a}{(\omega_{0a}^2 - \omega)^2 + \omega^2\gamma_a^2}\right] \qquad (2.41)$$

where A and B depend on the charges and their masses and γ_e and γ_a are dissipative parameters. The electronic polarizability arising from the displacement of the electrons in the atoms relative to the positive nuclei has a natural frequency of about 10^{-15} seconds, whereas the value of this quantity for the atomic polarizability lies in the range 10^{-12} to 10^{-14} seconds. As can be seen in Figure 2.4, resonance absorption will appear at the frequency of the ultraviolet light in the former case and at the frequency of the infrared in the latter. For a polar dielectric medium, n^2 is equal to the dielectric constant ϵ_∞ measured at a frequency so high that the permanent dipoles cannot contribute to the polarization of the medium. Therefore, ϵ_∞ is defined by the equation

$$M(n^2 - 1)/(n^2 + 2)\rho \cong P_e + P_a = (M/\rho)(\epsilon_\infty - 1)/(\epsilon_\infty + 2) \qquad (2.42)$$

where P_e and P_a are, respectively, the electronic and atomic molar polarizations.

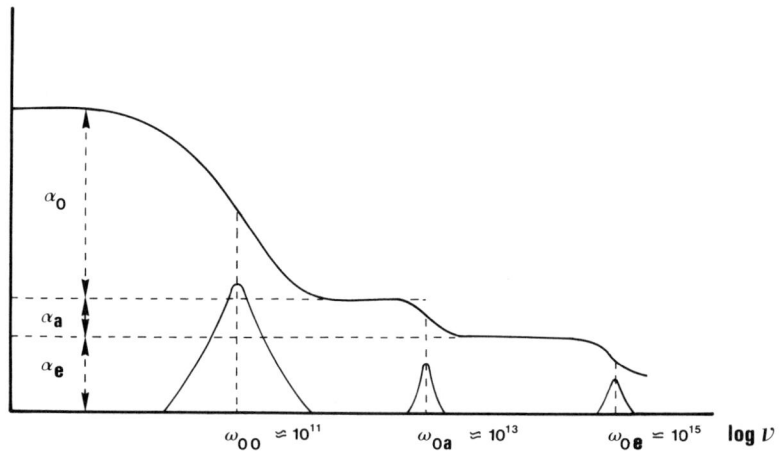

Figure 2.4 Variation of polarizability with frequency.

POLARIZATION OF A LIQUID: THE ONSAGER EQUATION

The limitations of the Debye equation to determine dipole moments of liquids from their static dielectric constants led Onsager to reexamine the effect of the internal electric field for polar molecules, which are spherical in form. Onsager [11] considered an additional polarization caused by the reaction field that acts upon the dipole as a result of the electric displacement caused by its own presence. Actually, a dipole located in the center of a spherical cavity (Fig. 2.5) polarizes the surrounding matter of permittivity ϵ, so that the inhomogenous polarization of the environment gives rise to a field at the dipole, which is called the *reaction field* **R**. Reasons of symmetry suggest that

$$\mathbf{R} = f\boldsymbol{\mu} \tag{2.43}$$

where f is called the factor of the reaction field.

The potential in the cavity, caused by both the dipole and the interaction between the dipole with the surrounding dielectric, can be obtained by taking the dipole as the origin of a coordinate system and choosing the direction of the Z axis along the dipole vector. There are no free charges inside the dielectric, and the Laplace equation

$$\nabla^2\phi = 0 \tag{2.44}$$

gives the distribution of the field in the cavity. Owing to the spherical symmetry of the cavity, the distribution of the field given by Eq. (2.44) takes the form

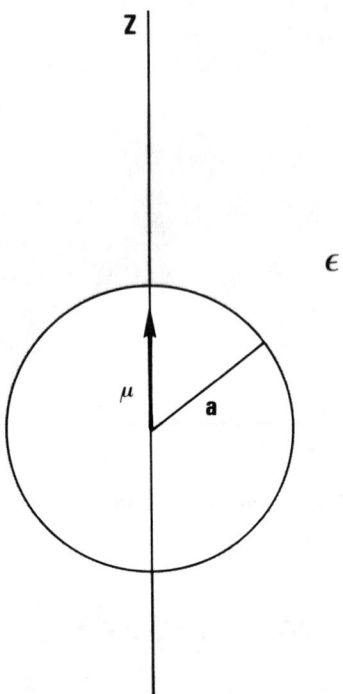

Figure 2.5 Dipole located inside a spherical cavity of radius a.

$$\nabla^2\phi = \frac{1}{r^2}\left[\frac{\delta}{\delta r}\left(r^2\frac{\delta\phi}{\delta r}\right)\right] + \frac{1}{r^2\sin^2\theta}\left(\frac{\delta^2\phi}{\delta\theta^2}\right)$$

$$+ \frac{1}{r^2\sin\theta}\frac{\delta}{\delta\theta}\left[\sin\theta\left(\frac{\delta\phi}{\delta\theta}\right)\right]$$

$$= \left(\frac{\delta^2\phi}{\delta r^2}\right) + \frac{2}{r}\left(\frac{\delta\phi}{\delta r}\right) + \frac{1}{r^2}\left[\cotan\theta\left(\frac{\delta\phi}{\delta\theta}\right)\right] = 0 \qquad (2.45)$$

The general solution of this equation is

$$\phi_1 = (Ar + Br^{-2})\cos\theta \qquad (2.46)$$

$$\phi_2 = (Cr + D\epsilon^{-1}r^{-2})\cos\theta \qquad (2.47)$$

where ϕ_1 and ϕ_2 represent, respectively, the potential inside and outside of a spheric cavity of radius a. The boundary conditions are

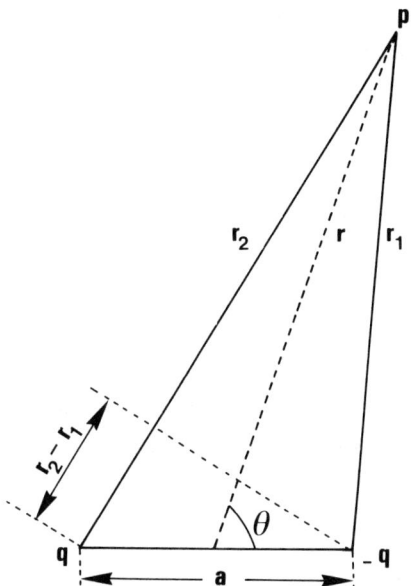

Figure 2.6 Potential produced by the dipole in point p in a vacuum.

$$[\phi_2]_{r\to\infty} = 0 \tag{2.48}$$

$$[\phi_1]_{r=a-\delta} = [\phi_2]_{r=a+\delta} \tag{2.49}$$

$$\left[\frac{d\phi_1}{dr}\right]_{r=a-\delta} = \epsilon\left[\frac{d\phi_2}{dr}\right]_{r=a+\delta} \tag{2.50}$$

The term B in Eq. (2.46) corresponds to the potential due to an ideal dipole along the z axis. As can be seen in Figure 2.6, the potential produced by the dipole in point p in a vacuum is

$$\phi = \frac{q}{r_1} - \frac{q}{r_2} = q\left(\frac{r_1 - r_2}{r_1 r_2}\right) \approx q\left(\frac{r_1 - r_2}{r^2}\right) \tag{2.51}$$

Because $q(r_1 - r_2) = qa\cos\theta = \mu\cos\theta$, the potential due to the ideal dipole is given by

$$\phi = (\mu/r^2)\cos\theta \tag{2.52}$$

The first boundary condition implies that $C = 0$ in Eq. 2.47. Accordingly,

$$\phi_1 = (\mu/r^2)\cos\theta + Ar\cos\theta \tag{2.53}$$

$$\phi_2 = (D/\epsilon r^2)\cos\theta \tag{2.54}$$

The second and third boundary conditions in conjunction with Eqs. (2.53) and (2.54) lead to

$$(D/\epsilon a^2) = (\mu/a^2) + Aa \tag{2.55}$$

$$-2(D/a^3) = -(2\mu/a^3) + A \tag{2.56}$$

Hence,

$$D = \frac{3\epsilon}{2\epsilon + 1} \mu \tag{2.57}$$

$$A = \frac{-2(\epsilon - 1)}{2\epsilon + 1} \frac{\mu}{a^3} \tag{2.58}$$

Coefficient D is called the *external moment of the immersed dipole* and it is usually represented by μ^*. It determines the force, modified by the intervening medium, that the dipole will exert upon a distant charge in the dielectric.

Potentials ϕ_1 and ϕ_2 inside and outside, respectively, the cavity can then be written as

$$\phi_1 = \frac{\mu}{r^2} \cos \theta - \frac{2(\epsilon - 1)}{2\epsilon + 1} \frac{\mu}{a^3} r \cos \theta \tag{2.59}$$

$$\phi_2 = \frac{3}{2\epsilon + 1} \frac{\mu}{r^2} \cos \theta \tag{2.60}$$

By subtracting the potential produced by the dipole in a vacuum, Eq. (2.52), to ϕ_1 and ϕ_2, the potentials ϕ_1' and ϕ_2' caused by an apparent surface charge on the surface cavity are obtained:

$$\phi_1' = -\frac{2(\epsilon - 1)}{2\epsilon + 1} \frac{\mu}{a^3} r \cos \theta \tag{2.61}$$

$$\phi_2' = -\frac{2(\epsilon - 1)}{2\epsilon + 1} \frac{\mu}{r^2} \cos \theta \tag{2.62}$$

Equation (2.59) suggests that the field in the cavity is the superposition of the field in a vacuum and a uniform field **R** given by

$$\mathbf{R} = \frac{1}{a^3} \frac{2(\epsilon - 1)}{2\epsilon + 1} \mu \tag{2.63}$$

By comparing this equation with Eq. (2.43), the factor of the reaction field f can be written as

$$f = \frac{1}{a^3} \frac{2(\epsilon - 1)}{2\epsilon + 1} \tag{2.64}$$

The expression for f is dependent on the model used for the cavity. For an ellipsoidal cavity, the reaction field can be calculated from the polarization of a homogeneous ellipsoid with dielectric constant ϵ_2 immersed in a dielectric with dielectric constant ϵ_1. The reader interested in details can find them in reference [37].

For polar molecules under the effect of an external field E, the internal field $\mathbf{E}_i$ can be built up from the cavity field $\mathbf{E}_c$ and the reaction field $\mathbf{R}$. The evaluation of $\mathbf{E}_c$ can be carried out by considering the modification of a homogeneous field E by action of an empty spherical cavity. This action can be calculated from the Laplace equation using the general solutions:

$$\phi_1 = -Gr \cos \theta \tag{2.65}$$

$$\phi_2 = -Er \cos \theta - (M/r^2) \cos \theta \tag{2.66}$$

where ϕ_1 and ϕ_2 refer to the potentials inside and outside the spherical cavity, respectively. By using the second and third boundary conditions, Eqs. (2.49) and (2.50), one obtains

$$Ga = Ea + M/a^2 \tag{2.67}$$
$$G = \epsilon E - 2\epsilon M/a^3 \tag{2.68}$$

whereby the values of G and M are

$$M = \frac{\epsilon - 1}{2\epsilon + 1} E a^3 \tag{2.69}$$

$$G = \frac{3\epsilon}{2\epsilon + 1} E \tag{2.70}$$

and where G represents the homogeneous cavity field $\mathbf{E}_c$ in the dielectric. The field, which acts upon a molecule in a polarized dielectric, involves $\mathbf{E}_c$ and the reaction field $\mathbf{R}$. The part of the field tending to direct the permanent dipole moments is called the *directing field* $\mathbf{E}_d$. In older theories, $\mathbf{E}_d$ was taken to be equal to the internal field $\mathbf{E}_i$, but Onsager showed that this approach is only satisfactory for gases at low pressures, because, in general, only a part of the internal field influences the direction of the permanent dipoles. The reaction field $\mathbf{R}$ does not contribute to $\mathbf{E}_d$ because it does not influence the direction of the dipole under consideration. However, the reaction field contributes to $\mathbf{E}_i$ because it polarizes the molecule. Because the reaction field $\mathbf{R}$ belongs to one particular orientation of the molecule, the difference between $\mathbf{E}_i$ and $\mathbf{E}_d$ will amount to the reaction field averaged over all orientations of the molecule. In other words,

$$\mathbf{E}_i = \mathbf{E}_d + \langle \mathbf{R} \rangle \tag{2.71}$$

The evaluation of $\mathbf{E}_d$ implies the elimination of the reaction field from $\mathbf{E}_i$ and this can be done by using the following procedure: Remove the permanent dipole moment from the molecule without changing its polarizability and let the surrounding dielectric adapt itself to the new situation; then fix the charge distribution of the surroundings and remove the central molecule. Had the polarizability of the molecule been removed before fixing the charge distribution of the surroundings, a physical cavity would be formed in the dielectric. In the evaluation of the directing field $\mathbf{E}_d$, the polarizability of the molecule must now be taken into account. The field $\mathbf{E}_d$ gives rise to a dipole $\alpha\mathbf{E}_d$ with reaction field $f\alpha\mathbf{E}_d$, and $\mathbf{E}_d$ is expressed by

$$\mathbf{E}_d = \mathbf{E}_c + f\alpha\mathbf{E}_d \tag{2.72}$$

Because $\mathbf{E}_c$ is given by G in Eq. (2.70), $\mathbf{E}_d$ can be written as

$$\mathbf{E}_d = \frac{1}{1 - f\alpha} \frac{3\epsilon}{2\epsilon + 1} \mathbf{E} \tag{2.73}$$

The calculation of the internal field $\mathbf{E}_i$ from Eq. (2.71) requires knowing $\mathbf{R}$. For a polarizable permanent dipole, having an average polarizability α, the reaction field induces a dipole $\alpha\cdot\mathbf{R}$, and according to Eq. (2.43), the reaction field $\mathbf{R}$ is

$$\mathbf{R} = f(\boldsymbol{\mu} + \alpha\mathbf{R}) \tag{2.74}$$

where $\boldsymbol{\mu}$ is the permanent dipole, and f is the reaction-field factor given by Eq. (2.64). Therefore, $\mathbf{R}$ is related to the permanent dipole moment by

$$\langle\mathbf{R}\rangle = \frac{f}{1 - f\alpha}\langle\boldsymbol{\mu}\rangle \tag{2.75}$$

By considering that $\langle\boldsymbol{\mu}\rangle = \mu\langle\cos\theta\rangle$, the average value of $\mathbf{R}$ required in Eq. (2.71) is

$$\langle\mathbf{R}\rangle = \frac{f}{1 - f\alpha} \frac{\mu^2}{3k_BT} \mathbf{E}_d \tag{2.76}$$

Hence, the final expression for the internal field is

$$\mathbf{E}_i = \mathbf{E}_d + \frac{f}{1 - f\alpha} \frac{\mu^2}{3k_BT}\mathbf{E}_d = \left(1 + \frac{f}{1 - f\alpha} \frac{\mu^2}{3k_BT}\right)\frac{1}{1 - f\alpha} \frac{3\epsilon}{2\epsilon + 1}\mathbf{E} \tag{2.77}$$

The polarization $\mathbf{P}$, which includes the orientation and inductive polarization, is related to the electric field $\mathbf{E}$ by Eq. (2.12):

$$\mathbf{P} = \frac{\epsilon - 1}{4\pi} \mathbf{E}$$

By expressing the total polarization in terms of $\mathbf{E}_i$ and $\mathbf{E}_d$, the fundamental relationship

$$\frac{\epsilon - 1}{4\pi} \mathbf{E} = N\left(\alpha\mathbf{E}_i + \frac{\mu^2}{3k_BT}\mathbf{E}_d\right) \tag{2.78}$$

is obtained for a system of N molecules per cubic centimeter. By substituting Eqs. (2.73) and (2.77) into Eq. (2.78), one finds that

$$\frac{(\epsilon - 1)(2\epsilon + 1)}{12\pi\epsilon} = N\frac{1}{1 - f\alpha}\left(\alpha + \frac{1}{3k_BT} \frac{\mu^2}{1 - f\alpha}\right) \tag{2.79}$$

where f is given by Eq. (2.64). In the evaluation of f, the radius of the spherical cavity intervenes. Onsager assumed that the volume of the cavity is equal to the volume available to each molecule, so that

$$(\tfrac{4}{3})\pi N a^3 = 1 \tag{2.80}$$

The value of a can approximately be obtained from Eqs. (2.33) and (2.80), making the assumption that $n^2 = \epsilon_\infty$. Accordingly,

$$\frac{\alpha}{a^3} = \frac{\epsilon_\infty - 1}{\epsilon_\infty + 2} \tag{2.81}$$

This expression, in conjunction with Eq. (2.64), gives

$$\frac{1}{1 - f\alpha} = \frac{(\epsilon_\infty + 2)(2\epsilon + 1)}{3(2\epsilon + \epsilon_\infty)} \tag{2.82}$$

Hence, substitution of Eq. (2.82) into Eq. (2.79) gives

$$\frac{\epsilon - 1}{4\pi} = \frac{3\epsilon(\epsilon_\infty - 1)}{4\pi(2\epsilon + \epsilon_\infty)} + \frac{(\epsilon_\infty + 2)^2(2\epsilon + 1)\epsilon}{(2\epsilon + \epsilon_\infty)^2}\frac{N\mu^2}{9k_BT} \tag{2.83}$$

where Eq. (2.33) and the approximation $n^2 = \epsilon_\infty$ were used to evaluate α. This expression, called the Onsager equation, can also be written as

$$\mu^2 = \frac{9k_bTM}{4\pi\rho N_A}\frac{(\epsilon - \epsilon_\infty)(2\epsilon + \epsilon_\infty)}{\epsilon(\epsilon_\infty + 2)^2} \tag{2.84}$$

by considering that $N = \rho N_A/M$ in Eq. (2.83), where M is the molecular weight of the dielectric, ρ is the density, and N_A is Avogadro's number. Rearrangements of this equation give

$$\frac{\epsilon - 1}{\epsilon + 2}\frac{M}{\rho} - \frac{\epsilon_\infty - 1}{\epsilon_\infty + 2}\frac{M}{\rho} = \frac{3\epsilon(\epsilon_\infty + 2)}{(2\epsilon + \epsilon_\infty)(\epsilon + 2)}\frac{4\pi N_A\mu^2}{9k_BT} \tag{2.85}$$

The Onsager equation differs from the Debye equation in the term $[3\epsilon(\epsilon_\infty + 2)/(2\epsilon + \epsilon_\infty)(\epsilon + 2)]$ on the right side of Eq. (2.85).

The Onsager equation converges to the Debye equation when ϵ approaches ϵ_∞, as occurs with gases at atmospheric pressure or below. Kirkwood [12, 38] generalized the Onsager theory by eliminating the approximation of a uniform dielectric constant identical with the dielectric constant of the medium, obtaining

$$\frac{(\epsilon - 1)(2\epsilon + 1)}{9\epsilon} = \frac{4\pi N_A\rho}{3M}\left(\alpha_d + \frac{g\mu^2}{3k_BT}\right) \tag{2.86}$$

where $g\mu^2$ is the product of the molecular dipole moment μ in the liquid and $\tilde{\mu}$, the latter being the sum of μ and the moment induced as result of hindered rotation in the spherical region surrounding the molecule. The Kirkwood equation [12, 38] represents an advance beyond the Onsager theory because it takes into consideration the hindrance of molecular orientation by molecular interaction. A more elaborate approach performed by Frölich [39] leads to

$$\frac{\epsilon - 1}{\epsilon + 2} - \frac{\epsilon_\infty - 1}{\epsilon_\infty + 2} = \frac{3\epsilon(\epsilon_\infty + 2)}{(2\epsilon + \epsilon_\infty)(\epsilon + 2)} \frac{4\pi\rho N_A g \mu^2}{9k_B M T} \tag{2.87}$$

which becomes the Onsager equation when $g = 1$. This parameter is defined as

$$g = 1 + \sum_{i,j} \cos \gamma_{ij} \tag{2.88}$$

where $\cos \gamma_{ij}$ represents the average of the cosine of the angle that forms the i molecule with the j molecule extended to all orientations. In other words, g is a measure of hindered relative molecular orientation arising from short-range intermolecular interactions, and its value is close to unity for normal liquids and significantly lower than 1 for associated liquids. In general, the results at hand indicate that the values of the dipole moment calculated for normal liquids by both the Onsager and the Kirkwood equations are in very good agreement with those obtained from measurements in the vapor phase [40].

Theories that treat intermolecular interactions explicitly at all stages, avoiding the introduction of cavities, have also been reported. The reader will find details of these theories in reference [41].

DETERMINATION OF DIPOLE MOMENTS FROM DIELECTRIC MEASUREMENTS IN SOLUTION

The Debye equation can only be used to determine the dipole moment of polar molecules in the vapor phase. However, if the molecules are sufficiently separated from one another by nonpolar molecules that reduce the interaction among their permanent dipole moments, the system would resemble the dielectric behavior of a gaseous condition. This approach is used to determine the dipole moments of long molecular chains. It should be stressed that whereas simple molecules have permanent dipole moments similar for all of them, long molecules are continuously changing from spatial conformation, and because the dipole moment associated with each conformation is generally different, the dipole moments that are measured are average values. The term μ_p^2 in the Debye equation should then be replaced by $\langle \mu_p^2 \rangle$.

In a solution containing n_1 molecules of solvent and n_2 molecules of solute of molecular weights M_1 and M_2, respectively, Eq. (2.30a) gives

$$\frac{4}{3}\pi N_A \alpha = \frac{\epsilon - 1}{\epsilon + 2} \frac{x_1 M_1 + x_2 M_2}{\rho} \tag{2.89}$$

where the average molecular weight M of the solution was considered to be

$$M = x_1 M_1 + x_2 M_2 \tag{2.90}$$

and where x_1 and x_2, representing the molar fraction of solvent and solute, respectively, are given by

$$x_1 = \frac{n_1}{n_1 + n_2} \quad \text{and} \quad x_2 = \frac{n_2}{n_1 + n_2} \tag{2.91}$$

and ϵ in Eq. (2.89) is the static dielectric constant of the solution.

For very dilute solutions ($x_2 \to 0$), the intermolecular interactions between solute molecules would be negligible, and α in Eq. (2.89) would be the average of the total polarizabilities of solute α_2 and solvent α_1:

$$\alpha = x_1 \alpha_1 + x_2 \alpha_2 \tag{2.92}$$

where the total polarization of the solvent is given by

$$\frac{4}{3} \pi N_A \alpha_1 = \frac{\epsilon_1 - 1}{\epsilon_1 + 2} \frac{M_1}{\rho_1} \tag{2.93}$$

By combining Eqs. (2.89), (2.92), and (2.93), one obtains

$$\frac{4}{3} \pi N_A \alpha_2 = \frac{4}{3} \pi N_A \left(\frac{\alpha}{x_2} - \frac{x_1}{x_2} \alpha_1 \right)$$

$$= \frac{\epsilon - 1}{\epsilon + 2} \left(\frac{x_1 M_1}{\rho x_2} + \frac{M_2}{\rho} \right) - \frac{x_1 (\epsilon - 1) M_1}{x_2 (\epsilon + 2) \rho_1} \tag{2.94}$$

Moreover, for very low concentrations, ϵ and ρ can be expanded in series:

$$\epsilon = \epsilon_1 + \frac{\delta \epsilon}{\delta w_2} w_2 = \epsilon_1 + \beta w_2 \tag{2.95}$$

$$\rho = \rho_1 + \frac{\delta \rho}{\delta w_2} w_2 = \rho_1 + \beta' w_2 \tag{2.96}$$

where w_2 is the weight fraction of solute, and ϵ_1 and ρ_1 are, respectively, the dielectric constant and density of the solvent. Both w_2 and $w_1 = 1 - w_2$ (the weight fraction of solvent) are related to x_2 and x_1 by the general relationship $x_i = (M/M_i) w_i$ and, consequently,

$$\frac{x_1}{x_2} = \frac{w_1}{w_2} \frac{M_2}{M_1} \tag{2.97}$$

Substitution of Eqs. (2.95), (2.96) and (2.97) into Eq. (2.94) gives the following relationship for the total polarization of the solute, P_2:

$$P_2 = \frac{4}{3} \pi N_A \alpha_2 = \frac{\epsilon_1 - 1}{\epsilon_1 + 2} \frac{M_2}{\rho_1} + \frac{M_2}{\rho_1} \frac{3\beta - (\epsilon_1 + 2)(\epsilon_1 - 1)(\beta'/\rho_1)}{(\epsilon_1 + 2)^2} \tag{2.98}$$

where the quadratic terms in w_2^2 were neglected and it was assumed that $\rho \to \rho_1$, $w_1 \to 1$, and $\epsilon \to \epsilon_1$ when $w_2 \to 0$. After regrouping terms in Eq. (2.98) and considering that $\beta' = \delta\rho/\delta w_2 = -(1/v_1^2)\delta v/\delta w_2$, the expression

$$P_2 = \frac{4}{3}\pi N_A \alpha_2 = M_2 \frac{3v_1}{(\epsilon_1 + 2)^2} \frac{\delta\epsilon}{\delta w_2} + M_2\left(v_1 + \frac{\delta v}{\delta w_2}\right)\frac{\epsilon_1 - 1}{\epsilon_1 + 2} \qquad (2.99)$$

is obtained, where v and v_1 represent the specific volumes of the solution and solvent, respectively. This relationship is known as the Halverstadt and Kumler equation [20].

Since the orientation polarizability α_{02} is given by

$$\alpha_{02} = \alpha_2 - \alpha_{a2} - \alpha_{e2} \qquad (2.100)$$

α_{e2} can be obtained from Eq. (2.99) by assuming that $\epsilon_1 = n_1^2$ and $\epsilon = n^2$, where n_1 and n represent the refractive indices of solvent and solute, respectively. Accordingly,

$$P_{e2} = \frac{4}{3}\pi N_A \alpha_{e2} = M_2 \frac{6v_1 n_1}{(n_1^2 + 2)^2}\frac{\delta n}{\delta w_2} + M_2\left(v_1 + \frac{\delta v}{\delta w_2}\right)\frac{n_1^2 - 1}{n_1^2 + 2} \qquad (2.101)$$

It follows from Eq. (2.24) that

$$\langle\mu_p^2\rangle = 3k_B T\alpha_{02} = \frac{9k_B T}{4\pi N_A}P_{02} \qquad (2.102)$$

where $P_{02} = P_2 - P_{e2} - P_{a2}$. In most systems, the molar atomic polarization amounts to only 5 to 10% of P_{e2}, and, therefore, it is often neglected in the calculations of dipole moments.

EQUATION OF GUGGENHEIM AND SMITH

By considering that the molar volume of component i in the solution is

$$V_i = M_i/\rho_i = M_i v_i \qquad (2.103)$$

and the molar volume of the solution can be expressed as

$$V = x_1 V_1 + x_2 V_2 \qquad (2.104)$$

Equation (2.94) can be rewritten for very dilute solutions in the following way:

$$\frac{\epsilon - 1}{\epsilon + 2}V = \frac{\epsilon - 1}{\epsilon + 2}V_1 x_1 + \frac{4}{3}\pi N_A\left(\alpha_{a2} + \alpha_{e2} + \frac{\langle\mu_p^2\rangle}{3k_B T}\right)x_2 \qquad (2.105)$$

which at high frequencies becomes

$$\frac{n^2 - 1}{n^2 + 2}V = \frac{n_1^2 - 1}{n_1^2 + 2}V_1 x_1 + \frac{4}{3}\pi N_A \alpha_{e2} x_2 \qquad (2.106)$$

A fictitious atomic polarizability α'_{a2} can be defined for the solute by means of the equation

$$\alpha'_{a2} = \alpha_{a1}(V_2/V_1) \tag{2.107}$$

where α_{a1} is the atomic polarizability of the solvent, and V_2 and V_1 are the molar volumes of the solute and solvent, respectively. Moreover, because orientation polarizability is not exhibited by nonpolar solvents, the Debye equation applied to the solvent gives

$$\frac{4}{3}\pi N_A(\alpha_{a1} + \alpha_{e1}) = \frac{\epsilon_1 - 1}{\epsilon_1 + 2}\frac{M_1}{\rho_1} \tag{2.108}$$

and

$$\frac{4}{3}\pi N_A\alpha_{e1} = \frac{n_1^2 - 1}{n_1^2 + 2}\frac{M_1}{\rho_1} \tag{2.109}$$

From Eqs. (2.107), (2.108), and (2.109), one obtains

$$\frac{4}{3}\pi N_A\alpha'_{a2} = \left(\frac{\epsilon_1 - 1}{\epsilon_1 + 2} - \frac{n_1^2 - 1}{n_1^2 + 2}\right)V_2 \tag{2.110}$$

By substituting Eq. (2.110) into the difference of Eqs. (2.105) and (2.106), the following relationship is obtained

$$\frac{\epsilon - 1}{\epsilon + 2} - \frac{n^2 - 1}{n_2 + 2} = \left(\frac{\epsilon_1 - 1}{\epsilon_1 + 2} - \frac{n_1^2 - 1}{n_1^2 + 2}\right)$$
$$+ \frac{4\pi N_A\rho}{3M_2}\left(\alpha_{a2} - \alpha'_{a2} + \frac{\langle\mu_p^2\rangle}{3k_BT}\right)w_2 \tag{2.111}$$

where use of the equation $x_i = w_i(M/M_i)$ was made. A close inspection of Eq. (2.111) reveals that if its first member,

$$\frac{\epsilon - 1}{\epsilon + 2} - \frac{n^2 - 1}{n^2 + 2} \tag{2.112a}$$

is plotted against w_2, the slope at $w_2 \to 0$ would give

$$\frac{4\pi N_A\rho}{3M_2}\left(\alpha_{a2} - \alpha'_{a2} + \frac{\langle\mu_p^2\rangle}{3k_BT}\right) \tag{2.112b}$$

Finally, because the derivative of Eq. 2.112a with respect to w_2, in the limit $w_2 \to 0$, has the form

$$\frac{\delta}{\delta w_2}\left(\frac{\epsilon - 1}{\epsilon + 2} - \frac{n^2 - 1}{n^2 + 2}\right) = \frac{3}{(\epsilon_1 + 2)^2}\frac{\delta\epsilon}{\delta w_2} - \frac{3}{(n_1^2 + 2)^2}\frac{\delta n^2}{\delta w_2} \tag{2.113}$$

by equating expression (2.112b) to the second member of Eq. (2.113), the so-called Guggenheim [21] and Smith [22] equation is obtained:

$$\frac{4}{3}\pi N_A(\alpha_{a2} - \alpha'_{a2}) + \frac{\langle \mu_p^2 \rangle}{3k_BT}$$

$$= 3M_2v_1\left(\frac{1}{(\epsilon_1 + 2)^2}\frac{\delta\epsilon}{\delta w_2} - \frac{1}{(n_1^2 + 2)^2}\frac{\delta n^2}{\delta w_2}\right) \qquad (2.114)$$

The only unknown term in Eq. (2.114) is the atomic polarization P_{a2}. For nonpolar substances, μ_p is negligible and both the atomic and electronic polarizations can be determined by means of Eqs. (2.107), (2.108), and (2.109). Careful experiments carried out in this way showed that P_{a2} may be one-tenth and even less of P_{e2}. There is no reason to assume that P_{a2} in polar substances is larger than in nonpolar ones. Moreover, P_{e2} for polar molecules is usually much smaller than P_{02}, and, consequently, P_{a2} turns out to be negligible in comparison with this term. In view of these circumstances, the assumption $P_{a2} = P'_{a2}$ was made, which implies that the ratio between the atomic polarization of solute and solvent should be equal to their molar volumes ratio; see Eq. (2.107). Therefore, Eq. (2.114) can be written in the more useful form

$$\langle \mu_p^2 \rangle = \frac{27k_BTM_2v_1}{4\pi N_A(\epsilon_1 + 2)^2}\left(\frac{\delta\epsilon}{\delta w_2} - \frac{\delta n^2}{\delta w_2}\right) \qquad (2.115)$$

where the approximation $n_1^2 \approx \epsilon_1$ is made and the derivatives are taken in the limit $w_2 \to 0$.

COMMENTS ON THE DETERMINATION OF DIPOLE MOMENTS

The major difficulty arising in the determination of dipole moments from permittivity measurements lies in how to deal with the internal field problem. Actually, as a consequence of the condensed nature of the liquid state, each dipole experiences not the externally applied field, but the field modified by neighboring dipoles. Among the approaches used to remedy this situation, the evaluation of dipole moments from permittivity measurements in solution stands out. In this method, use is made of solvents of such low permittivity that sample dipoles effectively experience the external field. In this case, methods based on the Debye equation [12] outlined by Helverstadt and Kumler [20] and Guggenheim [21] and Smith [22] become operative. A shortcoming of the Debye-based equations is that although progressive dilution eliminates intermolecular dipole–dipole interactions, intramolecular dipole–dipole interactions that may be important in polymers are not considered. Models such as those developed by Kirkwood [38] and Frölich [39], which allow for the interaction of surrounding dipoles by the correlation-function treatment, would be then most appropriate. However, their application introduces difficult computations that are often rather arbitrary. On the other hand, the Onsager theory, which is frequently favored to obtain the dipole moment of polymer chains, is for spherical symmetry and may be also unappropriate for polymers.

Many dipole moments obtained for oligomers and polymers by using treatments based on the Debye equation such as the Guggenheim and Smith equation show consistency among them, presumably because intramolecular dipole–dipole interactions in flexible chains fade away among dipoles separated by four or more skeletal bonds. As a consequence, the Guggenheim and Smith equation is one of the most reliable methods that can be used to evaluate the dipole moments of isolated molecular chains.

EXCLUDED VOLUME EFFECTS

As shown in Fig. 2.7, dipoles in molecular chains can be fixed to the chain skeleton in such a way that orientation of these dipoles requires movement of the molecular backbone, or they may be attached to flexible side chains. In the former case, dipoles may be classified geometrically into two types: (a) parallel to the chain direction (type A) and (b) rigidly attached to the backbone but bisecting a skeletal bond angle (type B). The last case, that in which dipoles are in the side chains, is called type C by Stockmayer [30, 42]. Although there is no example of polymers of only type A, there are polymers, such as polypropylene oxide, in which the repeat unit cannot be chosen to have a plane of symmetry and its dipole moment must have a component parallel and another one perpendicular to the chain contour. Therefore, these chains are of types A and B. The resultant value μ_n of n parallel vectors without reversal of direction sense must be correlated completely with the displacement vector r_n of the sequence, and for all the conformations, one has [30]

$$\langle \mu_n \cdot r_n \rangle = \text{constant} \times \langle r_n^2 \rangle \qquad (2.116)$$

The implication of this equation is that the mean-square dipole moment $\langle \mu_n^2 \rangle$ is subject to the same excluded volume effects as $\langle r_n^2 \rangle$ and the value of $\langle \mu_n^2 \rangle$ per skeletal bond increases with molecular weight except in θ conditions.

As far as molecular chains of type B are concerned, Marchal and Benoit [43] showed, based on a crude model without rigorous theoretical reasoning, that the excluded volume effects should vanish. This conclusion was also formulated by Stockmayer [30] on the grounds that a fixed displacement length r_n is compatible with many different values of μ_n (Figure 2.8). For these polymers,

$$\langle \mu_n \cdot r_n \rangle = 0 \qquad (2.117)$$

that is, there is no correlation between the vector sum of such type B components and the displacement vector. Nagai and Ishikawa [44] investigated this problem for linear polymer chains in a more rigorous way by using a first-order perturbation method. Following Fixman's application [45] of the multivariate Gaussian distribution for r and r_{ij} for the purpose of evaluating the expansion coefficient α_r^2, they

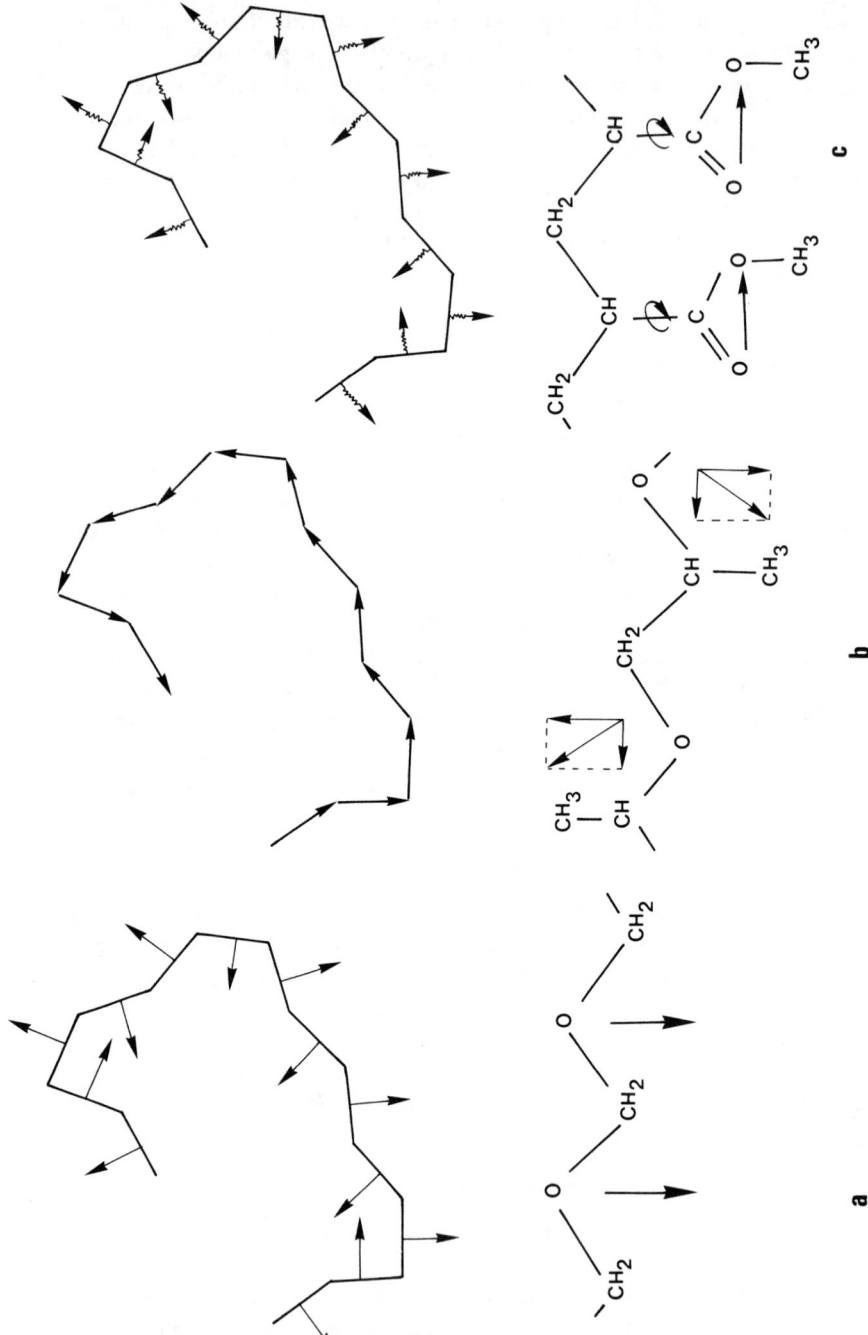

Figure 2.7 Schematic representation of chains in which dipoles are perpendicular (type B) and parallel (type A) to the chain contour. Dipoles located in flexible side chains are called type C dipoles. Examples of chains with dipoles of type B and C are shown in (a) and (c), respectively. A

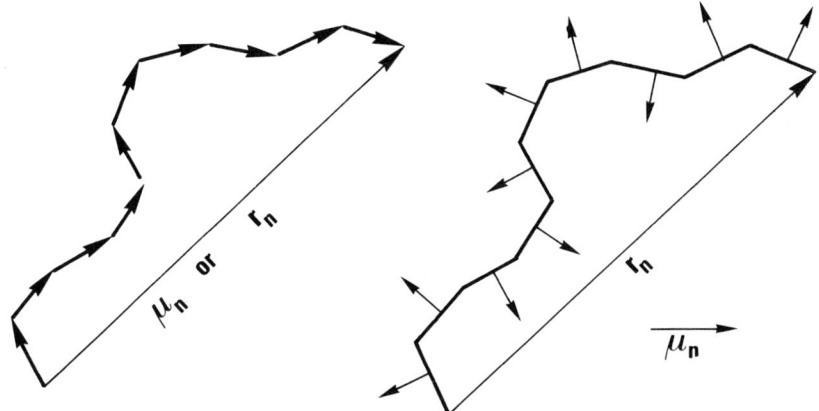

Figure 2.8　Scheme showing that correlation between dipole μ_n and displacement vector r_n only occurs when dipole components are parallel to the chain contour.

employed the multivariate Gauss...n distribution for μ and r_{ij} to calculate α_μ^2. As was indicated before, μ and r represent the dipole vector and the end-to-end vector for a specified conformation, respectively, whereas r_{ij} is the vector from segment i to segment j in that conformation; $\alpha_\mu^2 = \langle\mu^2\rangle/\langle\mu^2\rangle_0$ and $\alpha_r^2 = \langle r^2\rangle/\langle r^2\rangle_0$, where the averages without and with the subscript zero refer to the mean-square dipole moment and mean-square end-to-end distance in the presence and in the absence of the excluded volume, respectively. The expansion factors α_μ^2 for the dipole moment of a chain of infinite length, up to the term z^3, was found to be

$$\alpha_\mu^2 = 1 + (C_1 z + C_2 z^2 + C_3 z^3 + \cdots) \, \langle r \cdot \mu \rangle_0^2 / [\langle r^2 \rangle_0 \langle \mu^2 \rangle_0] \quad (2.118)$$

where z and C's are familiar quantities in the similar expression for the end-to-end distance, that is,

$$\alpha_r^2 = 1 + C_1 z + C_2 z^2 + C_3 z^3 + \cdots \quad (2.119)$$

The result obtained by Nagai and Ishikawa from the combination of Eqs. (2.118) and (2.119) is

$$\alpha_\mu^2 - 1 = X(\alpha_r^2 - 1) \quad (2.120)$$

$$X = \langle r \cdot \mu \rangle_0^2 \, [\langle r^2 \rangle_0 \langle \mu^2 \rangle_0]^{-1} \quad (2.121)$$

where the subscript zero denotes the ensemble unperturbed by long-range interactions. If $\langle r \cdot \mu \rangle_0 = 0$, then $\alpha_\mu^2 = 1$ irrespective of z, as was indicated a few years earlier by Stockmayer [30], without rigorous proof. If r is parallel to μ, $X = 1$ and $\alpha_\mu^2 = \alpha_r^2$. Because for each conformation, $0 \leq (r \cdot \mu)^2/(r^2 \mu^2) \leq 1$, it is then expected that

$$0 \le \frac{\langle \mathbf{r} \cdot \boldsymbol{\mu} \rangle_0^2}{\langle \mu^2 \rangle_0 \langle r^2 \rangle_0} \le 1 \qquad (2.122)$$

and from Eqs. (2.120) and (2.121), one finds

$$\alpha_\mu^2 - 1 \le \alpha_r^2 - 1 \qquad (2.123)$$

For a chain of N skeletal bonds $\langle r \cdot \boldsymbol{\mu} \rangle = 0$ whenever the chain in planar conformation has either (a) $N - 1$ symmetry planes, (b) $N - 1$ twofold symmetry axes, and (c) $N - 1$ symmetry points. Actually, $\boldsymbol{\mu}$ is perpendicular to $\mathbf{r}$ in cases (a) and (b), whereas it is zero in case (c). Most of conventional polar polymers have at least one of these symmetries. For example, polyamides from α,ω-dicarboxilic-n-alkanes and α,ω-diamino-n-alkanes have the symmetry planes (and the symmetry axes) or the symmetry points, depending on whether the number of carbons in dicarboxilic acids (or in diamines) is odd or even. The same can be said of polyesters derived from α,ω-dicarboxilic-n-alkanes and α,ω-dihydroxy-n-alkanes, polyethers, poly(vinyl halide)(s), and poly(acrylonitrile). The arguments are also valid for polymers such as poly(methyl methacrylate), poly(methyl acrylate), halo-substituted polystyrenes, etc., where the symmetries in the side groups are preserved by a proper disposition of side groups with respect to the main chain. The symmetry characteristics prevent the presence of a parallel dipole component in these polymers and, according to theory, excluded volume effects should be negligible.

For polymer chains with $\langle \mathbf{r} \cdot \boldsymbol{\mu} \rangle_0$ different from zero, Eq. (2.120) suggests that $\alpha_\mu^2 > 1$ if $\alpha_r^2 > 1$. Because X in Eq. (2.121) cannot be negative, no chain can have $\alpha_\mu^2 < 1 < \alpha_r^2$. Among the polymers with $\langle \mathbf{r} \cdot \boldsymbol{\mu} \rangle_0 \ne 0$ are polyesters and polyamides (and polypeptides) derived from α,ω-hydroxyacids and α,ω-aminoacids, respectively, poly(propylene oxide), poly(propylenesulfide), cis-polyisoprene, etc. According to the Nagai and Ishikawa [44] treatment, the dipole moments of these chains should present the excluded volume effects, and the unperturbed values should be obtained in nonpolar theta solvents. However, even in these polymers, the excluded volume seems to have no serious effect because the polar groups are almost perpendicular to the chain direction, and, hence, $\mathbf{r} \cdot \boldsymbol{\mu} \approx 0$.

Although there is a wealth of experimental work dealing with the dipole moments of polymers with different structures, many of the results were obtained on polymers of relatively low molecular weight, where long-range interactions are not important. However, the experimental dipole moments were found to be little affected by the excluded volume for many polymers of high molecular weight such as poly(vinyl chloride), poly(vinyl acetate), poly(methyl acrylate), etc., all of them of atactic structure [43,46–48].

Experimental testing of the validity of Eq. (2.120) would be required to determine the dipole moment of polymers for which $\langle \mathbf{r} \cdot \boldsymbol{\mu} \rangle_0 / \langle r^2 \rangle_0 \mu^2 \rangle_0$ departs significantly from zero. Because it is difficult to find a polymer soluble in nonpolar solvents with these characteristics, the plausibility of the Nagai and Ishikawa treatment has not yet been checked. Doi [49] has argued that Eq. (2.120) is valid for any order of perturbation, provided that $\langle \mu^2 \rangle$ is not independent of N.

Mattice and co-workers [50,51] have examined adherence to Eq. (2.120) for several model chains, containing atoms that behave as hard spheres, via Monte Carlo methods. The principal conclusion of this study is that adherence to the condition $\langle \mathbf{r} \cdot \boldsymbol{\mu} \rangle = 0$ does not guarantee that the mean-square dipole moment is independent of chain expansion. The combination $\alpha_\mu^2 \neq 1$ and $\langle \mathbf{r} \cdot \boldsymbol{\mu} \rangle = 0$ holds whenever the dipole moment vector assigned to bond i has a component perpendicular to the planes of bonds i and $i - 1$. Thus, patterns for the dipole moment vectors are found for which $1 < \alpha_\mu^2 < \alpha_r^2$ and $\alpha_\mu^2 < 1 < \alpha_r^2$ hold. Further simulations [51] performed on symmetric chains in which the dipole moment of the all transconformation is zero suggest that $\alpha_\mu^2 < 1 < \alpha_r^2$, even though $\langle \mathbf{r} \cdot \boldsymbol{\mu} \rangle = 0$. However, excluded volume effects were not detected for chains of this kind in many systems. For example, the experimental values of the dipole ratio $\langle \mu^2 \rangle / nm^2$ of hydroxyl terminated poly(oxytetramethylene) [52], where nm^2 is the mean-square dipole moment of the chain in the idealization that all the skeletal bonds are freely jointed, amounted to 0.54 and 0.52 at 20°C for fractions of number-average molecular weight 2500 and 408,000, respectively. The somewhat higher value of the low molecular-weight fraction is attributed to end-groups effect. Similar constancy of the dipole moment ratio with molecular weight was found for polydimethylsiloxane [53]. Some unexpected results reported by Burshtein and Stepanova [54] for the dipole moment ratio of poly(p-chlorostyrene) suggest that the excluded volume could affect the dipole moment of asymmetric chains. For example, assuming that the dipole moment of the repeat unit is 1.68 D, the dipole moment ratio at 30°C was found to be 0.77 and 0.71 when measured in toluene and p-xylene, respectively, but only 0.53 when measured in isopropyl benzene. Verification of these results [55], carried out later on samples of apparently the same stereochemical composition, showed that $\langle \mu^2 \rangle / nm^2$ amounts to 0.736 and 0.732 in toluene and isopropylbenzene, respectively. Therefore, excluded volume effects were not supported by the latter experiments.

Excluded volume interactions on the dipole moments of most polymers were found to be negligible, even in some polymer chains in which $\langle \mathbf{r} \cdot \boldsymbol{\mu} \rangle_0 \neq 0$. For example, the values at 30°C of the dipole moment ratio of two fractions of poly(propylene sulfide) of number-average molecular weight 6000 and 500,000, determined in benzene solutions, amounted to 0.45 and 0.44, respectively [56, 57]. The fact that the values of $\langle \mu^2 \rangle / nm^2$ show no dependence at all on the molecular weight indicates that excluded volume interactions have no significant effect on the dipole moment of this chain molecule.

The determination of excluded volume interactions carried out by comparing dipole moments determined in different nonpolar solvents may be misleading. Thus, studies concerned with the determination of dipole moment of low molecular-weight compounds in solution showed that the values obtained for some molecules were consistently different from those measured in the vapor state and varied from one solvent to another. They behave as if any moments induced in the solvent molecules by the solute dipoles were resolved with the permanent moment of the solute molecule into an apparent moment, which is then taken as the solution moment value. The difference between the values of the dipole moment in solution and in the vapor phase is the so-called solvent effect.

3
Experimental Methods

For frequencies lying in the range of 10^{-4} to 10^8 Hz, the experimental techniques used in dielectric experiments are based on the measurement of the equivalent capacitance and resistance at a given frequency [58]. Thus, a polymer sample may be considered as being electrically equivalent to a capacitance C_x in parallel with a resistance R_x, as shown in Figure 3.1. Under the effect of a voltage $V(t) = V_0 \exp(i\omega t)$, the admittance Y_x of the circuit can be written as

$$Y_x = 1/Z_x = i\omega C_x + 1/R_x \qquad (3.1)$$

where Z_x is the impedance of the circuit. The total current I is given by

$$I(t) = V(t)/Z_x \qquad (3.2)$$

The capacitance I_c and loss I_x components of the current I are

$$I_c(t) = \omega C_x V(t) \qquad (3.3)$$
$$I_x(t) = V(t)/R_x \qquad (3.4)$$

Dielectric experiments are usually performed by placing the samples between parallel-plate capacitors in the case of solid samples and high-viscosity liquids and concentric cylindrical capacitors in the case of low-viscosity liquids. Schemes of

31

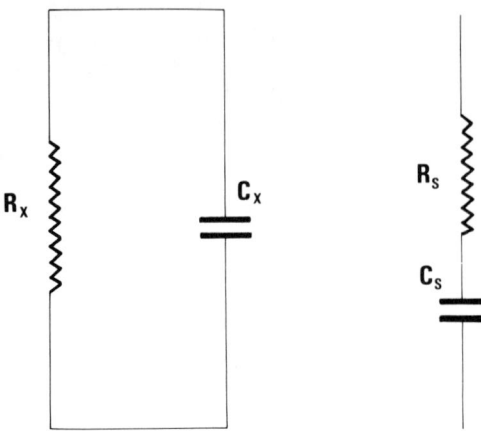

R_x C_x R_s C_s

a

b

Figure 3.1 Parallel and series equivalent circuits to the polymer samples.

these capacitors [59] are given in Figure 3.2. For vacuum-filled parallel-plate capacitors, the capacitance is given by

$$C_0 = \epsilon_0 A/d \tag{3.5}$$

where A is the area of each plate and d is the distance between them. The capacitance for a cylindrical capacitor in a vacuum is

$$C_0 = 2\pi\epsilon_0 \frac{l}{\ln (b/a)} \tag{3.6}$$

where l is the length of the electrodes, and b and a are, respectively, the radii of the outer and inner cylinders.

Replacement of the vacuum by a sample gives rise to a complex capacitance C^*, which is related to C_0 by

$$C^* = C_0\epsilon^* \tag{3.7}$$

where $\epsilon^* = \epsilon' - i\epsilon''$ is the complex permitivity.

The total current $I(t)$ passing through a capacitor under the action of a voltage $V(t) = V_0 \exp(i\omega t)$ is

$$I(t) = \left[\frac{dq(t)}{dt}\right] = \left\{\frac{d[C^*V(t)]}{dt}\right\} \tag{3.8}$$

Substitution of Eq. (3.7) into Eq. (3.8) gives

$$I(t) = (i\epsilon'\omega + \epsilon''\omega)C_0V(t) \tag{3.9}$$

and the capacitance and loss components of the current, I_c and I_x, are

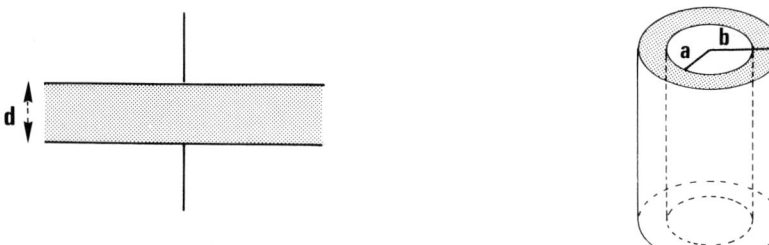

Figure 3.2 Schematic representation of parallel-plate and cylindrical condensers.

$$I_c = \epsilon'\omega C_0 V(t) \tag{3.10}$$

$$I_x = \epsilon''\omega C_0 V(t) \tag{3.11}$$

By comparing Eqs. (3.3) and (3.10) as well as Eqs. (3.4) and (3.11), the following expressions for the real ϵ' and loss component ϵ'' of the complex permittivity are obtained:

$$\epsilon' = C_x/C_0 \tag{3.12}$$

$$\epsilon'' = 1/R_x\omega C_0 \tag{3.13}$$

Hence, the loss tan δ_ϵ can be written as

$$\tan\delta_\epsilon = \epsilon''/\epsilon' = 1/R_x C_x \omega \tag{3.14}$$

There are techniques in which the measurements essentially correspond to the measurement of the equivalent series capacitance and resistance of the sample, as indicated in Figure 3.1. The impedance of the circuit, Z_s, is given by

$$Z_s = R_s - i/\omega C_s \tag{3.15}$$

The total current $I(t)$ in the circuit can be written as

$$I(t) = \frac{V(t)}{Z_s} = \left(\frac{\omega^2 C_s R_s}{1 + R_s^2\omega^2 C_s^2} + \frac{\omega i}{1 + R_s^2\omega^2 C_s^2}\right)V(t) \tag{3.16}$$

The expressions obtained for ϵ' and ϵ'' from Eqs. (3.9) and (3.16) are

$$\epsilon' = \frac{C_s}{C_0(1 + R_s^2\omega^2 C_s^2)} \tag{3.17}$$

$$\epsilon'' = \frac{R_s\omega C_s^2}{C_0(1 + R_s^2\omega^2 C_s^2)} \tag{3.18}$$

and the loss tan δ is given by

$$\tan\delta = R_s\omega C_s \tag{3.19}$$

By substituting Eq. (3.19) into Eq. (3.17) and (3.18), the real and loss components of the complex permittivity can also be written as

$$\epsilon' = \frac{C_s}{C_0(1 + \tan^2 \delta)} \tag{3.20}$$

$$\epsilon'' = \frac{C_s \tan \delta}{C_0(1 + \tan^2 \delta)} \tag{3.21}$$

The corresponding vector diagrams for the parallel and series circuits of resistance and capacitance are shown in Figure 3.3.

By comparing Eqs. (3.12) and (3.20), the following relationship between C_s and C_x is obtained

$$C_s/C_x = 1 + \tan^2 \delta \tag{3.22}$$

In the same way, the ratio R_x/R_s can be determined from Eqs. (3.19) and (3.14) as

$$\frac{R_x}{R_s} = \frac{1 + \tan^2 \delta}{\tan^2 \delta} \tag{3.23}$$

Since a sample may be considered equivalent to a resistance in parallel with capacitance C_x, the resistance can then be assumed to be made up of two contributions, R_0 and R_1: the first contribution arises from the dc conductivity of the specimen and the second from the non-dc resistance. Accordingly, R_0 is independent of frequency. Because R_0 and R_1 are in parallel,

$$1/R_x = 1/R_1 + 1/R_0 \tag{3.24}$$

the loss component of the dielectric permittivity can be expressed, after Eq. (3.13), by

$$\epsilon''(\omega) = 1/R_0\omega C_0 + 1/R_1\omega C_0 = \epsilon_0''(\omega) + \epsilon_1''(\omega) \tag{3.25}$$

where $\epsilon_0''(\omega)$, which represents the dc contribution to the loss, is given by

$$\epsilon_0''(\omega) = 1/R_0\omega C_0 = G_0/\omega C_0 \tag{3.26}$$

In this equation, G_0 represents the conductance of the sample. It is obvious that the product $\omega\epsilon_0''(\omega)$ is independent of frequency and, for a given value of R_0, $\epsilon_0''(\omega)$ shows a strong dependence on ω in the low-frequency region in the sense that it rapidly increases as ω decreases. $\epsilon_0''(\omega)$ can easily be evaluated in practice from Eq. (3.26) by measuring R_0 by means of a simple direct-current method. The loss $\epsilon_1''(\omega)$ due to nonconductivity effects can be obtained by subtracting $\epsilon_0''(\omega)$ from $\epsilon''(\omega)$.

At low frequencies, the dependence of $\epsilon_1''(\omega)$ on ω is rather small and Eq. (3.25) can be written as

$$\epsilon''(\omega) = k/\omega + \epsilon_1''(\omega) \tag{3.27}$$

where k is proportional to the specific ionic conductance. Accordingly,

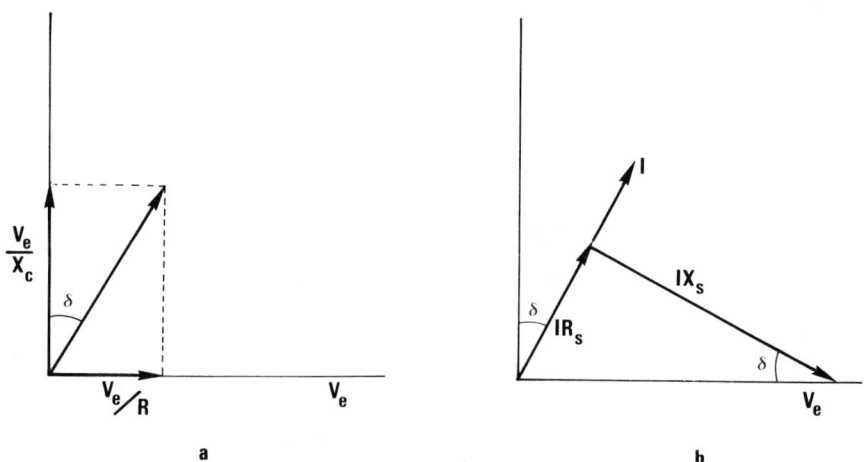

Figure 3.3 Vector diagram of the (a) parallel and (b) series equivalent circuit of Figure 3.1. R, X, and V represent the resistance, capacitance, and effective voltage, respectively.

$$\omega\epsilon''(\omega) = k + \omega\epsilon_1''(\omega) \tag{3.28}$$

The simple extrapolation of $\omega\epsilon''(\omega)$ versus ω gives the value of k.

A large part of the work on the dielectric behavior of polymers in the liquid, crystalline, and glassy states has been carried out in the frequency range of 10^2 to 10^5 Hz. The determination of the static dielectric constant of diluted polymer solutions is usually performed at frequencies below 10^6 Hz. This is because dilute solutions of most flexible polymers present relaxation phenomena at frequencies above 10^6 Hz, so that $\epsilon \approx \epsilon'(\omega)$ for $\omega < 10^6$ Hz. A simple method to measure C_x and R_x for a polymer sample is schematically shown in Figure 3.4. It consists [58] in a capacitance bridge made up of two variable capacitors, C_1 and C_3, and two fixed capacitors, C_2 and C_4. Capacitors C_3 and C_4 are in parallel with resistances R_3 and R_4, respectively. The capacitor with the sample is connected in parallel with variable capacitor C_1. For the balanced circuit, in which the capacitor containing the sample is unconnected, it holds that

$$(Z_1 Z_3)_u = (Z_2 Z_4)_u \tag{3.29}$$

If the capacitor with the sample is connected to the circuit, the relationship

$$(Z_1 Z_3)_c = (Z_2 Z_4)_c \tag{3.30}$$

also holds at balance.

Hence, at balance,

$$(Z_1 Z_3)_u = (Z_2 Z_4)_u = (Z_1 Z_3)_c = (Z_2 Z_4)_c \tag{3.31}$$

The values of impedances Z_1 and Z_3 are

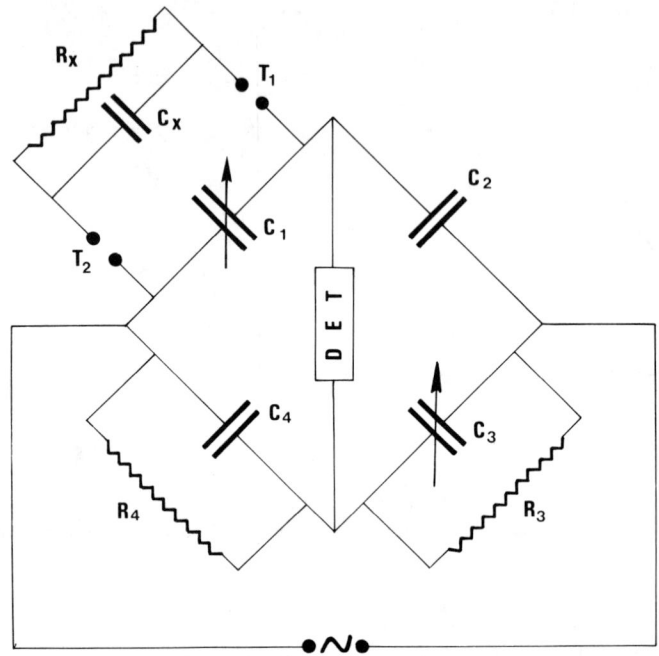

Figure 3.4 Capacitance bridge.

$$(Z_1)_u = \frac{1}{(i\omega C_1)_u} \tag{3.32}$$

$$(Z_1)_c = \frac{R_x}{1 + [C_x + (C_1)_c]i\omega R_x} \tag{3.33}$$

$$(Z_3)_u = \frac{R_3}{1 + i\omega R_3(C_3)_u} \tag{3.34}$$

$$(Z_3)_c = \frac{R_3}{1 + i\omega R_3(C_3)_c} \tag{3.35}$$

By substituting Eqs. (3.32) to (3.35) into the expression

$$(Z_1 Z_3)_u = (Z_1 Z_3)_c \tag{3.36}$$

of Eq. (3.31), one obtains

$$\frac{R_x}{\{1 + [C_x + (C_1)_c]i\omega R_x\}[1 + (C_3)_c i\omega R_3]} = \frac{1}{i\omega(C_1)_u[1 + (C_3)_u i\omega R_3]} \tag{3.37}$$

Because the respective real and imaginary parts of the two members of this equation must be equal, C_x, R_x, and tan δ_x can easily be obtained. The expressions for these quantities are

$$C_x = \frac{(C_1)_u - (C_1)_c - R_3^2\omega^2(C_3)_c[(C_1)_c(C_3)_c - (C_1)_u(C_3)_u]}{1 + [(C_3)_c\omega R_3]^2} \tag{3.38}$$

$$R_x = \frac{1 + [(C_3)_c\omega R_3]^2}{(C_1)_u\omega^2 R_3[(C_3)_c - (C_3)_u]} \tag{3.39}$$

$$\tan \delta_x = \frac{1}{R_x\omega C_x} = \frac{(C_1)_u\omega R_3[(C_3)_c - (C_3)_u]}{C_x\{1 + [(C_3)_c\omega R_3]^2\}} \tag{3.40}$$

A Wagner earth is often used to ensure that at balance, potentials X and Y are at earth potential so that noise and stray pickup are at a minimum. Highly accurate values for ϵ' and ϵ'' are obtained by this method, and the uncertainties in the measurements are due because the sample dimensions are not known to the accuracy that C_1 and C_3 changes can be determined. Detailed information on different techniques to measure C_x and R_x at different ranges of frequencies are given in reference [58].

Dieletric measurements on polymer solutions and on low-viscosity polymer liquids, in the range of 10^2 to 10^6 Hz, can be carried out in a two-terminal cell, whose schematic is shown in Figure 3.5(a). It consists of a container for the liquid, a group of concentric cylinders comprising the electrodes, and an insulated mechanical support assembly and terminals. With the terminals of the instrument connected to a bridge or other electrical measuring instrument, capacitance C_{xy} and resistance R_{xy} include not only those of the sample to be measured, C_x and R_x, but also extraneous and residual effects, C_y and R_y, as indicated in the equivalent circuit of

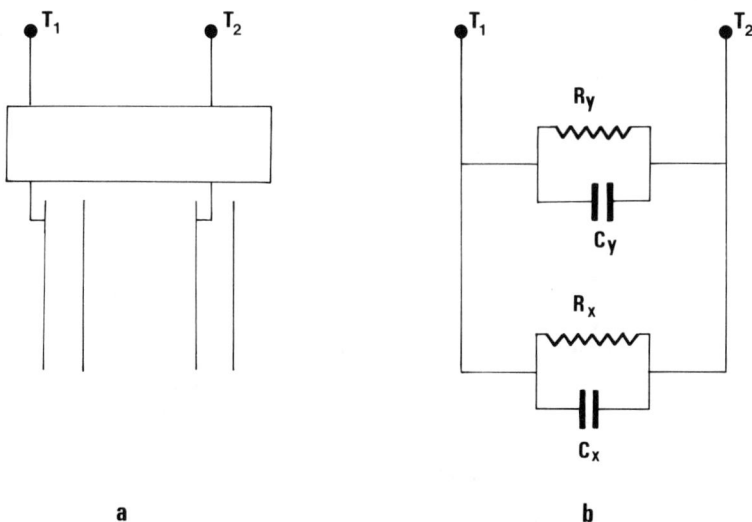

a   b

Figure 3.5 Schematic of (a) a cylindrical two-terminal cell and (b) equivalent circuit including residual impedance effects.

Figure 3.5(b). These effects arise from the leads from bridge to cell, the electrical insulation used to support the cell electrodes, surface leakages, and residual effects due to capacitive coupling to other metallic surfaces such as ovens, etc. It is advisable to ground one terminal of the cell to fix the extraneous effects so that they may be either eliminated or evaluated. The capacitance of the empty cell, C_0, can be determined from two measurements. The first measurement is carried out in the cell containing a nonpolar liquid. In this case,

$$(C_{xy})_{11} = C_{0x}\epsilon_{11} + C_y \tag{3.41}$$

where ϵ_{11} is the dielectric constant of the liquid at the temperature of measurement. Another measurement is made with another nonpolar liquid of dielectric constant ϵ_{12}. If the extraneous effects remain constant,

$$(C_{xy})_{12} = C_{0x}\epsilon_{12} + C_y \tag{3.42}$$

Hence,

$$C_{0x} = \frac{(C_{xy})_{11} - (C_{xy})_{12}}{\epsilon_{11} - \epsilon_{12}} \tag{3.43}$$

Because $C_{0x} \approx C_{ax}$, where C_{ax} represents the capacitance of the cell with dry air as the dielectric medium, the capacitance of the cell containing a liquid of permittivity ϵ_1 can also be written as

$$(C_{xy})_1 = C_{ax}\epsilon_1 + C_y \tag{3.44}$$

On the other hand,

$$(C_{xy})_a = C_{ax} + C_y \tag{3.45}$$

Then from Eqs. (3.44) and (3.45),

$$C_{ax} = \frac{(C_{xy})_1 - (C_{xy})_a}{\epsilon_1 - 1} \tag{3.46}$$

$$C_y = (C_{xy})_a - C_{ax} \tag{3.47}$$

In this way, the capacitance of the cell in a vacuum and the residual capacitances can also be determined from two measurements: one of these measurements is made with the cell empty and the other with a liquid of known dielectric constant.

It is very important that C_y remain constant and, therefore, it is advisable to use rigid leads in test procedures. Extraneous and residual effects also affect the loss tan δ. For two dielectrics in parallel, as shown in Figure 3.5, having individual capacitances C_x and C_y and dissipation factors D_x and D_y, the capacitance C_{xy} and dissipation factor D_{xy} of the combination are

$$C_{xy} = C_x + C_y \tag{3.48}$$

$$C_{xy}D_{xy} = C_xD_x + C_yD_y \tag{3.49}$$

Hence,

$$D_x = (C_{xy}D_{xy} - C_yD_y)/C_x \tag{3.50}$$

If the cell is thoroughly clean and empty,

$$(C_{xy})_a (D_{xy})_a = C_{ax}D_{ax} + C_yD_y \tag{3.51}$$

and because D_{ax} is negligible or zero, D_y can be written as

$$D_y = (D_{xy})_a(C_{ax}/C_y) + 1 \tag{3.52}$$

In order to reduce as far as practicable extraneous or residual effects, three-terminal cells are commonly used. A cell of this kind consists of the two-terminal cell described previously and the addition of a guard, or third terminal. This guard, or third terminal, T_3 is schematically shown in Figure 3.6(a), and it consists of a coaxial tube connected to guard lead or terminal T_2. The equivalent circuit is shown in Figure 3.6(b), where the sample being measured is represented by R_x and C_x as in the two-terminal cell. In certain systems, the guard electrode may also be used to control the electrostatic field distribution. In general, the techniques of measurement are designed in such a way that C_x and R_x can be measured independently of impedances Z_{HG} and Z_{LG} indicated in Figure 3.6. For example, these impedances, in the bridge described before, can be eliminated by using a Wagner earthing device.

Dielectric measurements of high-viscosity liquids and of both crystalline and

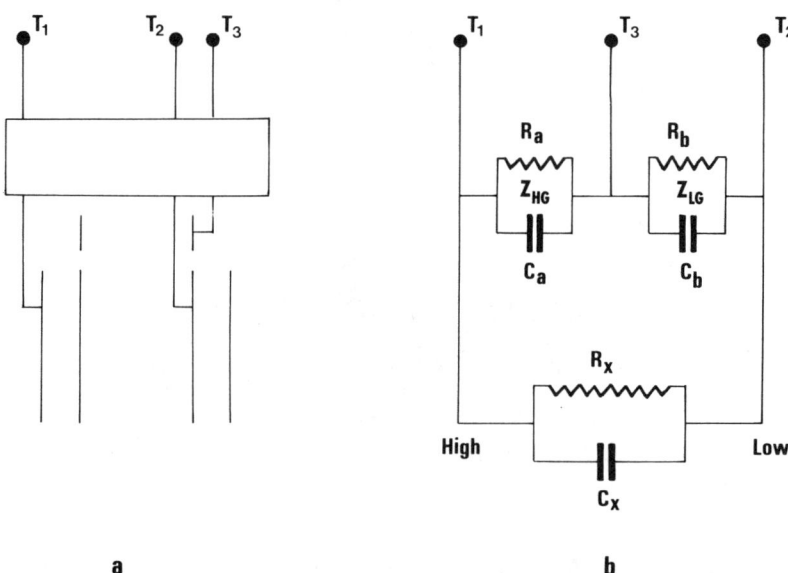

Figure 3.6 Schematic of (a) a three-terminal cell and (b) equivalent circuit including residual impedance effects.

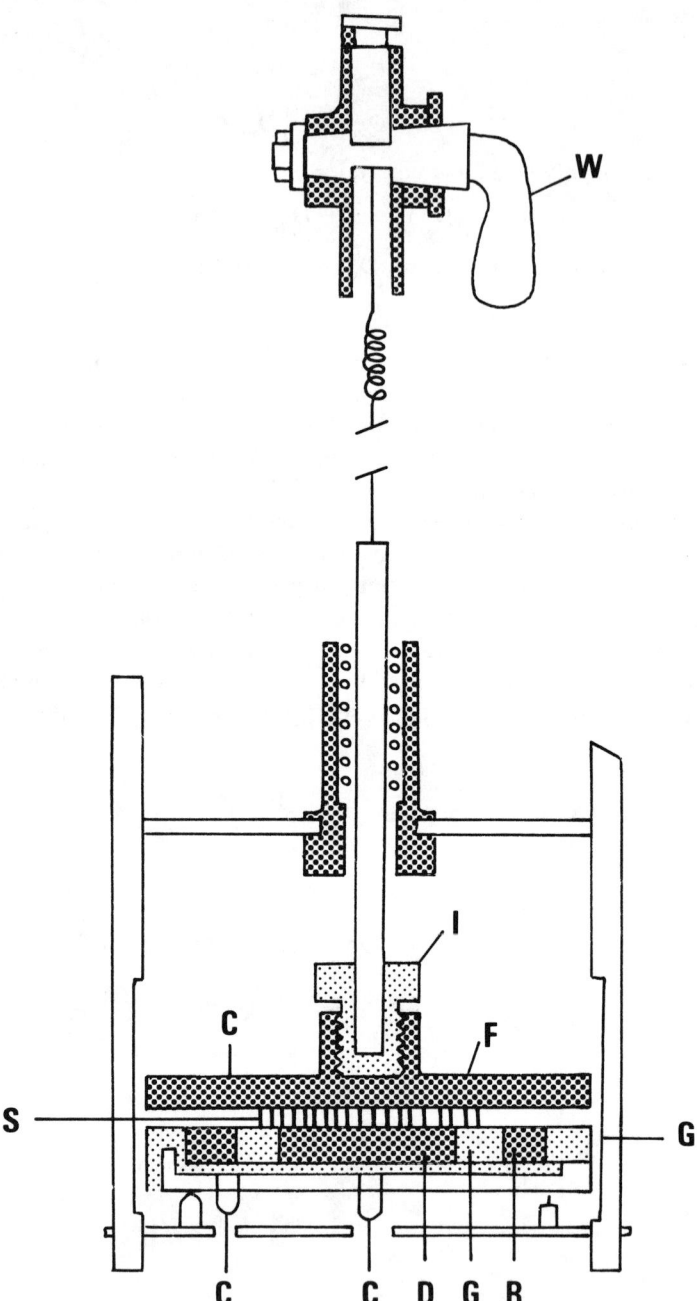

Figure 3.7 Parallel-plate capacitor: (F) floating electrode, (R) ring electrode, (D) disk electrode, (G) guard rings; (S) specimen, (W) windlass; (C) electrical connectors, and (I) insulating bushing. (Redrawn from reference [60] by courtesy of the American Institute of Physics.)

glassy polymers are carried out by using parallel-plate capacitors. Designs of either the two-terminal type, in which the effective capacitance is not well defined, or of the three-terminal type contain no means by which dimensional changes of the specimen can be monitored. By adding a third active electrode to a three-terminal cell, Work and co-workers [60] could obtain the dielectric permittivity without knowledge of the sample thickness. The cell is shown in Figure 3.7. In the lower section of the cell, there are two active electrodes D and R, in disk and ring form, respectively, that are fastened in guard block G but are electrically isolated from it. The disk electrode *sees* nothing but the sample, whereas the ring electrode *sees* nothing but the surrounding atmosphere. If the measurements are performed in a vacuum, the disk and ring capacitances, C_D and C_R, respectively, are

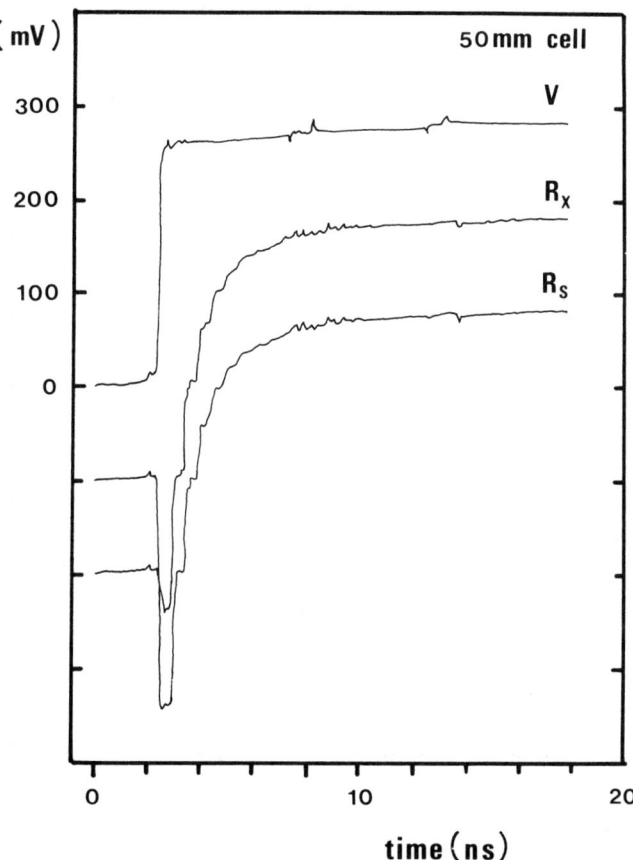

Figure 3.8 Time-domain signals of V, R_s, and R_x for 4.8% (weight/weight) solution of PMMA in benzene at 20°C. (Redrawn from reference [64] by courtesy of the Physical Society of Japan and the Japan Society of Applied Physics.)

$$C_D = \epsilon_0 \epsilon' A_D / d \qquad (3.53)$$

$$C_R = \epsilon_0 A_R / d \qquad (3.54)$$

where ϵ' is the real part of the complex permittivity corresponding to the specimen; ϵ_0 is the dielectric constant of the vacuum; A_D and A_R are the effective areas of the disk and ring electrodes, respectively; and d is the electrode separation, that is, the sample thickness. Then the value of ϵ' is

$$\epsilon' = (C_D / C_R)(A_R / A_D) \qquad (3.55)$$

The ratio A_R / A_D is constant over all temperatures if the cell is in thermal equilibrium and can be determined once and for all. In order to overcome the

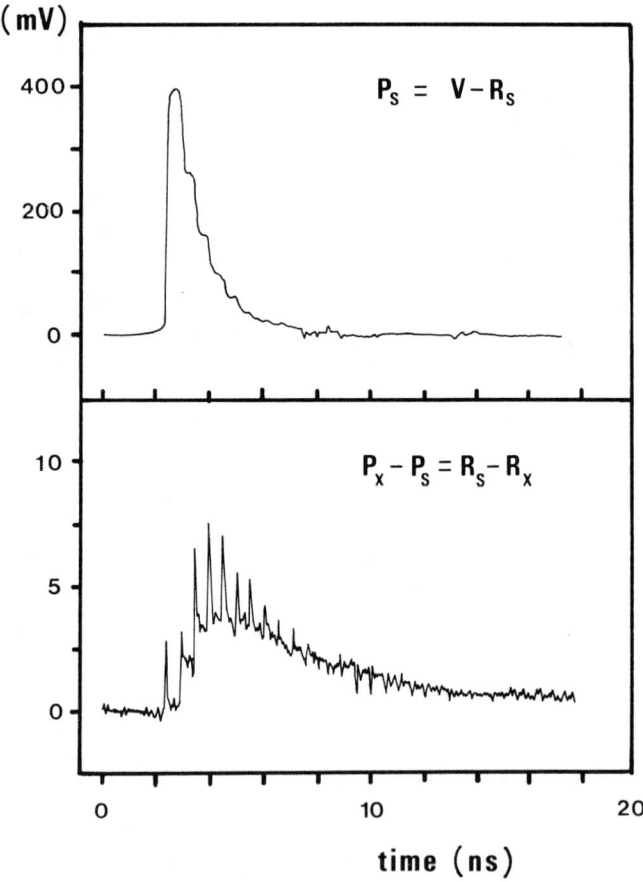

Figure 3.9   Time-domain signals of $P_s = V - R_s$ and $P_x - P_s = R_x - R_s$ of the solution of Figure 3.8. (Redrawn from reference [64] by courtesy of the Physical Society of Japan and the Japan Society of Applied Physics.)

mechanical instability of viscoelastic materials, the upper electrode can be partially supported by means of a suspension wire containing a spring of variable tension. Details of the ensamblage are given in Figure 3.7. Good electrical contact is essential to obtain correct values of C_D (or C_x) and R_x, because an air gap between the electrodes' face and the sample gives rise to a capacitance in series with the sample, leading to an appreciable error in the measurement of C_D and R_x. Good contact is ensured by evaporating a metal film on the sample surface. A less satisfactory method of ensuring a good electrical contact is to attach metal foil to the surface with the aid of silicone grease.

Dielectric relaxations of polymer solutions in the high-frequency region above 10 MHz are caused by the conformational change and internal motion of local chain molecules. Precise observations of the dielectric spectra of macromolecules are important to understand the dynamics of molecular chains. The technique of time-domain refractometry (TDR) has successfully been applied to several polymer solu-

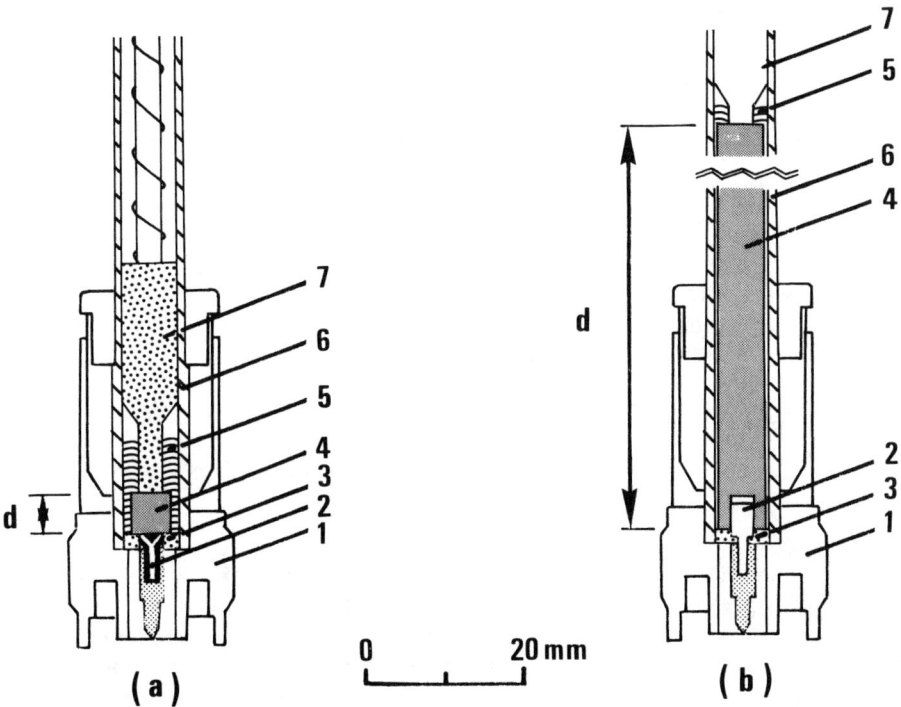

Figure 3.10   Sample cells for (a) high frequencies and (b) low frequencies: (1) N-type connector (plug); (2) center pin; (3) insulating washer (Kel-F); (4) inner conductor; (5) liquid sample; (6) outer conductor; (7) insulating block (Kel-F). d indicates the length of the inner conductor. (Redrawn from reference [64] by courtesy of the Physical Society of Japan and the Japan Society of Applied Physics.)

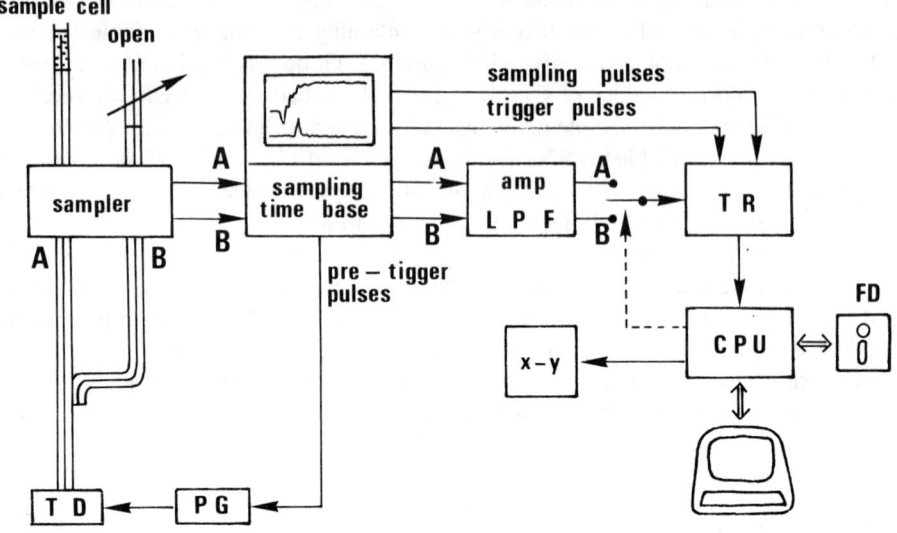

Figure 3.11 Block diagram of the time-domain reflectometry (TDR) measuring system: (TD) tunnel diode, (PG) pulse generator; (LPF) low-pass filter; (TR) transient recorder; (CPU) minicomputer; (X–Y) X–Y recorder and display, (FD) floppy-disk system. (Redrawn from reference [64] by courtesy of the Physical Society of Japan and the Japan Society of Applied Physics.)

tions using the difference method. The technique is based on the relation between an input signal $V(t)$ to the open-circuit sample and its reflection $R(t)$, which in the frequency domain is given by [61–63]:

$$p(s) = v(s) - r(s) = \frac{[1 - \rho(s)]\{1 - \exp[-2ds(\epsilon^*)^{1/2}/C]\}}{1 + \rho(s)\exp[-2ds(\epsilon^*)^{1/2}/C]} \tag{3.56}$$

$$\rho(s) = \frac{1 - \gamma(\epsilon^*)^{1/2}}{1 + \gamma(\epsilon^*)^{1/2}} \tag{3.57}$$

where $s = i\omega$; $v(s)$ and $r(s)$ are Fourier transforms of $V(t)$ and $R(t)$, respectively; ϵ^* is the complex permittivity; ρ is the complex reflection cofficient; d is the cell length; c is the velocity in a vacuum; ω is the angular frequency; and γ is the ratio of the vacuum capacitance per unit length of the cell to the capacitance per unit length of the matched coaxial cable. The difference method measures the differences $P_s(t)$ and $P_x(t)$ for the solvent and solution, respectively, where $P(t) = V(t) - R(t)$. Examples of $V(t)$, $R_s(t)$, and $R_x(t)$ are shown [64] in Figure 3.8, and $P_s(t)$ and $P_x(t) - P_s(t)$ are shown in Figure 3.9.

Cole [62] rewrote Eq. (3.56) for the frequency-domain analysis in the much simpler form:

$$\epsilon^*(\omega) = \frac{c}{i\omega\gamma d} \frac{p(\omega)/sv(\omega)}{1 - p(\omega)/2v(\omega)} f(z) \qquad (3.58)$$

$$f(z) = z \cot z \qquad (3.59)$$

where $z = d\omega(\epsilon^*)^{1/2}/C$. By applying Eq. (3.58) to the difference method, the relationship between the complex permittivity of solution ϵ_x^* and that of solvent ϵ_s^* is given by

$$\epsilon_x^* = \epsilon_s^* \frac{(v - r_s) + (r_s - r_x)}{(v - r_s) - i\omega(\gamma d/cf_s)(r_s - r_x)} \frac{f_x}{f_s} \qquad (3.60)$$

Here v, r_s, and r_x are the Laplace transforms of the incident pulse, that of the reflected pulse from the known solvent, and that of the reflected pulse from the unknown solution, respectively. The product γd represents the effective length of the conductor. Schematics of the cells used for high- and low-frequency measurements are shown in Figure 3.10. Equation (3.60) places a definite restriction on the choice of cell length, because f diverges at $|z| = \pi/2$ and this equation is hence useful for $|z| < 1$. The block diagram of the measuring system is shown in Figure 3.11. The diagram is described in detail in reference [64].

4
Dielectric Equilibrium Properties of Polymer Chains

CONFORMATIONAL ENERGIES

The knowledge of the energy effects accompanying the changes from one conformation to another in combination with the structural characteristics of the chains is paramount to interpret and predict the physical properties of polymeric materials. The evaluation of conformation-dependent properties is often carried out by means of the rotational isomeric state (RIS) [19] model, which assumes that each skeletal bond can adopt a small number of discrete states. Intramolecular interactions are accounted for in the model by assigning statistical weights to pairs of conformations about two consecutive skeletal bonds. The statistical weights are expressed as Boltzmann factors:

$$\sigma_i = \exp(-E_{\alpha\beta}/RT) \tag{4.1}$$

where $E_{\alpha\beta}$ represents the conformational energy of the rotational state β of bond i provided that bond $i - 1$ is in the α state. For a chain of n skeletal bonds, the conformational partition function Z can be generated by a serial multiplication of matrices $\mathbf{U}_i$ that embody the statistical weights for the states of each bond i relative to a reference state. Following the scheme commonly used, statistical-weight parameters σ, α, β, γ, and τ represent first-order interactions (between groups separated for three bonds), whereas ω, ω', and ω'' represent second-order interactions (between groups separated by four bonds). In the case of a symmetric chain with three rotational states (t, g^+, and g^-), the statistical-weight matrix for any bond i can be expressed in a generalized way by [19]

$$\mathbf{U}_i = \begin{bmatrix} 1 & \sigma & \sigma \\ 1 & \sigma\omega & \sigma\omega' \\ 1 & \sigma\omega' & \sigma\omega \end{bmatrix} \tag{4.2}$$

with the rows and columns being indexed in the order t, g^+, and g^-. It should be pointed out that in most symmetric chains, second-order interactions associated with $g^{\pm}g^{\pm}$ conformations are not important, and, consequently, $\omega \approx 1$. However, $g^{\pm}g^{\mp}$ conformations give rise to pentane-type interactions between atoms or atomic groups separated by four bonds, and, therefore, the values of ω' are lower than 1. As is discussed in what follows, the statistical-weight matrices are not symmetrical for chains with side groups in their structure.

The conformational partition function Z may be written as

$$Z = \prod_{i=1}^{n} \mathbf{U}_i \tag{4.3}$$

where U_1 and $\mathbf{U}_n$ are row and column matrices, respectively, to effect the required sum of terms.

Standard matrix-multiplication methods were developed by Flory and co-workers [19, 65] (see the Appendix "Matrix Multiplication Scheme") to calculate the average of any equilibrium conformation-dependent property a, such as the mean-square dipole moment $\langle \mu^2 \rangle$, mean-square end-to-end distance $\langle r^2 \rangle$, etc. The method involves the serial multiplication of matrices $\mathbf{A}_i$, one for each skeletal bond, followed by division by the conformational partition function:

$$\langle a \rangle = Z^{-1} \prod_{i=1}^{n} \mathbf{A}_i \tag{4.4}$$

where $\mathbf{A}_i$ is a supermatrix that contains the appropriate statistical-weight matrix $\mathbf{U}_i$, a coordinate transformation matrix $\mathbf{T}_i$ that depends on both the skeletal bond angle θ_i and the allowed rotational angle ϕ, and the specific contribution, either vectorial or tensorial, of bond i to property a. Here, $\mathbf{A}_1$ and $\mathbf{A}_n$ are row and column matrices, respectively, to yield the sum of the pondered contribution of each configuration to property a.

The location of the rotational states and the establishment of their relative energies are obtained, whenever possible, by direct measurements, either spectroscopic or thermodynamic, on small molecules having structural characteristics similar to the chain molecules under investigation [17,66,67]. When information of this kind is not available, precise values of conformational energies could be obtained by using a complete ab initio quantum mechanical approach [68]. However, the size of the associated secular equation is so large that the time involved in its solution becomes prohibitively long even for the faster computer. Simplified energy calculations were then developed to account for the interaction between nonbonded atoms trying to combine the sophistication of the potential functions with the minimization of the time required to solve them [69]. It is obvious that the level of sophistication of the potential functions is inversely proportional to the size of the molecules under consideration.

By considering an allowed conformation as that one in which all the variable nonbonded interaction distances are equal or greater than those corresponding to the sum of the van der Waals radii, all the sterically allowed conformations of a macromolecular chain can, in principle, be calculated [2]. The distances between bonded atoms, usually obtained from the distances between atomic centers of atoms in various crystals, were found to be relatively constant and, consequently, additive for the same pairs of bond atoms regardless of the structural environment. This has permitted definition of the atomic radii, which predict with great accuracy the interatomic distances [70].

The interatomic distances between pairs of nonbonded atoms are somewhat dependent on the structural environment, and the average values of the atomic radii in this case, known as the van der Waals radii, are much larger than the atomic radii obtained from interatomic distances of bonded atoms. In spite of the fact that the van der Waals radii are only roughly constant and additive, they have proved to be useful in crystal-structure analysis [71]. The van der Waals radii of different atoms that usually enter in the constitution of molecular chains are shown in Table 4.1. Because the van der Waals radii of nonbonded interactions are not strictly additive,

TABLE 4.1 Van der Waals Radii r_i,
Effective Number of Outer Shell Electrons
N_{eff}, and Atomic Polarizability α for Various
Atoms and Groups of Atoms [73, 74]

Atom or Group	r_i (Å)	α (Å^3)	N_{eff}
H	1.3	0.42	0.9
C	1.8	0.93	5
CH$_3$	2.0	1.77	7
O	1.6	0.70	7
C (carbonyl)	1.8	1.23	5
O (carbonyl)	1.6	0.84	7
N (sp^2)	1.35	1.15	6
F	1.40	1.93	8.5
Cl	1.80	2.45	14
Br	2.05	3.34	23
P	1.75	2.30	13
S	1.9	2.39	13.5

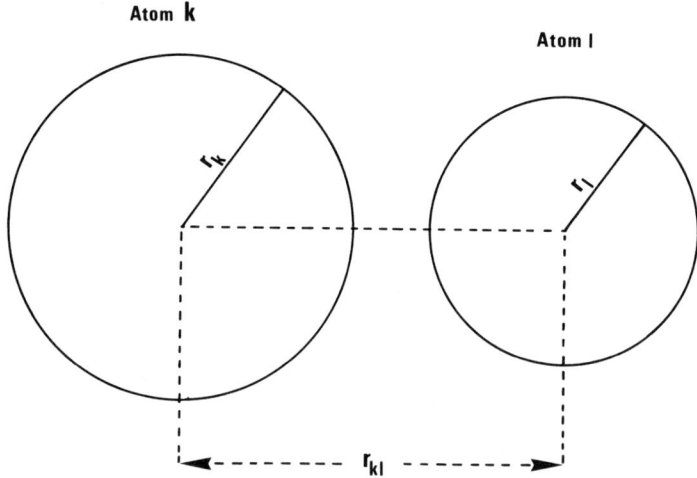

Figure 4.1 Distinction between the van der Waals radii (r_k, r_l) and the contact distance for the k,l pair of atoms.

contact distances have been used instead of van der Waals radii in conformational analysis. A contact distance is defined as the minimum distance between the centers of two nonbonded atoms, and, in fact, the value of this quantity is not equal to the sum of the van der Waals radii, but it is slightly larger (Figure 4.1). The contact distance for a given pair of nonbonded atoms lies between two limits: the normal limit, defined as the arithmetic average distance in which all recorded distances obtained from crystal structures for the pair of atoms are included, and the outer limit, which is the shortest contact distance observed in the solid state [2].

When two molecules come into proximity, induced-dipole–induced-dipole interactions occur. Actually, although the molecules have no permanent dipoles, their electron clouds are continuously fluctuating so that the molecules can be considered as having a permanent dipole that is constantly changing in magnitude and direction. A molecule with an instantaneous dipole μ_1^* will polarize the other molecule and will induce in it an instantaneous dipole μ_2^*, giving rise to an attractive interaction between them that does not average to zero. It is obvious that the interaction will depend on the polarizabilities of the two molecules, because the size of the instantaneous dipoles μ_1^* and μ_2^* is dependent on the looseness of the nuclear potential's control on the outer electrons in the two molecules. The potential associated with this interaction was rigorously derived for a pair of identical spherically symmetrical and chemically saturated molecules, finding that it varies as R^{-6}, where R is the distance between the two molecules [1]. It should be pointed out that induced quadrupoles and octupoles are neglected in the derivation of intermolecular interactions. By analogy, the attractive term for the potential of two nonbonded atoms or atomic groups could be written as $-A_{ij}/r_{ij}^{-6}$, where r_{ij} is the distance between them. The attractive term coefficient A_{ij} can be determined by using the Slater–Kirkwood [72] equation as modified by Scott and Scheraga [73]:

$$A_{ij} = \frac{3eh\alpha_i\alpha_j}{4\pi m^{1/2}C_{ij}} \tag{4.5}$$

where

$$C_{ij} = [\alpha/N_{\text{eff}}]_i^{1/2} + [\alpha/N_{\text{eff}}]_j^{1/2} \tag{4.6}$$

In Eq. (4.5), e is the electronic charge; h is Planck's constant; m is the electron mass; α_i is the atomic polarizability; and N_{eff} is a correcting factor that takes into account the nature, strengths, and orientation of the chemical bonds to which the pair of atoms belong. N_{eff} can be considered as the number of effective electrons surrounding the atoms in the chemical bonds. Values of α_i and N_{eff} for some atoms and groups of atoms are given in Table 4.1.

As the distance between two molecules or atomic groups decreases, a distance is reached at which the electronic and nuclear repulsions begin to dominate the attractive forces, and the repulsive interactions rise very steeply with decreasing separation. The overall curve representing the potential is reasonably well approximated by the equation

$$V_{ij}(r) = B_{ij}/r_{ij}^n - A_{ij}/r_{ij}^6 \tag{4.7}$$

where n is a large integer. When $n = 12$, Eq. (4.7) is called the Lennard–Jones potential. It should be pointed out that the repulsion term is a useful approximation that describes the shape of the repulsion. For small values of r_{ij}, r_{ij}^{12} is much smaller than r_{ij}^6 and the positive term B_{ij}/r_{ij}^n dominates the negative A_{ij}/r_{ij}^6. The positive-term coefficient B_{ij} is obtained by assuming that the potential function V_{ij} has a minimum at the normal constant-interaction distance R_0.

$$\left[\frac{dV_{ij}(r)}{dr}\right]_{r=R_0} = \frac{6A_{ij}}{R_0^7} - \frac{12B_{ij}}{R_0^{13}} = 0 \tag{4.8}$$

so that

$$B_{ij} = \tfrac{1}{2}A_{ij}R_0^6 \tag{4.9}$$

As was indicated before, the attractive term arises from the attraction between mutually induced fluctuating dipoles produced by the distortion of electron orbitals of two freely rotating molecules. Therefore, in the application of Eq. (4.7) to the study of interactions between pairs of nonbonded atoms or groups of atoms, it is assumed that the possible distortion of spherical symmetry will only affect the depth and position of the minimum in energy for the interaction.

Another useful function in the evaluation of molecular conformations is the Buckingham potential function [19]:

$$V_{ij}(r) = B_{ij} \exp[-C_{ij}r_{ij}] - A_{ij}/r_{ij}^6 \qquad (4.10)$$

which, as can be seen, differs from the Lennard–Jones potential in only the repulsive term. Although it would be expected that the presence of an additional parameter in the repulsive term of Eq. (4.10) should make it more specific than Eq. (4.7), the fact that accurate values of C_{ij} are not known is a shortcoming in the use of the Buckingham potential function. It is believed that the value of C_{ij} lies in the interval $4.35 < C_{ij} < 4.70$, and the value $C_{ij} = 4.60$ is often used for all types of pair interactions. The values of A_{ij} and B_{ij} are calculated following the method described before for the Lennard–Jones potential function.

The potential curve for an ethane molecule includes, in addition to the non-bonded interactions between the hydrogen atoms, a contribution arising from orbital–orbital interactions of the electrons centered about the two carbons. This latter interaction is a quantum mechanics effect caused by the geometry of the orbitals in the molecule. The orbital–orbital interactions for rotations about perfect single bonds is zero due to the cylindrical symmetry of these bonds. Because most of the bonds corresponding to the skeletal bonds are not pure single bonds, a torsional potential appears whose value increases as the deviation from pure single bonds increases. For near-double and near-triple bonds, the orbital–orbital interaction terms clearly dominate, so that rotations about these bonds become highly hindered. For single bonds, the torsional potential can be written as [1, 2, 19]:

$$E(\phi) = \tfrac{1}{2}E^*(1 - \cos n\phi) \qquad (4.11)$$

where E^* is the barrier height, n is the periodicity of the function, and ϕ is the rotational angle. The value of E^* is usually obtained from microwave spectroscopy and the temperature dependence of NMR spectra corresponding to low molecular-weight compounds with structural characteristics similar to those of the bonds under investigation.

Polar bonds in the chains give rise to dipole–dipole interactions. The residual charges $\delta+$ and $\delta-$ associated with the atoms of the chains can be determined either from the values of the dipoles or by using semiempirical quantum mechanics methods. Therefore, the coulombic contribution can be expressed by

$$E_c = \sum \frac{\delta_i \delta_j}{\epsilon r_{ij}} \qquad (4.12)$$

where ϵ is the dielectric constant of the medium. In these calculations, it is assumed that the dielectric medium is homogeneous with respect to polymer and solvent, an assumption that does not affect the results whenever ionic groups are not present in the chains [2].

By considering the atoms as spheres, conformational energies arising from rotations about two consecutive skeletal bonds can be estimated by using the standard equation

$$E(\phi_i,\phi_j) = \tfrac{1}{2}[E_i^*(1 - \cos n\phi_i) + E_j^*(1 - \cos n\phi_j)]$$
$$+ \sum\sum \left(\frac{B_{kl}}{r_{kl}^{12}} - \frac{A_{kl}}{r_{kl}^{12}} + \frac{\delta_k\delta_l}{\epsilon r_{kl}}\right) \tag{4.13}$$

which includes the torsional, van der Waals, and coulombic contributions to the potential. In this equation, ϕ_i and ϕ_j are the rotational angles of the skeletal bonds i and j measured from the trans state. The average values of the conformational energies in the potential wells can be obtained by the expression

$$\langle E \rangle = Z^{-1} \sum\sum \{E(\phi_i,\phi_j) \exp[-E(\phi_i,\phi_j)/RT]\} \tag{4.14}$$

with the rotational partition function Z defined as

$$Z = \sum\sum \exp[-E(\phi_i,\phi_j)/RT] \tag{4.15}$$

In the same way, the rotational angles average at the minimum of potential can be written as

$$\langle \phi_i \rangle = Z^{-1} \sum\sum \{\phi_i \exp[-E(\phi_i,\phi_j)/RT]\} \tag{4.16}$$

Semiempirical potential functions provide information on the overall conformational-energy surface as a function of skeletal rotational angles. However, the reliability of the conformational energies obtained by means of Eq. (4.13) or by means of semiempirical quantum mechanics methods must be carefully tested against appropriate experimental data such as dipole moments, unperturbed dimensions, optical properties, etc., of molecular chains. In the case of n alkanes, for example, the experimental values of the mean-square end-to-end distance $\langle r^2 \rangle_0$ and its temperature coefficient $d[\ln(\langle r^2 \rangle_0)]/dT$ are reproduced by using conformational energies obtained [75] from Eq. (4.13). However, semiempirical potential functions do not always provide quantitative information on the values of conformational energies, especially when oxygen, sulfur, or halogen atoms are present in the molecular chains.

Vibrational spectra of dimethoxymethane CH_3O—CH_2—OCH_3 and dimethoxyethane CH_3OCH_2—CH_2OCH_3 suggest that gauche states about both the central CH_2—O bonds in the former compound and CH_2—CH_2 in the latter are preferred over the alternative trans states [76,77]. However, semiempirical potential calculations, carried out using Eq. (4.13), failed to predict lower energy for the gauche states as experimentally found [78]. Quantum mechanics studies on dimethoxymethane, based on the CNDO/2 method, predict that the $g^\pm g^\pm$ conformation of the central pair of bonds is the most stable one [79]. This quantum mechanics approach, however, overestimates the dipole moment by about 40% as compared to the experimental result. Other attempts at a quantum mechanics calculation of the conformational energies of low molecular-weight compound analogues of poly(methylene oxide) (PMO) and poly(ethylene oxide) (PEO) gave equally discouraging results [80, 81]. Consequently, a number of semiquantitative explanations have been postulated to explain the unusual steric attractions or *gauche effects* that have been detected in these and a wide variety of molecules.

Equilibrium properties of the chains such as the mean-square end-to-end distance $\langle r^2 \rangle_0$ and the dipole moment $\langle \mu^2 \rangle_0$ as unperturbed by long-range interactions, as well as their temperature coefficients $d[\ln(\langle r^2 \rangle_0)]/dT$ and $d \ln (\langle \mu^2 \rangle_0)/dT$, can be used to obtain information on the location and relative energies of the conformations that appear along the chains [19, 69]. Other configuration-dependent properties suitable for this purpose, when available, are the equilibrium cyclization constant K_x, the stress–optical coefficient C, and the molar Kerr constant $_mK$. Whereas the mean-square end-to-end distance is sensitive to excluded volume effects, both the mean-square dipole moment and the molar Kerr constant are independent of chain expansion whenever the resulting dipole moment of each repeating unit directs along the bisector of a skeleton bond angle, or, more generally, whenever the chains have symmetry planes, twofold symmetry axes, or symmetry points. Moreover, both $\langle \mu^2 \rangle_0$ and $_mK$ can be measured for chains of any length, in contrast with what occurs with the unperturbed dimensions that only in the cases of relatively long chains can be determined in a reliable way. On the other hand, because the skeletal bonds change more in polarity than they do in length, $\langle \mu^2 \rangle_0$ is in general more sensitive to the structure than is $\langle r^2 \rangle_0$.

The unperturbed dimensions and the dipole moments of molecular chains are usually expressed in terms of the characteristic ratio $\langle r^2 \rangle_0/nl^2$ and the dipole moment ratio $\langle \mu^2 \rangle_0/nm^2$, where nl^2 and nm^2 represent, respectively, the mean-square end-to-end distance and the mean-square dipole moment of a freely jointed chain; here n is the number of skeletal bonds of the chains, each having an average length l and an average dipole moment m. Therefore, the characteristic ratio and the dipole moment ratio are the factors by which the mean-square unperturbed dimensions and the mean-square dipole moments differ from those of a random-flight chain. For vinyl chains, the dipole moment ratio is better expressed as $\langle \mu^2 \rangle_0/x\mu_0^2$, where x is the degree of polymerization, and μ_0 is the dipole moment associated with each repeating unit.

RELATIONSHIP BETWEEN DIPOLE MOMENT AND STRUCTURE FOR SOME POLYMER CHAINS

α,ω-Dibromealkanes as Models of Polyethylene

Polyethylene is a relatively simple polymer of negligible polarity whose conformational characteristics have been obtained from the critical interpretation of both the unperturbed dimensions of high molecular-weight chains [19] and the dipole moments of α,ω-dibromealkanes [82] $Br(CH_2)_{n-1}Br$. In the latter case, significant polarity is introduced through the two terminal bonds. Experimental values [83] of $\langle \mu^2 \rangle^{1/2}$ for α,ω-dibromealkanes with n ranging from 4 through 11 are shown in

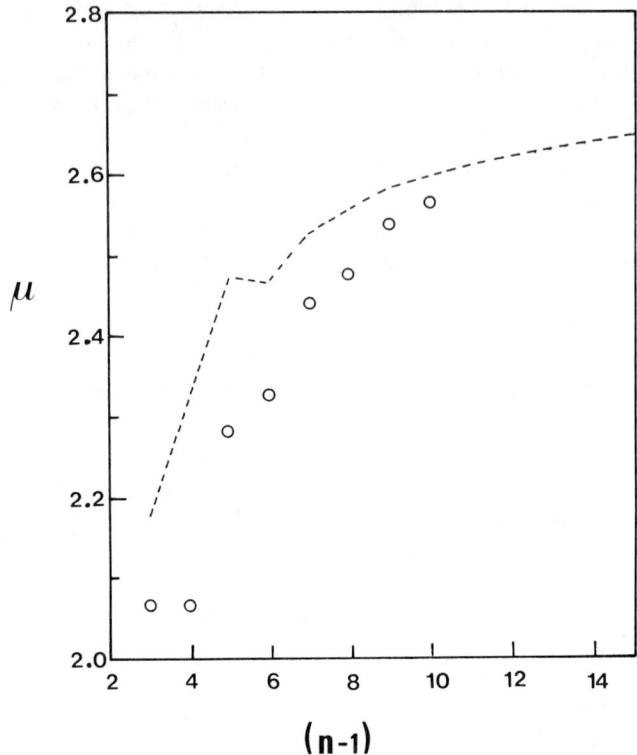

Figure 4.2 Experimental values (○) of the dipole moment for $Br\text{-}(CH_2)_{n-1}$ at 25°C in benzene. The dashed line represents the values of μ calculated by the rotational isomeric state (RIS) model. (Redrawn from reference [82] by courtesy of the American Institute of Physics.)

Figure 4.2. For the lower members of the series, significant correlations between the polar end groups exist so that the dipole moments of these compounds are strongly dependent on the relative conformational energies of the rotational states of the backbone.

Theoretical calculations of $\langle \mu^2 \rangle$ by using the rotational isomeric state model requires knowledge of the conformational energies associated with the rotational states about CH_2CH_2—CH_2CH_2 and CH_2CH_2—CH_2Br bonds. In the former case, conformational energy calculations using semiempirical potential functions suggest that the rotational states about CH_2—CH_2 bonds in n-alkanes are located at $\phi = 0$, $\pm 115°$. These states are called trans (t, $\phi = 0°$), gauche positive (g^+, $\phi = 115°$), and gauche negative (g^-, $\phi = -115°$). Moreover, gauche states have an energy E_σ approximately 0.5 kcal mol^{-1} higher than trans states, and pairs of gauche states $g^\pm g^\mp$ of opposite sign, which give rise to pentane-type interferences between two

CH_2 groups, are almost completely suppressed. Similar calculations about CH_2CH_2—CH_2Br bonds indicate that the minima are located at $\phi = 0, \pm120°$, and gauche states have an energy 0.6 kcal mol^{-1} above that of the corresponding trans states. Theoretical values of the mean-square dipole moment were calculated by assuming that the dipole moment associated with the CH_2—Br bond has a value of 1.90 D. As can be seen in Figure 4.2, values of $\langle\mu^2\rangle$ thus calculated are in satisfactory agreement with the experimental results, thus giving strong support to the basic features of the RIS model.

Polyoxides

Polyoxides with the repeating unit $-(CH_2)_x$—O— have a relatively simple structure and, consequently, are suitable to analyze the relationship between structure and properties in polymer chains. The dielectric properties of polyoxides with number of methylene groups in the repeating unit $x = 2, 3, 4, 6$, and 10 were measured in nonpolar solvents. The results [52, 84–89], expressed in terms of the dipole moment ratio $\langle\mu^2\rangle/nm^2$, where nm^2 represents the mean-square dipole moment of the chains in the idealization that all the skeletal bonds are freely oriented, are shown in Table 4.2. It was assumed that the dipole moment associated with C—O bonds m_{C-O} has a value of 1.07 D, whereas the corresponding to C—C bonds m_{C-C} is zero.

Information on the polarity of the first member of the series ($x = 1$) poly(methylene oxide) is scarce. Actually, PMO is a crystalline polymer that exhibits a high melting point ($\approx210°C$) and, therefore, it is insoluble in most solvents. The dipole moment ratio $\langle\mu^2\rangle/nm^2$ of PMO, obtained from dielectric results on molten PMO [84], was estimated to be about 0.3. Owing to the difficulties involved in the determination of $\langle\mu^2\rangle$ from dielectric measurements performed on the bulk, the dipole moment measurements on two oligomers ($CH_3O[CH_2$—O—$]_xCH_3$; $x = 1$ and 2) by Uchida, Kurita, and Kubo [89] have been considered as the most reliable source for the conformational energies. Three minima appear in the potential curves about C—O bonds, located at $0, \pm115°$, respectively. The statistical-weight matrices associated with the CH_2—O and O—CH_2 bonds of the repeating unit have the form [69]:

$$U(CH_2—O) = \begin{bmatrix} 1 & \sigma' & \sigma' \\ 1 & \sigma' & 0 \\ 1 & 0 & \sigma' \end{bmatrix} \qquad U(O—CH_2) = \begin{bmatrix} 1 & \sigma' & \sigma' \\ 1 & \sigma' & \sigma'\omega \\ 1 & \sigma'\omega & \sigma' \end{bmatrix} \qquad (4.17)$$

The columns and rows in the statistical-weight matrices refer, respectively, to the rotational states (t, g^+, or g^-) of bonds i and $i - 1$. Here, $\sigma' = \exp(-E_{\sigma'}/RT)$, where $E_{\sigma'}$ is the conformational energy associated with gauche states about both CH_2—O and O—CH_2 bonds, which give rise to first-order $CH_2 \cdots O$ interactions, taking the alternative trans states as reference, that is, $E_{\sigma'} = E_{g^\pm} - E_t$.

TABLE 4.2 Experimental Values of the Dipole Moment Ratio
$D_x = \langle \mu^2 \rangle / nm^2$ and Its Temperature Coefficient TC $= 10^3 d[\ln (D_x)]/dT$
of Polyoxides ($m_{C—O} = 1.07$ D, $m_{C—C} = 0$)

Polymer	D_x	TC	Reference
Polymethylene oxide			
$CH_3(CH_2O)_xCH_3$			
$x = 1$	0.1	6.0	[89]
$x = 2$	0.29		[89]
$x = \infty$	0.30		[84]
Poly(ethylene oxide)			
—$(CH_2)_2O$—	0.51	2.6	[69, 85–87]
Poly(trimethylene oxide)			
—$(CH_2)_3O$—	0.41	1.5	[88]
Poly(tetramethylene oxide)			
—$(CH_2)_4O$—	0.50, 0.53	2.7, 1.8	[69, 52]
Poly(hexamethylene oxide)			
—$(CH_2)_6O$—	0.54	1.2	[87]
Poly(decamethylene oxide)			
—$(CH_2)_{103}O$—	0.54	1.8	[52]
Isotactic poly(propylene oxide)			
—$CH(CH_3)CH_2O$—	0.50, 0.45	—	[91, 92]
Poly(3,3-dimethyloxetane)			
—$CH_2C(CH_3)_2CH_2O$—	0.20, 0.25	2.5, 4.5	[95, 96]
Poly(2-methyl oxetane)			
—$OCH(CH_3)CH_2CH_2$—	0.35	4.5	[93]
Poly(3-methyl oxetane)			
—$OCH_2CH(CH_3)CH_2$—	0.35	2.7	[94]
Poly(3-methyltetrahydrofuran)			
—$OCH_2CH(CH_3)(CH_2)_2$—	0.53	1.6	[87]

Semiempirical potential calculations carried out by using Buckingham potentials suggest that $E_{\sigma'} \approx -0.3$ kcal mol^{-1}. Second-order interactions about CH_2—O and O—CH_2 bonds cause, respectively, $CH_2 \cdots CH_2$ interactions of high energy ($E_\omega' \rightarrow \infty$) and $CH_2 \cdots$ O interactions whose energy $E_\omega = 1.5$ kcal mol^{-1}. Evaluation of $\langle \mu^2 \rangle / nm^2$ was performed as a function of $E_{\sigma'}$ assuming that the dipole associated with CH_2—O bonds lies along the bonds and its value is 1.07 D. The results obtained, shown in Figure 4.3, indicate that the dipole moment ratio significantly increases as $E_{\sigma'}$ increases because trans conformations become favored and the dipoles are essentially parallel in the all-trans conformation. For $E_{\sigma'} = -0.3$ kcal mol^{-1}, the theoretical values of $\langle \mu^2 \rangle / nm^2$ for the oligomers of PMO are much larger than the experimental results. An inspection of the curves [69] of Figure 4.3 suggests that only values of $E_{\sigma'}$ lying in the range -0.8 to -1.4 kcal mol^{-1} reproduce satisfactorily the experimental results for the oligomers of PMO with $x = 1$ and $x = 2$. This study suggests that $g^+g^+ \ldots g^+$ and $g^-g^- \ldots g^-$ sequences, occasionally separated by tt conformations, occur along PMO chains and, hence, result in the relatively high unperturbed dimensions and low polarity exhib-

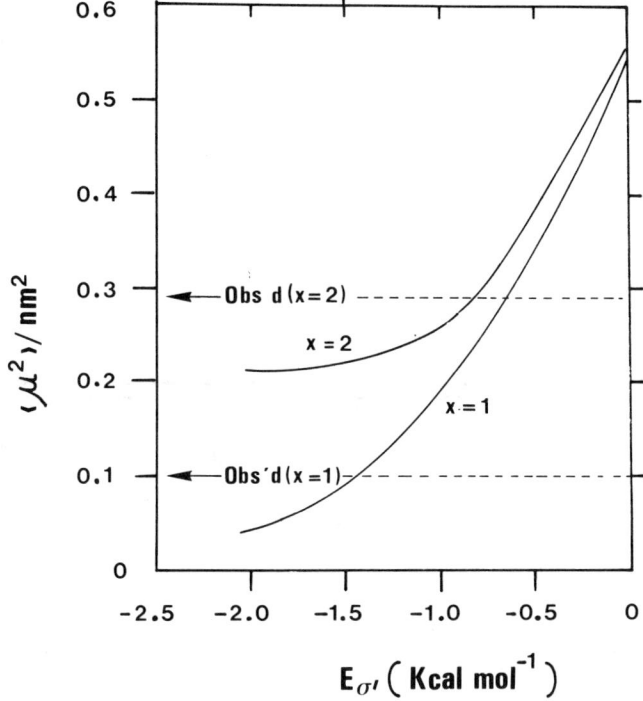

Figure 4.3 Values of the dipole moment ratio for dimers ($x = 1$) and trimers ($x = 2$) of $CH_3O(CH_2{-}O{-})_xCH_3$. The calculated results were obtained as a function of $E_{\sigma'}$ assuming $E_\omega = 1.5$ kcal mol^{-1}. (Redrawn from reference [69] by courtesy of the American Chemical Society.)

ited by this polymer (Figure 4.4). An increase in temperature should increase the fraction of *tt* conformations in which the dipole moments associated with the CH_2OCH_2 groups are parallel, according to Scheme 4.1, and hence the dipole moment of PMO should exhibit an unusually high temperature coefficient.

Another useful polymer in the study of the relationship between structure and properties is poly(ethylene oxide) (PEO), whose repeating unit is shown in Scheme

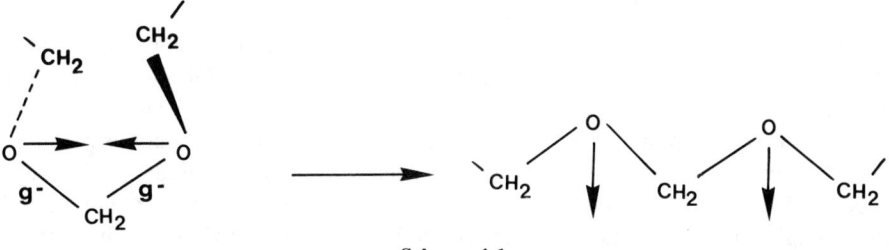

Scheme 4.1

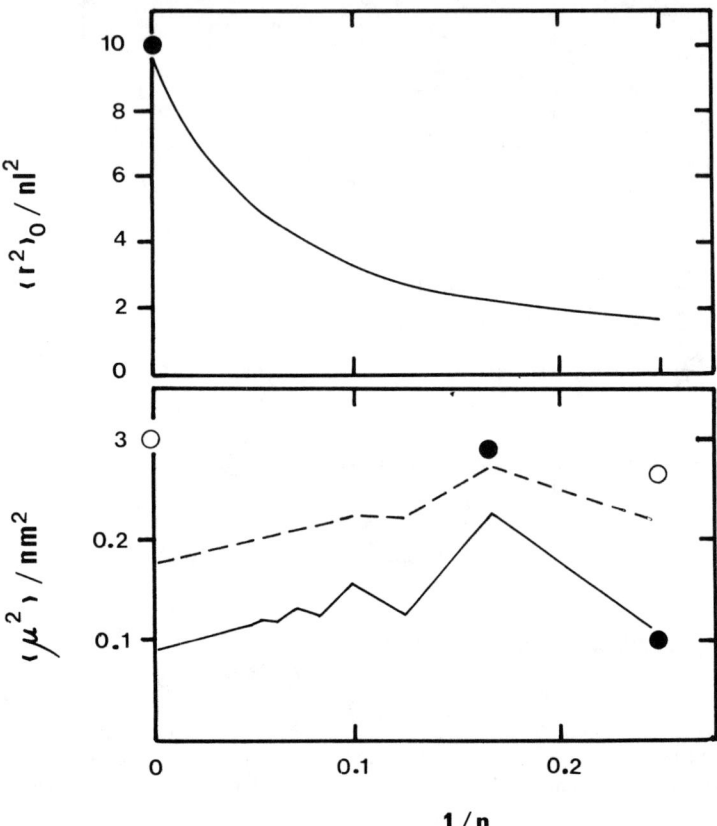

Figure 4.4 Statistical properties at 20°C of poly(methylene oxide) (PMO) chains shown as a function of the reciprocal of the number of skeletal bonds. The upper part represents the calculated (solid curve) and experimental values (filled circles) at 25°C of the characteristic ratio $\langle r^2 \rangle_0 / nl^2$. Calculated and experimental values of $\langle \mu^2 \rangle_0 / nm^2$ are shown in the lower portion. The results at 25°C are represented by the solid curve and filled circles and at 200°C by the dashed-line curve and open circles. (Redrawn from reference [69] by courtesy of the American Chemical Society.)

4.2. Gauche rotations about bonds of type a and c produce first-order interactions between two methylene groups. Semiempirical potential calculations [69] predict for gauche conformations about these bonds an energy $E_{\sigma''}$ of 1.25 kcal mol^{-1} with respect to the corresponding trans states. Gauche states about C—C bonds give rise to first-order O · · · O interactions, and the energy $E_{\sigma'}$ of these bonds is calculated to be 0.55 kcal mol^{-1} above that of the alternative trans states, as consequence of the strong coulombic repulsions that occur between O · · · O atoms in gauche conformation. The calculations also suggest that rotations of different sign about ca

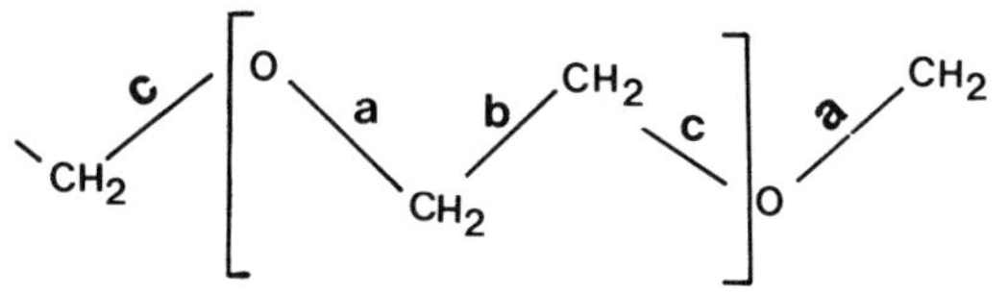

Scheme 4.2

bonds give rise to strong repulsions between two methylene groups and, consequently, $g^{\pm}g^{\pm}$ conformations are forbidden. These conformations are, however, permitted about ab and bc bonds, where second-order interactions $CH_2 \cdots O$ occur with an energy E_{ω} of 0.4 kcal mol^{-1}. Moreover, the rotational states are located at 0, $\pm 120°$ for C—C bonds and 0, $\pm 100°$ for C—O bonds.

The dipole moment ratio calculated with the energies given before is only 50% of the observed value [69]. As can be seen in Figure 4.5, the values of the characteristic ratio $\langle r^2 \rangle / nl^2$ close to the experimental results can only be reached for values of $E_{\sigma'}$ below -0.2 kcal mol^{-1}. The dipole moment slightly increases as $E_{\sigma'}$ decreases, but even for values of $E_{\sigma'} < -0.5$ kcal mol^{-1}, $\langle \mu^2 \rangle / nm^2$ only amounts to 60% of the experimental result. The dipole moment ratio of PEO chains is very sensitive to $E_{\sigma''}$ and the location of gauche states about C—O bonds. Thus, by assuming $E_{\sigma''} = 0.9$ kcal mol^{-1} and $\phi_g = \pm 110°$ for C—O and C—C bonds, the experimental values of the dipole moment ratio, the characteristic ratio, and their corresponding temperature coefficients are reproduced for $E_{\sigma'} = -0.4$ to -0.5 kcal mol^{-1}. These conformational-energy parameters also give a good account of the dipole moment and its temperature coefficient for oligomers of PEO of the general formula [90] $CH_3O(CH_2CH_2O)_xCH_3$. Thus, the theoretical values of both $\langle \mu^2 \rangle / nm^2$ and $10^3 \, d[\ln(\langle \mu^2 \rangle)]/dT$ for the oligomer with a single repeating unit ($x = 1$) at 25°C are 0.65 and 0.3 K^{-1}, respectively, in good agreement with the experimental values 0.64 and 0.6. For the oligomer with four repeating units ($x = 4$), the experimental data for the dipole moment ratio and its temperature coefficient are 0.58 and 2.0 K^{-1}, respectively, which are also close to the theoretical values 0.52 and 1.3 K^{-1}.

Poly(propylene oxide) (PPO) is an asymmetric polyoxide that can be schematically derived from PEO by substituting a hydrogen atom in the repeating unit of this polymer for a methyl group. This polymer, as in general for all polymers with asymmetric skeleton carbon atoms in their structure, can be considered a stereochemical copolymer of meso and racemic diads. As a rule, a diad is called meso when rotations of the same sign ($\pm 120°$) about the two CH_2—CHR skeletal bonds located immediately before the asymmetric carbon atoms of the diad place the side pendant group attached to these carbons in the plane that contains the skeletal bonds of the diad in all trans conformations. It is considered racemic if this is achieved by rotations of different sign about the same CH_2—CHR bonds. Accordingly, meso diads in which the asymmetric carbon atoms are separated by an even number of skeletal bonds have the side groups located at the same side of the plane, whereas

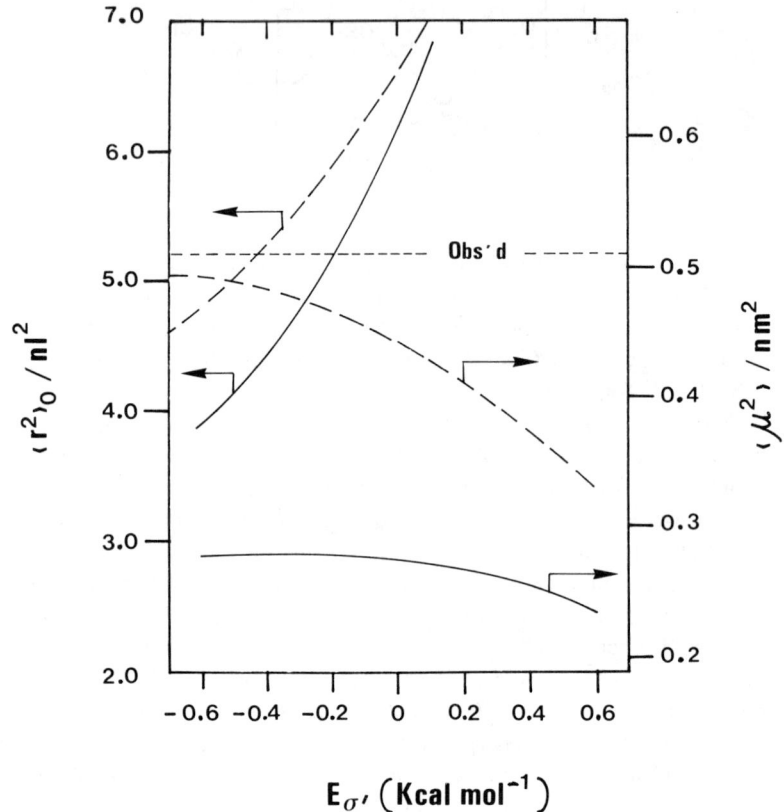

Figure 4.5 The characteristic ratio and the dipole moment ratio for PEO calculated at 30 and 20°C, respectively, as a function of the energy $E_{\sigma'}$ associated with gauche states about C—C bonds. The two solid curves were obtained using $E_{\sigma''} = 1.3$ kcal mol^{-1}, $E_{\omega} = 0.4$ kcal mol^{-1}, ϕ_g(C—O) $= \pm 100°$, and ϕ_g(C—C) $= \pm 120°$. The dashed-line curves were calculated using $E_{\sigma''} = 0.9$ kcal mol^{-1}, $E_{\omega} = 0.4$ kcal mol^{-1}, and ϕ_g(C—O) $= \phi_g$(C—C) $= 100°$. The horizontal line represents the experimental values for both ordinates. (Redrawn from reference [69] by courtesy of the American Chemical Society.)

racemic diads have the two pendant groups located above and below the plane. On the contrary, the side groups are located above and below the plane for meso diads in which the asymmetric carbon atoms are separated by an odd number of skeletal bonds and are at the same side of the plane for racemic diads. Polymer chains in which the molar fractions of meso diads are 1 and 0 (the fractions of racemic diads are 0 and 1) are called isotactic and syndiotactic chains, respectively.

An isotactic diad in all trans conformations for PPO is shown in Figure 4.6. Inductive effects produced by the methyl groups prevent the dipole associated with the (CH)$_3$CH—O—CH$_2$ group from bisecting the bond angle and, therefore, it has a

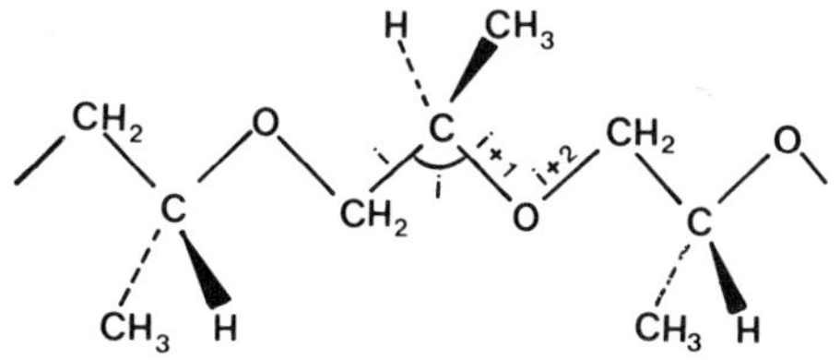

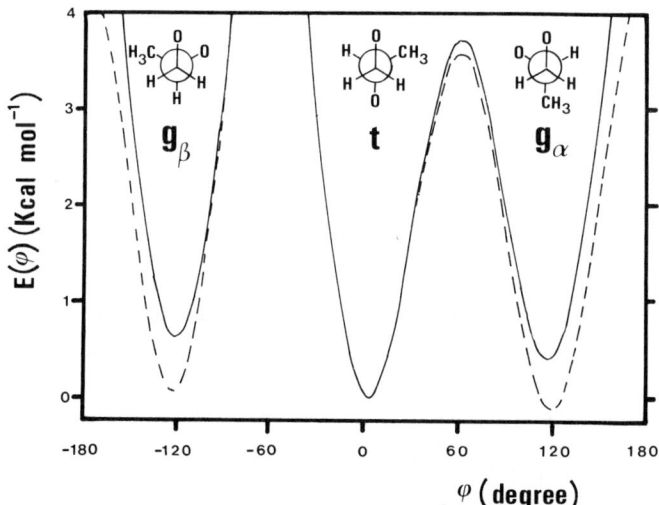

Figure 4.6 The isotactic poly(propylene oxide) chain in the all trans conformation.

component parallel to the chain contour. As consequence, the dipole moment of PPO chains should present excluded volume effects. Dielectric measurements carried out on low molecular-weight PPO gave for $\langle \mu^2 \rangle / nm^2$ the values of 0.54 in benzene and 0.49 in cyclohexane [91, 92]. Nonbonded interactions were computed by using the Buckingham 6-exp. energy function, and the resulting potential curves about bonds of type i in the repeating unit of the isotactic diad, represented in Figure 4.7, suggest that the rotational states about the bonds of the repeating unit are located at $(0, \pm 120)$. Similar calculations for bonds of type $i + 1$ and $i + 2$ predict that the rotational minima correspond to $(-20, 120, -100)$ and $(0, \pm 100)$, respectively.

A close inspection of Figures 4.6 and 4.7 suggests that g^+ states about bonds of type i have a lower energy than the alternative g^- states, though both states are disfavored with respect to the trans states. The conformational energies E_α and E_β of g^+ and g^- states, respectively, amount to 0.4 and 1.25 kcal mol^{-1}, respectively,

Figure 4.7 Conformational energies of 1,2-dimethoxypropane as a function of the rotational angle about the C—C bond, the adjoining O—C and C—O bonds being kept trans. (Redrawn from reference [92] by courtesy of the American Chemical Society.)

with respect to the alternative trans states. The trans state in bond i is not compatible with the g^+ rotation about bond $i - 1$ because the $i - 3$ CH and the i CH_3 groups' bonds are placed at a distance at which strong repulsive interactions take place. Similar strong repulsive interactions arise from $g^- g^-$ states about the $i - 1$ and i bonds. Accordingly, the statistical-weight matrix corresponding to i skeletal bonds can be written as [92]:

$$\mathbf{U}_i = \begin{bmatrix} 1 & \alpha & \beta \\ 0 & \alpha & \beta\omega \\ 1 & \alpha\omega & 0 \end{bmatrix} \tag{4.18}$$

Positive rotation about bonds of type $i + 1$ places the $i + 2$ methylene group and the i methyl group nearly in cis conformation, giving rise to a very unstable conformation due to the strong interactions between the two groups, whereas g^- rotations about these bonds give a conformation whose energy is similar to that of the alternative trans conformation. As for the $i + 2$ skeletal bonds, g^+ states have an energy 1.2 kcal mol^{-1} above that of the corresponding trans states, whereas g^- states cause strong interactions between the $i + 3$ C atom and the i CH_3 group and consequently are forbidden. Similar interactions occur between the $i + 3$ C atom and either the i CH_3 or the $i - 1$ CH_2 by $g^+ g^+$, and $g^{\pm} g^{\pm}$ rotations about $i + 1$ and $i + 2$ bonds, and the resulting conformations are also forbidden. The statistical-weight matrices associated with the $i + 1$ and $i + 2$ bonds are

$$\mathbf{U}_{i+1} = \begin{bmatrix} 1 & 0 & 1 \\ 1 & 0 & \omega \\ 1 & 0 & 1 \end{bmatrix} \qquad \mathbf{U}_{i+2} = \begin{bmatrix} 1 & \sigma & 0 \\ 1 & 0 & 0 \\ 1 & 0 & \sigma \end{bmatrix} \tag{4.19}$$

The corresponding statistics for mirror-image structures can be obtained by pre- and postmultiplying each of matrices $\mathbf{U}_i$, $\mathbf{U}_{i+1}$, and $\mathbf{U}_{i+2}$ by the elementary matrix $\mathbf{Q}$ defined as

$$\mathbf{Q} = \begin{bmatrix} 1 & 0 & 0 \\ 0 & 0 & 1 \\ 0 & 1 & 0 \end{bmatrix} \tag{4.20}$$

Theoretical calculations [92] of the dipole moment ratio and the unperturbed dimensions indicate that the experimental results can only be reproduced for values of $E_\alpha \approx -0.3$ kcal mol^{-1}. Here, as in PEO, semiempirical potential calculations overestimate the energies associated with gauche states about C—C bonds, which give rise to first-order O $\cdots$ O interactions.

The next member of the series in the family of polyoxides is poly(trimethylene oxide) (P3MO). This chain permits obtaining information on the conformational energy associated with gauche states about bonds, which gives rise to first-order $CH_2 \cdots$ O interactions. See Scheme 4.3.

Conformational-energy calculations suggest that the energy $E_{\sigma'}$ of gauche

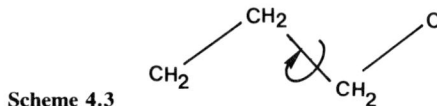

Scheme 4.3

states about CH_2CH_2—CH_2O bonds is about 0.1 kcal mol^{-1} lower than that corresponding to the trans states. Experimental values of $\langle\mu^2\rangle/nm^2$ and its temperature coefficient are reproduced by using this energy and assuming that the energy of gauche states about C—O bonds is 0.9 kcal mol^{-1} above that of the alternative trans states. However, a value of $E_{\sigma'} = -0.7$ kcal mol^{-1} is necessary to reproduce the observed value of the characteristic ratio (see Figure 4.8). An optimum agree-

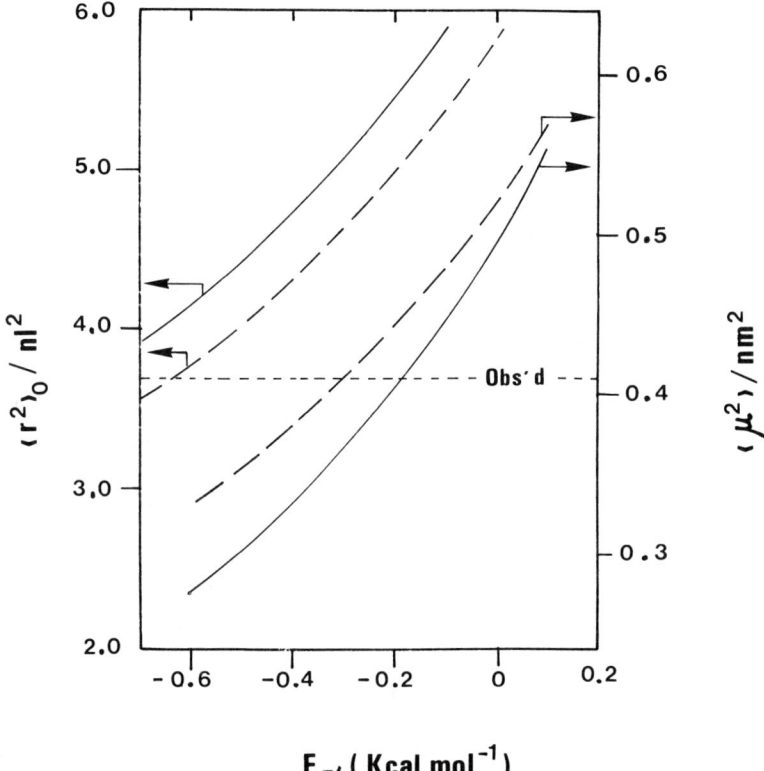

$$E_{\sigma'}\,(\text{Kcal mol}^{-1})$$

Figure 4.8 Values of $\langle r^2\rangle_0/nl^2$ and $\langle\mu^2\rangle_0/nm^2$ at 20°C for poly(trimethylene oxide) calculated as a function of the conformational energy $E_{\sigma'}$ about C—C bonds. The continuous lines represent the values calculated using $E_{\sigma''} = 1.2$ kcal mol^{-1}, $E_\omega = 0.9$ kcal mol^{-1}, $\phi_g(\text{C—O}) = \pm100°$, and $\phi_g(\text{C—C}) = \pm120°$. Dashed lines were obtained using $E_{\sigma''} = 0.9$ kcal mol^{-1}, $E_\omega = 0.6$ kcal mol^{-1}, $\phi_g(\text{C—O}) = \pm110°$, and $\phi_g(\text{C—C}) = \pm120°$. Observed values are specified by the horizontal line. (Redrawn from reference [69] by courtesy of the American Chemical Society.)

ment between theory and experiment can be reached [69, 88] by assuming $E_\sigma \approx$ -0.4 kcal mol^{-1} and by adjusting the second-order interaction parameter to $E_\omega \approx$ 0.6 kcal mol^{-1} for the conformation in which two oxygen atoms are brought into close contact as a consequence of rotations of different sign about OCH$_2$—CH$_2$—CH$_2$O bonds.

By schematically substituting one or more hydrogen atoms of the repeating unit of oxetane and poly(trimethylene oxide) for methyl groups, polymers are obtained whose dielectric conformational properties are dependent upon both the number and location of the methyl groups in the structural unit. Values of the dipole moment ratio and its temperature coefficient for poly(2-methyloxetane) (P2DO) [93], poly(3-methyloxetane) (P3DO) [94], and poly(3,3-dimethyloxetane) (PDMO) [95,96] are shown in Table 4.2, where it can be seen that the polarity of asymmetric substituted oxetanes is larger than that of symmetric ones.

Semiempirical potential calculations about C—O bonds in PDMO [95] show that the energy steeply increases as the rotational angle ϕ departs from zero, as consequence of the decreasing distance between the methylene $i + 1$ group in Figure 4.9 and the methyl group attached to the $i - 2$ C. Because of these strong repulsive interactions (see Figure 4.10), C—O bonds are always restricted to trans ($\phi = 0$) states.

As for the rotational states about C—C bonds, there is no major difference as far as the van der Waals interactions are concerned. Thus, the trans conformation places O atom i between the methyl groups attached to C atom number $i - 2$, whereas the gauche conformation places that oxygen between a methyl and a methylene ($i - 3$) group. However, the electrostatic attractions between oxygen and carbon atoms (numbers i and $i - 3$) are higher in the gauche conformation than in the alternative trans state. Figure 4.10 shows that as a result of these effects, gauche states have an energy $E_{\sigma'}$ of about 0.4 kcal mol^{-1} below that of the corresponding trans states. The statistical-weight matrices for the bonds included into a repeating unit can be written as [95, 96]

$$\mathbf{U}_{i+1} = [1] \qquad\qquad \mathbf{U}_{i+2} = [1 \quad \sigma' \quad \sigma'] \qquad (4.21)$$

$$\mathbf{U}_{i+3} = \begin{bmatrix} 1 & \sigma' & \sigma' \\ 1 & \sigma' & \sigma'\omega \\ 1 & \sigma'\omega & \sigma' \end{bmatrix} \qquad \mathbf{U}_{i+4} = \begin{bmatrix} 1 \\ 1 \\ 1 \end{bmatrix} \qquad (4.22)$$

Theoretical values of $\langle\mu^2\rangle/nm^2$ were obtained as a function of $E_{\sigma'}$ for different values of the second-order energy E_ω arising from rotations of different sign about two consecutive pair of bonds of type C—C. The results obtained, plotted in Figure 4.11, indicate that the experimental dipole moment ratio is reproduced for values of $E_{\sigma'} \approx -0.5$ to -0.7 kcal mol^{-1} and $E_\omega = 1.2$ kcal mol^{-1}. The critical analysis of other conformation-dependent properties such as the temperature coefficients of both the dipole moment $d(\ln \langle\mu^2\rangle)/dT$ and the unperturbed dimensions shows [96] that the best set of conformational energies that give a good account of the confor-

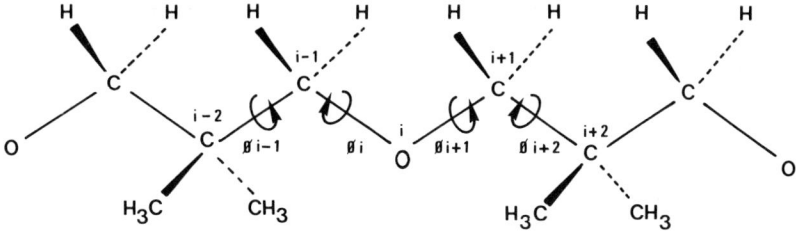

Figure 4.9 A segment of the poly(3,3 dimethyl oxetane) chain in its planar all-trans conformation.

mation-dependent properties of polymers are $E_{\sigma'} = -0.7$ and $E_\omega = 1.2$ kcal mol^{-1}.

The low polarity of PDMO in comparison with that of poly(trimethylene oxide) (P3MO) is explained on the one hand because gauche states over C—C bonds in the former polymer have lower energy, relative to trans, than similar bonds in the latter, and on the other hand because a noticeable fraction of rotational states

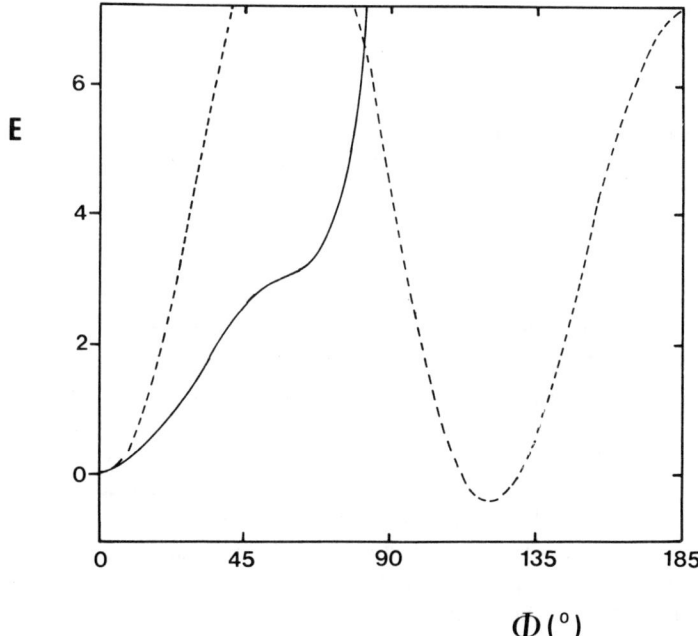

$\Phi(°)$

Figure 4.10 Calculated conformational energies (in kcal mol^{-1}) for poly(3,3-dimethyl oxetane) as functions of rotational angles ϕ about C—O (solid line) and C—C (dashed line) bonds. (Redrawn from reference [95] by courtesy of the American Institute of Physics.)

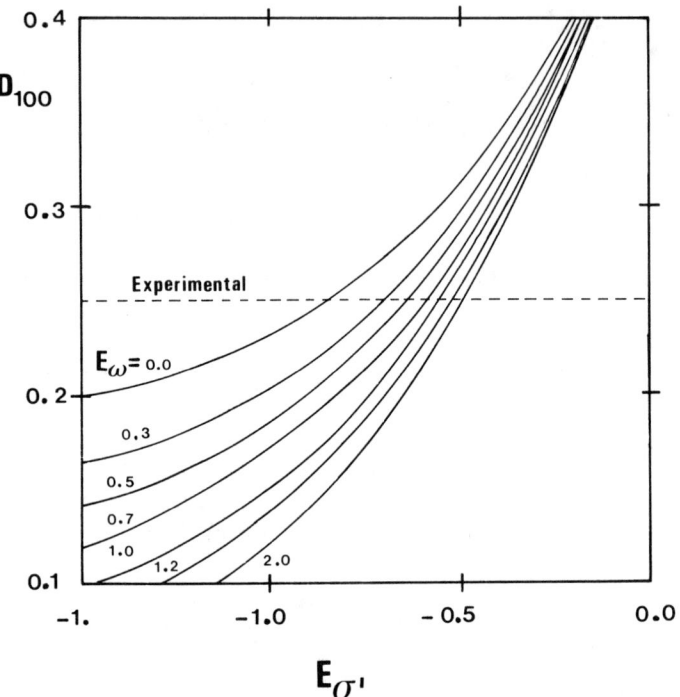

Figure 4.11   Calculated values at 20°C of the dipole moment ratio of poly(3,3-dimethyl oxetane) as a function of the conformational energy about C—C bonds. Energies are in kcal mol^{-1}. (Redrawn from reference [95] by courtesy of the American Institute of Physics.)

about C—O bonds are in gauche states in P3MO. As a consequence, the occurrence of $ttg^{\pm}g^{\pm}tt$ conformations about the sequence CH_2—O—CH_2—CR_2—CH_2—O—CH_2 (where R = H and CH_3 for P3MO and PDMO, respectively), which place pairs of bond dipoles in the nearly antiparallel direction, are larger in the latter polymer. As can be seen in Figure 4.12, an increase in temperature causes an increase of the fraction of $ttttt$ conformations in PDMO, which place the pair dipoles in a parallel direction and, hence, results in the positive temperature coefficient exhibited by the dipole moments of these chains. The positive change in polarity with temperature is not so large in P3MO because an increase in temperature also increases the gauche population about C—O bonds, which opposes the reduction of polarity of the chains.

The substitution of a single hydrogen atom for a methyl group in the central methylene group of the repeating unit of P3MO causes a significant change in the statistics of the chains. As a consequence of the presence of an asymmetric carbon in P3DO, the conformation-dependent properties of this polymer are dependent on

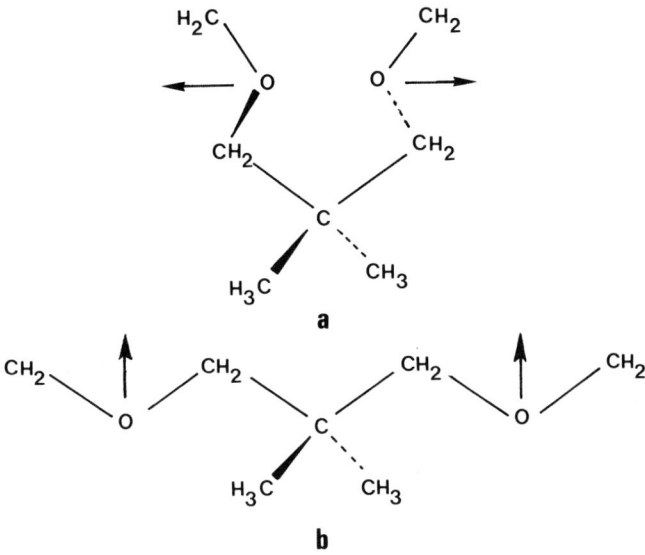

Figure 4.12 A segment of poly(3,3-dimethyl oxetane) in (a) its tg^+g^+t and (b) $tttt$ conformations. The dipole moments of the CH_2OCH_2 bonds are marked by arrows.

the stereoregularity of the chains. Taking as reference the diad shown in Figure 4.11, g^+ states about C—O bonds have an energy of about 1 kcal mol^{-1} above that of the alternative trans states, whereas the g^- states are forbidden. The energy interaction arising from g^+ rotations about $CH_2CH(CH_3)$—CH_2O bonds would be similar to that of the alternative trans states were it not for the coulombic interactions between the positively charged $i + 2$ carbon atoms ($\delta = 0.155$ in electron units) and the negatively charged oxygen atom ($\delta = -0.31$ in the same units), which amounts to -1.83 kcal mol^{-1}, a value significantly lower than that corresponding to trans states, which is only -1.41 kcal mol^{-1}. Therefore, an energy $E_\beta = -0.42$ kcal mol^{-1} was used for these states. Theoretical values of the dipole moment ratio were then calculated as a function of the energy E_α associated with gauche states in which the oxygen atom is placed between a methylene and a methyl group. The statistical-weight matrices corresponding to the skeletal bonds of the repeating unit of the isotactic diad indicated in Figure 4.13 are [94]

$$\mathbf{U}_i = \begin{bmatrix} 1 & \sigma'' & 0 \\ 0 & 0 & 0 \\ 1 & 0 & 0 \end{bmatrix} \qquad \mathbf{U}_{i+1} = \begin{bmatrix} 1 & \alpha & \beta \\ 1 & 0 & 0 \\ 0 & 0 & \beta \end{bmatrix} \tag{4.23}$$

$$\mathbf{U}_{i+2} = \begin{bmatrix} 1 & \beta & \alpha \\ 1 & \beta & \alpha\omega \\ 1 & \beta\omega & \alpha \end{bmatrix} \qquad \mathbf{U}_{i+3} = \begin{bmatrix} 1 & 0 & \sigma'' \\ 0 & \sigma'' & 0 \\ 0 & 0 & 0 \end{bmatrix} \tag{4.24}$$

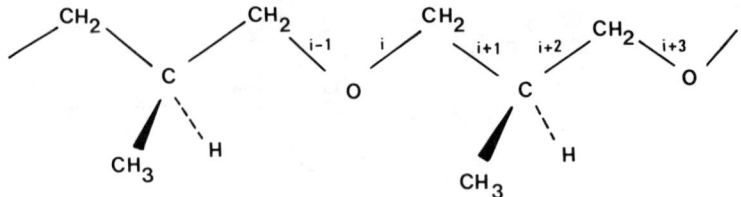

Figure 4.13   Isotactic diad of poly(3-methyl oxetane) in all-trans conformation.

where the statistical-weight zero applied to g^- and g^+ states in $\mathbf{U}_i$ and $\mathbf{U}_{i+3}$, respectively, reflect severe interactions between methyl and methylene groups separated by four bonds in these states. Values of 1.2 and 1.3 kcal mol^{-1}, respectively, were used for $E_{\sigma''}$ and E_ω. The corresponding matrices for mirror-image or enantiomorphic structure can be obtained by pre- and postmultiplying each of the matrices $\mathbf{U}_i$, $\mathbf{U}_{i+1}$, $\mathbf{U}_{i+2}$, and $\mathbf{U}_{i+3}$ by the elementary matrix $\mathbf{Q}$ defined before.

The results obtained for an atactic chain, shown in Figure 4.14, suggest that

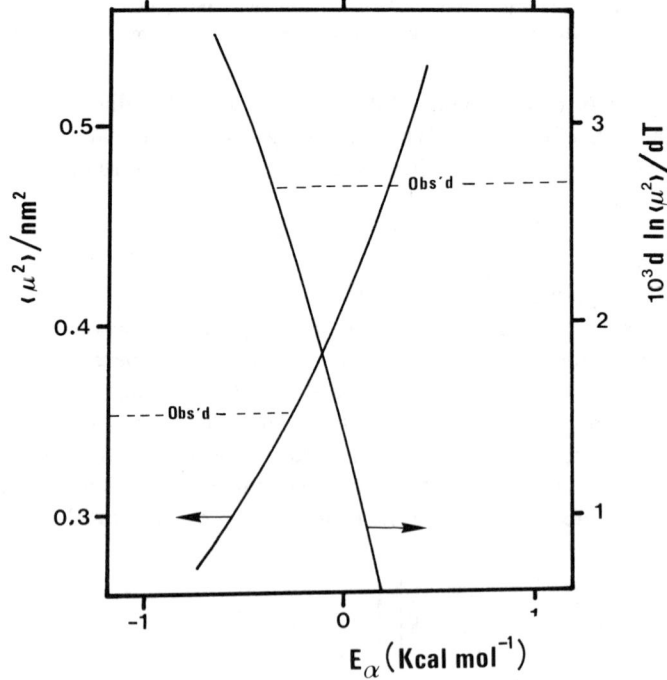

Figure 4.14   Dependence of the dipole moment ratio and its temperature coefficient on E_α for poly(3-methyl oxetane). See the text for details. (Redrawn from reference [94] by courtesy of the American Chemical Society.)

the experimental values of both $\langle \mu^2 \rangle / nm^2$ and $d[\ln (\langle \mu^2 \rangle)]/dT$ are satisfactorily reproduced for $E_\alpha \approx -0.3\,\text{kcal mol}^{-1}$. As occurs with other polymers in which the assymmetric carbon atoms are separated for more than three bonds, the conformation-dependent properties of the polymer do not show a significant dependence on the stereoregularity of the chains. The decrease in the fraction of $g^{\pm}g^{\pm}$ conformations about the central C—C bonds of P3DO as a consequence of the presence of a methyl group contributes to the higher polarity of the former polymer with respect to that of PDMO.

Because the dipole moment ratios of both P3DO and P2DO have the same value (see Table 4.2) suggests that the substitution of a single hydrogen atom for a CH_3 group in the first or second methylene group of P3MO does not alter the dielectric conformation-dependent properties of the resulting polymers. The theoretical analysis of the dipole moment of P2DO indicates [93] that gauche states about the C—C bonds are slightly disfavored with respect to similar states about the C—C bonds in P3DO.

For polyoxides in which $x > 3$, conformational calculations suggest that gauche states about both OCH_2—CH_2CH_2 and $CHCH_2$—CH_2CH_2 bonds, respectively, have energies of $E_{\sigma'}$ and E_σ whose values are 0 and 0.5 kcal mol^{-1}. However, by comparing the theoretical and experimental values of conformation-dependent properties such as unperturbed dimensions, dipole moments, and their respective temperature coefficients (Figure 4.15), one can see that the best set of conformational energies are [52, 69, 87] $E_{\sigma'} = -0.2$, $E_{\sigma''} = 0.9$, $E_\sigma = 0.5$, and $E_\omega = 0.56$ kcal mol^{-1}. This last quantity accounts for the conformational energy arising from rotations of different sign about two consecutive CH_2—CH_2 bonds that cause second-order $CH_2 \cdots O$ interactions.

The conformational energies involving oxygen atoms, calculated from semiempirical potential functions, are not always in agreement with the values obtained from the critical analysis of the dimensions and dipole moments. As can be seen in Table 4.3, the semiempirical calculations overestimate the energies of each of the anomalous interactions, the overestimation being higher the lower the number of methylene groups in the repeating unit of polyoxides.

TABLE 4.3 Calculated and Experimental
Conformational Energies in kcal mol^{-1}
Associated with Gauche States that Involve
Oxygen Atoms in Some Polyoxides [69]

Polymer	Bond	Calculated	Experimental
PMO	O—C—O—C	−0.3	−1.4, −1.2
PEO	O—C—C—O	0.6	−0.4
PMO3	O—C—C—C	−0.1	−0.4
PMO4	O—C—C—C	~0	−0.2

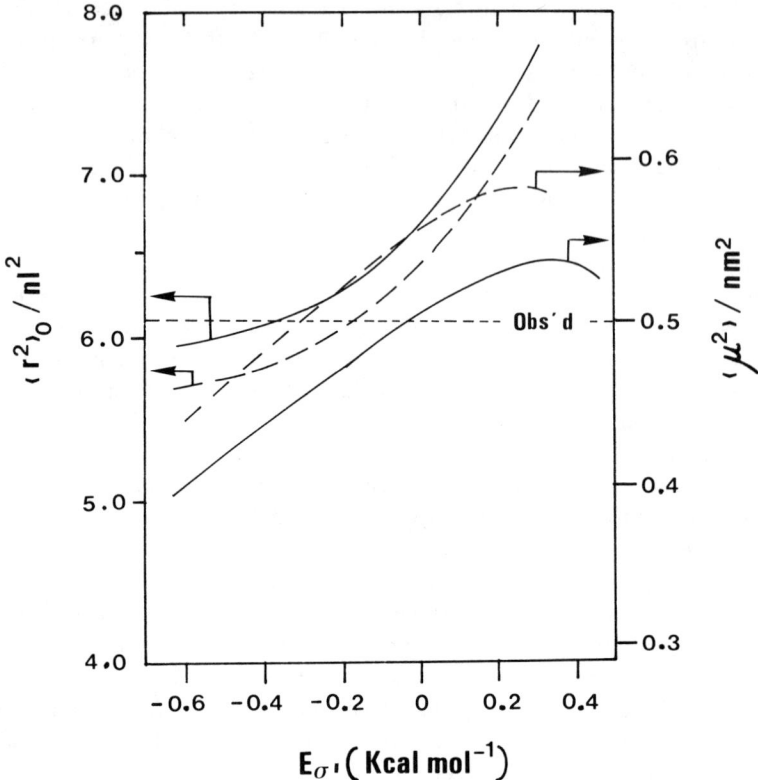

Figure 4.15 Values at 20°C of both the characteristic and the dipole moment ratios as a function of the conformational energy associated with gauche states about CC—CO bonds for poly(tetramethylene oxide). The two solid curves were calculated using $E_{\sigma''} = 1.2$, $E_\sigma(CC—CC) = 0.5$, $E_\omega = 0.6$, $\phi_g(CC—CO) = \pm120°$, $\phi_g(CC—CC) = \pm112.5°$, and $\phi_g(C—O) = \pm100°$. The two dashed-line curves were calculated using $E_{\sigma''} = 0.9$, $E_\sigma(CC—CC) = 0.5$, $E_\omega = 0.6$, $\phi_g(CC—CO) = \pm120°$, $\phi_g(CC—CC) = \pm112.5°$, and $\phi_g(C—O) = \pm110°$. The horizontal line represents the experimental values. (Redrawn from reference [69] by courtesy of the American Chemical Society.)

Alternating Copolymers: Polyformals

Linear polyformals are polyoxides that have the structural unit $O—CH_2—O—(CH_2)_y—$. Therefore, these polymers can be considered alternating copolymers of methylene oxide, $O—CH_2—$, and alkylene oxide, $O—(CH_2)_y—$. The physical properties of the members of the series are strongly dependent on the value of y. Thus, the first member of the series ($y = 1$), poly(methylene oxide), exhibits, as was indicated before, a high melting point and, therefore, it is insoluble in most common organic solvents. The second member of the series, poly(1,3-dioxolane)

(PXL; $y = 2$), has a relatively low melting point ($\approx 55°C$) and it is readily soluble in a number of organic solvents; low melting points and solubility in nonpolar solvents are also exhibited by the polyformals in which $y > 2$. Experimental values of the dipole moment ratio and its temperature coefficient for PXL [97], poly(1,3-dioxepane) (PXP; $y = 4$) [98], poly(1,3-dioxocane) (PXC; $y = 5$) [99], poly(1,3-dioxonane) (PXN; $y = 6$) [99], and an alicyclic polyformal [100] are given in Tables 4.4 and 4.5, respectively. A general feature of these polymers is that in spite of the high concentration of polar groups along the chains, they have relatively low polarity and a high temperature coefficient.

The repeating unit of poly(1,3-dioxolane), shown in Figure 4.16, suggests that this polymer can be considered an alternating copolymer of methylene oxide and ethylene oxide. The acetalic a and e bonds are similar to the bonds of PMO and, as a consequence, the critical interpretation of the dipole moments of PXL is useful to determine the conformational energy E_{σ_η} associated with gauche states about C—O bonds in the former polymer. The statistical-weight matrices corresponding to the skeletal bonds of the repeating unit of PXL are [97]

$$
\mathbf{U}_a = \begin{bmatrix} 1 & \sigma_\eta & \sigma_\eta \\ 1 & \sigma_\eta & 0 \\ 1 & 0 & \sigma_\eta \end{bmatrix} \qquad \mathbf{U}_b = \begin{bmatrix} 1 & \sigma'' & \sigma'' \\ 1 & \sigma'' & \sigma''\omega \\ 1 & \sigma''\omega & \sigma'' \end{bmatrix}
$$

$$
\mathbf{U}_c = \begin{bmatrix} 1 & \sigma' & \sigma' \\ 1 & \sigma' & \sigma'\omega \\ 1 & \sigma'\omega & \sigma' \end{bmatrix} \tag{4.25}
$$

$$
\mathbf{U}_d = \begin{bmatrix} 1 & \sigma'' & \sigma'' \\ 1 & \sigma'' & \sigma''\omega \\ 1 & \sigma''\omega & \sigma'' \end{bmatrix} \qquad \mathbf{U}_e = \begin{bmatrix} 1 & \sigma_\eta & \sigma_\eta \\ 1 & \sigma_\eta & \omega'\sigma_\eta \\ 1 & \omega'\sigma_\eta & \sigma_\eta \end{bmatrix}
$$

Values of the dipole moment ratio of PXL were calculated as a function if E_{σ_η} by using the values of -0.5, 0.9, and 0.56 kcal mol^{-1} for the conformational energies $E_{\sigma'}$, $E_{\sigma''}$, and E_ω, respectively, obtained in the conformational analysis of PEO. The results, shown in Figure 4.17, indicate that $\langle\mu^2\rangle/nm^2$ decreases markedly with an increase in the preference for gauche states about the acetalic bonds. This is because the $g^\pm g^\pm$ states in the bond pairs ea place pairs of bond dipoles in nearly antiparallel orientations (Scheme 4.1). As can be seen in Tables 4.4 and 4.5, coincidence between theoretical and experimental values of both the dipole moment ratio and its temperature coefficient is achieved for values of E_{σ_η} of approximately -1.2 kcal mol^{-1}, in very good agreement with the value of -1.4 kcal mol^{-1} obtained for this energy in the analysis of olygomers of PMO. As usual, the dipole moment ratio decreases as the chain length increases (see Figure 4.18).

A close inspection of the experimental values of $\langle\mu^2\rangle/nm^2$ (calculated with $E_{\sigma_\eta} = -1.2$ kcal mol^{-1}) for the members of the series of the polyformals indicates that the gauche population about the acetalic bonds conditions the polarity of these

TABLE 4.4 Experimental and Theoretical Values of the Dipole Moment Ratio, $D_x = \langle \mu^2 \rangle / nm^2$, at 30°C of Polyformals

Polymer	Structural Unit	D_x		
		Experimental	Theoretical	Reference
Poly(methylene oxide)	—CH$_2$O—	0.3		[84]
Poly(1.3-dioxolane)	—CH$_2$O(CH$_2$)$_2$O—	0.17	0.17	[97]
Poly(1.3-dioxepane)	—CH$_2$O(CH$_2$)$_4$O—	0.16	0.16	[98]
Poly(1.3-dioxocane)	—CH$_2$O(CH$_2$)$_5$O—	0.17	0.18	[99]
Poly(1.3-dioxonane)	—CH$_2$O(CH$_2$)$_6$O—	0.18	0.18	[99]
Poly(trans-1,4-cyclohexane dimethanol-alt-formaldehyde)	—CH$_2$OC$_6$H$_{10}$CH$_2$O	0.16	0.16	[100]

TABLE 4.5 Experimental and Theoretical Values of the Temperature Coefficient of the Dipole Moment Ratio of Polyformals (TC $= 10^3$ d ln $\langle \mu^2 \rangle / dT$)

Polymer	Structural Unit	TC		
		Experimental	Theoretical	Reference
Poly(1.3-dioxolane)	—CH$_2$O(CH$_2$)$_2$O—	6.0	6.1	[97]
Poly(1.3-dioxepane)	—CH$_2$O(CH$_2$)$_4$O—	5.4	5.8	[98]
Poly(1.3-dioxocane)	—CH$_2$O(CH$_2$)$_5$O—	5.4	4.9	[99]
Poly(1.3-dioxonane)	—CH$_2$O(CH$_2$)$_6$O—	5.5	4.9	[99]
Poly(trans-1,4-cyclohexane dimethanol-alt-formaldehyde)	—CH$_2$OC$_6$H$_{10}$CH$_2$O	5.0	5.5	[100]

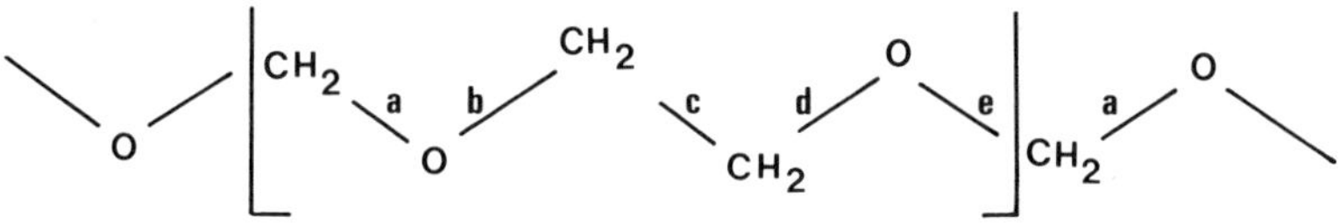

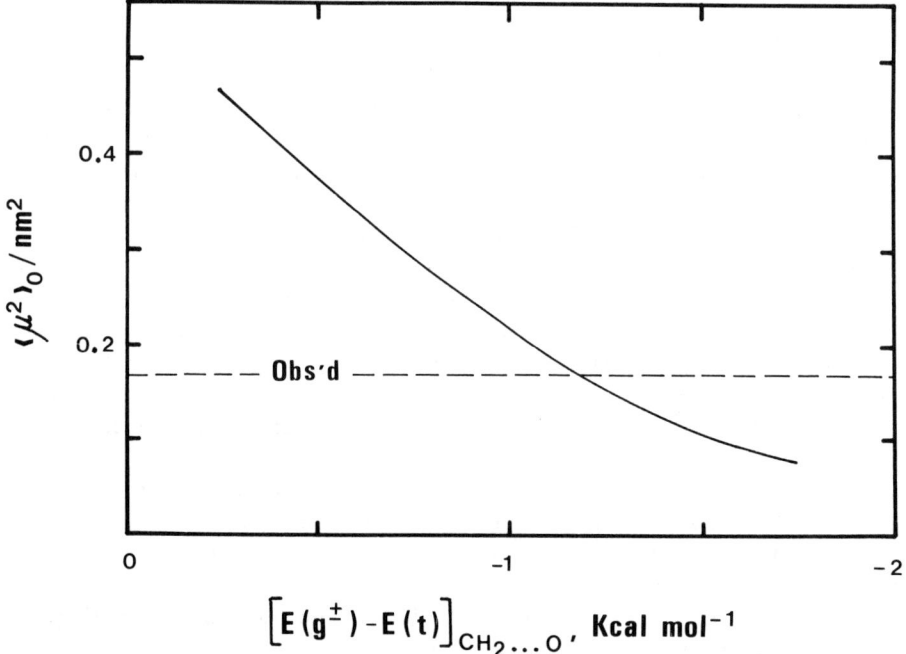

Figure 4.16 Repeating unit of poly(1,3-dioxolane) in the all-trans conformation.

polymers. In simple molecular terms, these chains have a very low dipole moment ratio largely because of their preference for gauche states of a low dipole moment. On the contrary, the temperature coefficient $d[\ln(\langle\mu^2\rangle)]/dT$ is very large because an increase in temperature increases the fraction of alternative *tt* conformations about the two acetalic bonds of higher energy and larger dipole moment. Comparison of the dipole moment ratio of the polyformals with that of their parent homopolymers indicates that as far as this conformation-dependent property is concerned, polyformals are much more similar to poly(methylene oxide) than they are to the other parent homopolymer. This comparison demonstrates that it is not possible in general

Figure 4.17 Dependence of the dipole moment ratio of poly(1,3-dioxolane) on the conformational energy about bonds of type *a* and *e* (Figure 4.16) in which the interacting species are CH_2 groups and O atoms (T = 30°C). (Redrawn from reference [97] by courtesy of the American Chemical Society.)

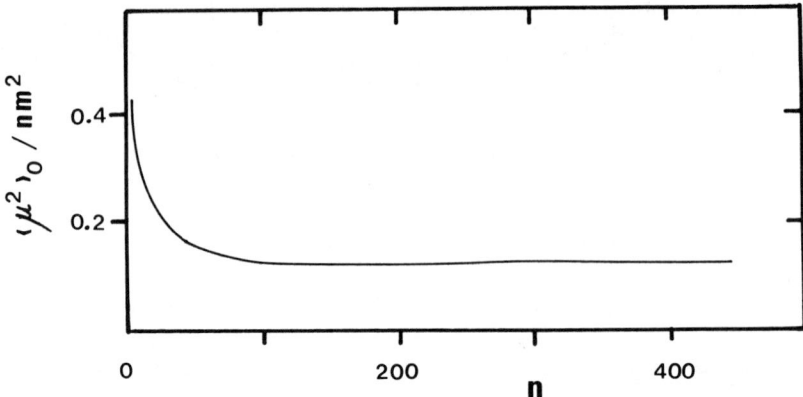

Figure 4.18   The dipole moment ratio of poly(1,3-dioxolane) at 25°C shown as a function of the number of skeletal bonds in the chain.

to predict the statistical properties of a copolymeric chain by simply averaging the values of the same property for the parent homopolymers.

Polysulfides

Some significant differences appear in melting points, crystalline-state configurations, and conformational preferences of poly(alkylene sulfides) and the analogous polyoxides. The differences are much more pronounced for the lower members of both series. For example, poly(ethylene sulfide) [101, 102] has a much higher melting point (216°C) than poly(ethylene oxide) (68°C). In general, the differences between the melting points of poly(alkylene sulfides) and poly(alkylene oxides) decrease as the number of methylene groups in the repeating unit decreases.

Intramolecular-energy calculations carried out to evaluate the conformational energies of the poly(alkylene sulfides) of structure $—S(CH_2)_x—$, with $x = 1, 2, 3,$ and 5, showed that the minimum conformational energy corresponds to the known crystalline-state configurations [103]. The polysulfides were found to be more flexible than their polyoxide counterparts, primarily because the C—S bond is considerably longer than the C—O bond (1.815 vs. 1.43 Å). The steric congestion about C—S bonds is less pronounced than about C—O bonds and as a result, gauche states about C—S bonds have an energy lying in the range of -0.1 to 0.1 kcal mol^{-1} with respect to the trans states.

Values of the dipole moment ratio for both symmetric and asymmetric polysulfides are shown in Table 4.6, where it can be seen that in most cases, the polarity of the polysulfides is significantly higher than that of their polyoxide counterparts as a consequence not only of the larger polarity of the C—S bond (1.21 D versus 1.07 D for the C—O bond), but also due to the differences that the oxygen and sulfur atoms stamp in the respective polymers.

TABLE 4.6 Experimental Values of the Dipole Moment Ratio (at 30°C) D_x and Its Temperature Coefficient TC = $10^3 \ d \ln (D_x)/dT$ for Some Polysulfides

Polymer	Repeating Unit	D_x	TC	Reference
Poly(trimethylene sulfide)	—S(CH$_2$)$_3$—	0.61	1.3	[112]
Poly(pentamethylene sulfide)	—S(CH$_2$)$_5$—	0.76	1.1	[104]
Alternating copolymer of methylene sulfide and pentamethylene sulfide	—SCH$_2$S(CH$_2$)$_5$—	0.26	4.9	[106]
Alternating copolymer of ethylene sulfide and pentamethylene sulfide	—S(CH$_2$)$_2$S(CH$_2$)$_5$—	0.64	0.9	[109]
Poly(propylene sulfide):	—S(CHCH$_3$)CH$_2$—			
atactic		0.44	1.5	[56, 57]
isotactic		0.39	2.0	[56, 57]
Poly(3,3-dimethylthietane)	—SCH$_2$C(CH$_3$)$_2$CH$_2$—	0.62	0.6	[113]
Poly(thiomethylene-1,4-trans-cyclohexylenemethylenethiomethylene	—SCH$_2$C$_6$H$_{10}$CH$_2$SCH$_2$—	0.18	4.3	[107]

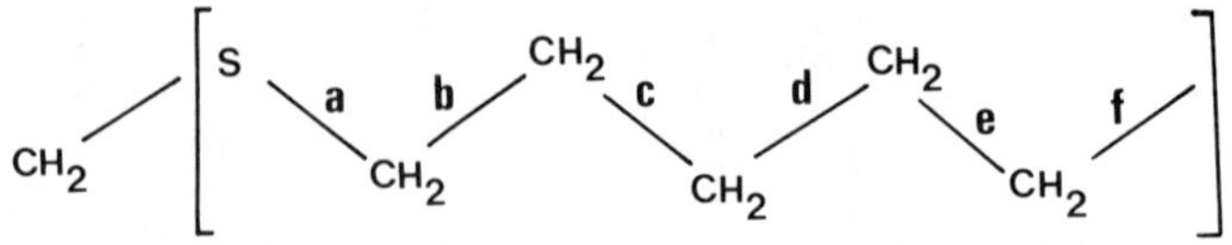

Figure 4.19   The repeating unit of poly(pentamethylene sulfide) in the all-trans conformation.

The repeating unit of a member of the series of polysulfides, poly(pentamethylene sulfide) (P5MS), is shown in Figure 4.19. Gauche states about bonds of type b and e bring S atoms and CH_2 groups into proximity; these states are located at $\phi = \pm 115°$ and have an energy of approximately 0.4 kcal mol^{-1} higher than the corresponding trans states [104]. Gauche states about bonds of type c and d bring two CH_2 groups into proximity in essentially the same interaction found in poly(ethylene). Here gauche states are located at $\phi = \pm 112°$ and have an energy of about 0.5 kcal mol^{-1} above that of the alternative trans states. The energies corresponding to $g^{\pm}g^{\mp}$ conformations were so high as to permit them to be ignored in the calculations of the dipole moments. By using these conformational energies, $\langle \mu^2 \rangle / nm^2 = 0.75$ was obtained [104], which is in excellent agreement with the experimental value of 0.76 corresponding to this quantity at 25°C.

From a conformational point of view, the first two members of the series of polysulfides are important to investigate whether gauche states about CH_2—S bonds in poly(methylene sulfide) (PMS) and CH_2—CH_2 bonds in poly(ethylene sulfide) (PES) are strongly favored over the alternative trans states, as occurs with their polyoxide counterparts, poly(methylene oxide) and poly(ethylene oxide), respectively. However, PMS has not been investigated with regard to its configuration-dependent properties largely because its rather high melting point (245°C) [105] and its instability in the molten state render it difficult to handle experimentally. However, information on the conformational properties of this polymer can be obtained from the analysis of the conformation-dependent properties of poly(1,3-dithiocane) (PDTC), whose repeating unit is given in Figure 4.20. This polymer can be considered an alternating copolymer of methylene sulfide and pentamethylene sulfide. Semiempirical potential-function calculations about S—CH_2 bonds of the CH_2S—CH_2S segment suggest that the rotational states are located at 0, $\pm 110°$ and

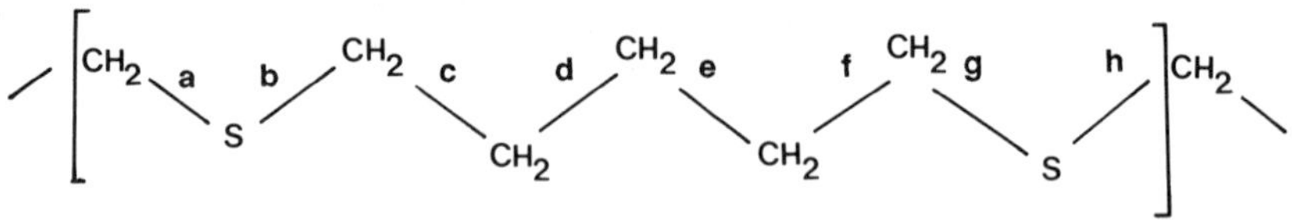

Figure 4.20   Repeating unit of the alternating copolymer of pentamethylene sulfide and methylene sulfide.

gauche states have an energy $E_{\sigma'}$ of about 0.36 kcal mol^{-1} below that of the corresponding trans states. By using this conformational energy, together with the energies discussed before for the pentamethylene sulfide moiety, a theoretical value for $\langle\mu^2\rangle/nm^2$ at 25°C is found (0.61) that is more than twice the experimental value (0.26) at the same temperature [106].

The dependence of the dipole moment ratio on $E_{\sigma''}$, shown in Figure 4.21, indicates that $\langle\mu^2\rangle/nm^2$ decreases as $E_{\sigma'}$ decreases, so that agreement between theory and experiment is found [106] for $E_{\sigma'} \approx -1.2$ kcal mol^{-1}. Utilization of this energy in the calculation of $d[\ln(\langle\mu^2\rangle)]/dT$ gives the value 4.6×10^{-3} K^{-1}, in excellent agreement with the experimental result 4.9×10^{-3} K^{-1}. In conclusion, these results suggest that gauche states about both the OCH$_2$—OCH$_2$ and SCH$_2$—SCH$_2$ segments have similar energies with respect to the corresponding trans states. It should be pointed out that this value of $E_{\sigma'}$ also satisfactorily reproduces the low dipole moment ratio and the high-temperature coefficient exhibited by polyalicyclic tioformals [107].

The analysis of the conformation-dependent properties of poly(ethylene sulfide) (PES) presents serious difficulties because this polymer is highly crystalline, melts above 200°C, and is unstable above the melting temperature. The evaluation of the conformational energy $E_{\sigma'}$ associated with gauche states about SCH$_2$—CH$_2$S bonds is important to explain the differences in the physical behavior of PES and its

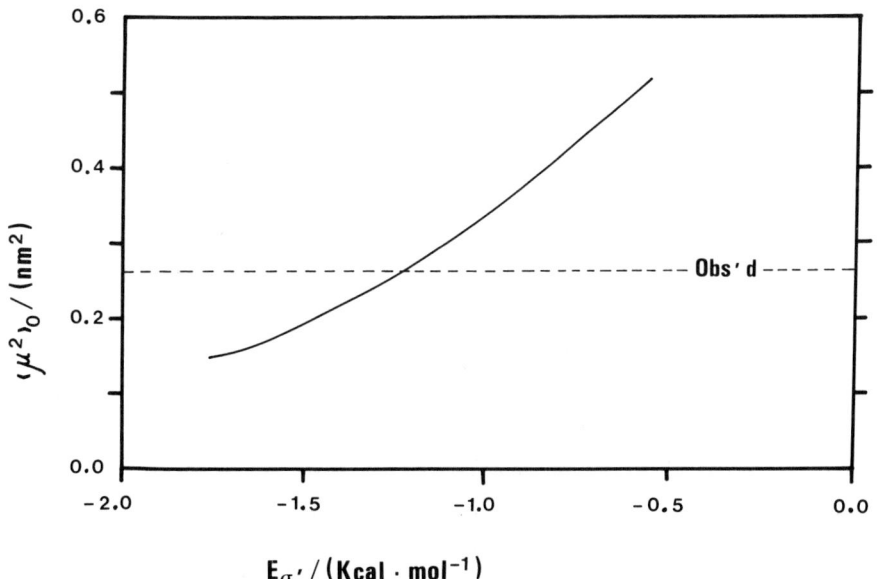

Figure 4.21 Influence of the conformational energy $E_{\sigma'}$ associated with gauche states about bonds of type a and h in Figure 4.20. (Redrawn from reference [106] by courtesy of Hüthig & Wepf Verlag.)

polyoxide counterpart PEO. Semiempirical potential calculations predict a value of nearly 0.4 kcal mol^{-1} for $E_{\sigma'}$ with respect to the energy of the alternative trans states [108]. Moreover, gauche states about CH_2S—CH_2CH_2 bonds have an energy $E_{\sigma''}$ of 0.1 kcal mol^{-1} below that of the trans states, whereas second-order interactions involved in the sequences CH_2S—CH_2—CH_2S and CH_2CH_2—S—CH_2CH_2 have conformational energies $E_{\omega'}$ and $E_{\omega''}$ of 1.1 and 0.4 kcal mol^{-1}, respectively.

In view of the insolubility of PES in common organic solvents, the experimental determination of $E_{\sigma'}$ for PES was obtained from the critical analysis of the dipole moment of both 1,2-bis(butylthio)ethane (BBT) and an alternating copolymer of ethylene sulfide and pentamethylene sulfide (PXS) [109]. The experimental values of $\langle \mu^2 \rangle /nm^2$ and its temperature coefficient for BBT and PXS are shown in Table 4.7 on page 82. Theoretical values of the dipole moment ratio calculated as a function of $E_{\sigma'} = [E(g^{\pm}) - E(t)]_{S \ldots S}$ for both BBT and PXS are shown in Figure 4.22. An increase in $E_{\sigma'}$ decreases the dipole moment ratio because of the increase in the all-trans conformation that places the dipoles corresponding to the CH_2SCH_2 groups in a nearly antiparallel direction. Agreement between theory

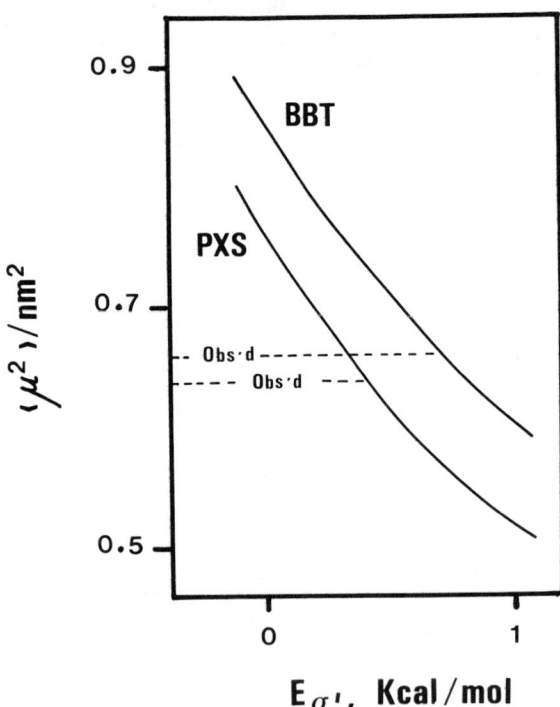

Figure 4.22 Variation of the dipole moment ratio with energy $E_{\sigma'}$ corresponding to gauche states about C—C bonds for BBT and PXS. See the text for details. (Redrawn from reference [109] by courtesy of the American Chemical Society.)

and experiment is found [109] for values of E_σ, lying in the range of 0.4 to 0.6 kcal mol^{-1}, which is in excellent agreement with the predictions from semiempirical potential-function calculations. As can be seen in Table 4.7, these conformational energies also give a good account of the experimental temperature coefficients of the dipole moments of BBT and PXS. According to this analysis, the dipole moment ratio of PES (0.43) is smaller than that of PEO (0.52) [69]. The flexibility of PES, as expressed by the conformational partition function z, is significantly higher than that of PEO, since the values of z per monomer residue are 10.2 and 4.3, respectively. The high melting point of PES in comparison with that of PEO, therefore, must be attributed to stronger van der Waals intermolecular interactions in the former compound, due partly to the inherently more attractive interatomic interactions associated with the highly polarizable sulfur atoms, and partly due to differences in the crystalline-state configuration [102].

The poly(propylene sulfide) chain [CH(CH$_3$)—S—CH$_2$—] (PPS), shown in Figure 4.23, is obviously related to PES, but it is much more tractable in experimental investigations because the melting point (52°C) [110] of the crystallizable (isotactic) modification is well below that of PES. The addition of the methyl group to the repeating unit gives the chain a stereochemically variable structure. The methyl group causes the appearance of a component of the dipole moment of the repeating unit parallel to the chain contour, and, consequently, excluded volume effects could affect the dipole moment in this polymer. The values of the experimental dipole moment ratio and its temperature coefficient for isotactic and atactic polymers in different solvents [56, 57] suggest that excluded volume effects do not play an important role in the polarity of PPS chains with different stereochemical compositions.

The fact that the C—S bond is considerably longer than the C—O bonds makes possible that conformations that are forbidden in PPO can exist in PPS chains. The statistical weight matrices for the skeletal bonds of the repeating unit can be written as [111]

$$\mathbf{U}_i = \begin{bmatrix} 1 & \alpha & \beta \\ \omega & \alpha & \beta\omega' \\ 1 & \alpha\omega' & \beta\omega \end{bmatrix} \qquad \mathbf{U}_{i+1} = \begin{bmatrix} 1 & \tau & \gamma \\ 1 & \tau & \gamma\omega' \\ 1 & \tau\omega' & \gamma \end{bmatrix} \qquad (4.26)$$

$$\mathbf{U}_{i+2} = \begin{bmatrix} 1 & \sigma & \sigma\omega'' \\ 1 & \sigma\omega'' & \sigma\omega'' \\ 1 & \sigma\omega'' & \sigma \end{bmatrix} \qquad (4.27)$$

The conformational energies $E_\alpha = -0.33$, $E_\beta = 1.3$, $E_\tau = 1.2$, $E_\gamma = 0.1$, $E_\omega = 1.5$, $E_{\omega'} = 1.1$, and $E_{\omega''} = 1.2$ kcal mol^{-1}, obtained from semiempirical potential-function calculations, give a good account of the experimental dielectric results.

Among the alkylene sulfide polymers, poly(trimethylene) sulfide (P3MS) [—S(CH$_2$)$_3$—]$_x$ represents a particularly interesting polymer for configurational

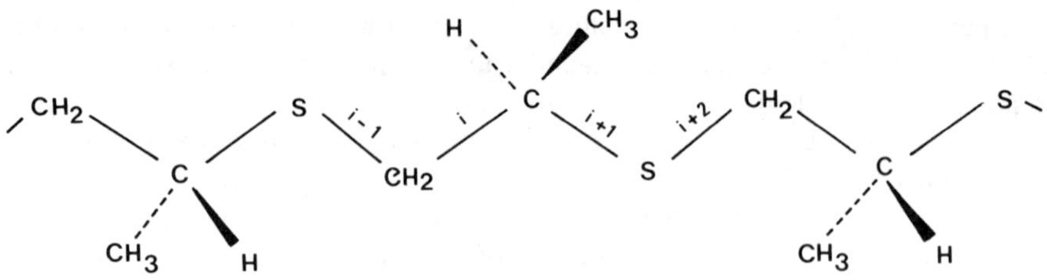

Figure 4.23   Isotactic segment of poly(propylene sulfide) in the all-trans conformation.

analysis. Despite its obvious similarity in structure to the analogous polyoxide poly(trimethylene oxide), the configurational properties of both polymers should be vastly different because all the skeletal bonds in the P3MS repeating unit should have gauche and trans states of very nearly the same energy, whereas the bonds in P3MO are much more constrained to either gauche or trans states. As a consequence, P3MS serves as an approximation to a freely rotating chain used as a conceptual idealization of real chains in many treatments dealing with the statistics of chain molecules [103, 112].

The experimental value of the dipole moment ratio for P3MS is shown in Table 4.6. In fact, the value of $\langle \mu^2 \rangle / nm^2$ at 25°C is nearly 50% higher than the value for its polyoxide counterpart [112]. Particular importance involves the conformational energy $E_{\sigma'}$ associated with gauche states about bonds of type b and c (Figure 4.24), that give rise to first-order interactions between a CH_2 group and a sulfur atom. It is obvious that except for favorable coulombic contributions, the value of $E_{\sigma'}$ should be the same as that for S $\cdots$ CH_2 interactions occurring in P5MS. As in this latter polymer, gauche states about b and c bonds are located at $\pm 113°$. Theoretical calculations of $\langle \mu^2 \rangle / nm^2$, carried out by using a three rotational states scheme, gave for this quantity a value of 0.69 at 25°C, which is somewhat larger than the experimental result of 0.61 at the same temperature. Better agreement between theory and experiment could be obtained by using a five rotational states scheme, which is described in detail in reference [112].

A segment of the polysulfide analogous of PDMO, poly(3,3-dimethyl thietane) (PDMS), is shown in Figure 4.25. Whereas C—O bonds in PDMO seem to be restricted to trans states, the energy associated with rotation about C—S bonds

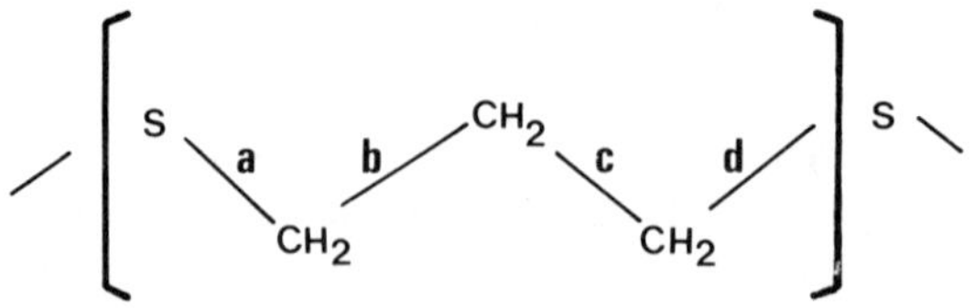

Figure 4.24   Repeating unit of poly(trimethylene sulfide) in the all-trans conformation.

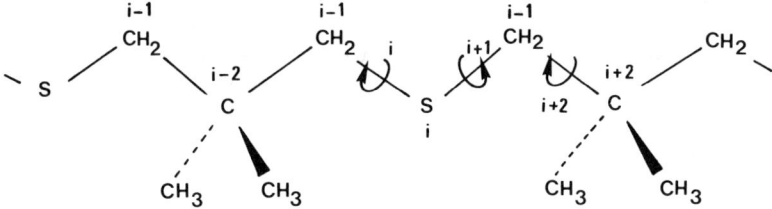

Figure 4.25   Segment of poly(3,3-dimethyl thietane) in the all-trans conformation.

increases as the rotational angle ϕ departs from zero. In the region $0 < \phi < 75°$, the variation of energy with ϕ, depicted in Figure 4.26, is mainly due to the rotational barrier about the C—S bond because all the interaction groups are placed far apart because of the length of the C—S bond and, as a result, nonbonded and electrostatic interactions are negligible. However, for $\phi > 75°$, the energy rises steeply because of severe interactions between the $i + 1$ methylene group and the methyl groups. As a consequence of this interaction, $g^{\pm}$ states are located at $\pm 75°$ and their energy E_τ is 1.8 kcal mol^{-1} above that of the alternative trans states. Gauche rotations about C—C bonds are stabilized mainly due to the attractive coulombic interactions between S atom i of the repeating unit and carbon atom $i - 3$. According to these

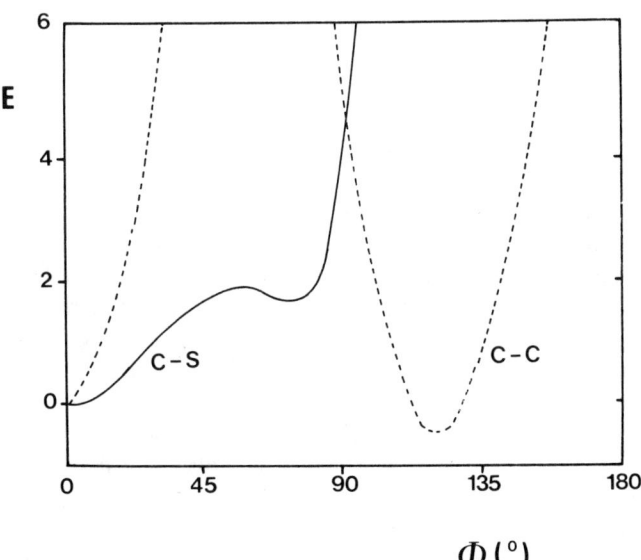

Figure 4.26   Calculated conformational energies in kcal mol^{-1} as a function of the rotational angle ϕ about C—S (solid line) and C—C (dashed line) bonds for poly(3,3-dimethyl thietane). (Redrawn from reference [113] by courtesy of the American Chemical Society.)

TABLE 4.7 Dipole Moment Results (at 30°C) for 1,2-Bis(butylthio)ethane (BBT) and an Alternating Copolymer of Poly(methylene Sulfide) and Ethylene Sulfide (PXS) Used as Model Compounds to Determine the Conformational Energy $E_{\sigma'}$ About SCH_2—CH_2S Bonds [109]

Model Compound	$D_x = \langle \mu^2 \rangle / nm^2$		$TC = 10^3 \, d \, [\ln(D_x)]/dT$	
	Experimental	Theoretical	Experimental	Theoretical[†]
BBT	0.656	0.696	1.2	0.7
PXS	0.635	0.604	1.7	0.9

[†]The theoretical values were calculated assuming $E_{\sigma'} = 0.6$ kcal mol^{-1}.

calculations, gauche states about C—C bonds are located at $\phi = \pm 120°$ and the energy $E_{\sigma'}$ associated with these states is 0.5 kcal mol^{-1} below that of the alternative trans states. Rotations of different sign about the two central C—C bonds of the repeating unit give rise to second-order interactions between two sulfur atoms whose energy E_ω amounts to 1.9 kcal mol^{-1}. The statistical weights required for evaluating the configurational properties of PDMS may be expressed in a 3 × 3 matrix scheme as [113]

$$\mathbf{U}_{i+1} = \mathbf{U}_{i+4} = \begin{bmatrix} 1 & \tau & \tau \\ 1 & \tau & \tau \\ 1 & \tau & \tau \end{bmatrix} \tag{4.28}$$

$$\mathbf{U}_{i+2} = \begin{bmatrix} 1 & \sigma' & \sigma' \\ 1 & \sigma' & \sigma' \\ 1 & \sigma' & \sigma' \end{bmatrix} \qquad \mathbf{U}_{i+3} = \begin{bmatrix} 1 & \sigma' & \sigma' \\ 1 & \sigma' & \sigma'\omega \\ 1 & \sigma'\omega & \sigma' \end{bmatrix} \tag{4.29}$$

Theoretical calculations of $\langle \mu^2 \rangle / nm^2$ carried out using the conformational energies indicated before give a value for this quantity significantly lower than the experimental result (≈ 0.62 at 30°C). The dependence of the dipole moment ratio and its temperature coefficient on E_τ, $E_{\sigma'}$, and E_ω, shown in Figures 4.27, 4.28, and 4.29, suggests that the set of conformational energies that gives a better account of the experimental results are $E_{\sigma'} = -0.1$ to -0.2, $E_\tau = 1.8$, and $E_\omega = 1.8$ kcal mol^{-1}. According to this analysis, the fraction of *tttt* conformations about the bonds of the repeating unit, which place the dipole in a nearly parallel direction, is higher in PDMS than in PDMO, and, hence, the polarity of the former polymer is higher than that of the latter.

    Comparison of the experimental values of the dipole moment ratios of polysulfides and some pertinent polyoxides shows that the polysulfides appear to have, on the average, larger dipole moment ratios than the analogous polyoxides. It is also interesting to note that in both cases, CH_2X—CH_2—XCH_2 linkages (where the two X's are both either S or O) result in very low values of the dipole moment ratio, probably because these sequences show a much higher preference for gauche than for trans states. Since pairs of $g^{\pm}$ states in such CH_2X—CH_2—XCH_2 sequences would be expected to give nearly antiparallel alignment of the bond dipoles, this would cause a corresponding decrease in $\langle\mu^2\rangle$.

The conformational analysis suggests strong preference for gauche states about the central bonds in the sequences

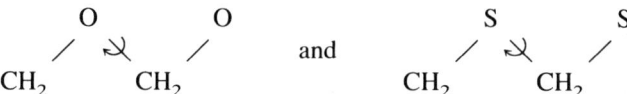

where the gauche population about C—O and C—S bonds seems to be similar. The alternation of sulfur and oxygen atoms in the $CH_2SCH_2OCH_2SCH_2$ moiety gives rise to the sequences

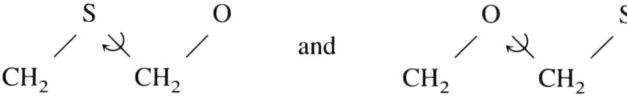

in which gauche rotations about the S—C and C—O bonds cause first-order $CH_2 \cdots O$ and $CH_2 \cdots S$ interactions. Potential curves reflecting these interac-

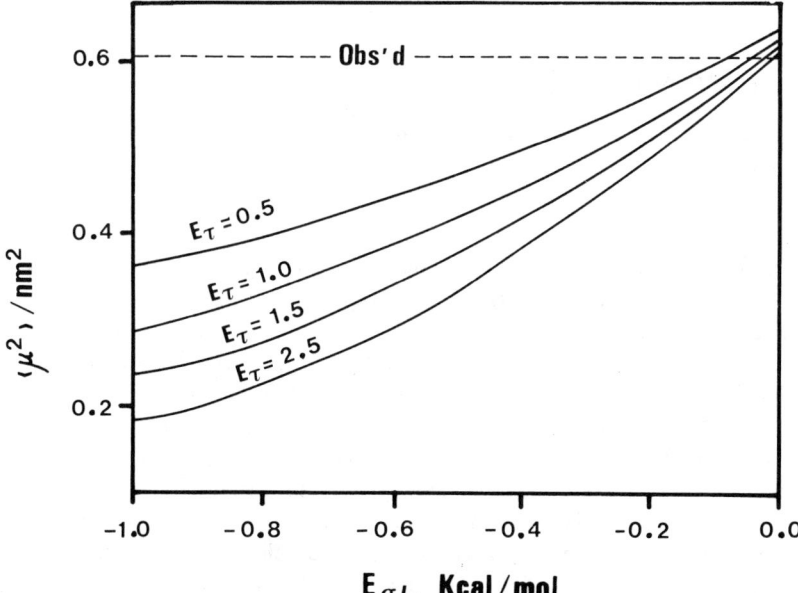

Figure 4.27 Calculated values of the dipole moment ratio of poly(3,3-dimethyl thietane) at 30°C as a function of $E_{\sigma'}$ for several values of E_T (in kcal mol^{-1}), taking E_ω = 1.8 kcal mol^{-1}. (Redrawn from reference [113] by courtesy of the American Chemical Society.)

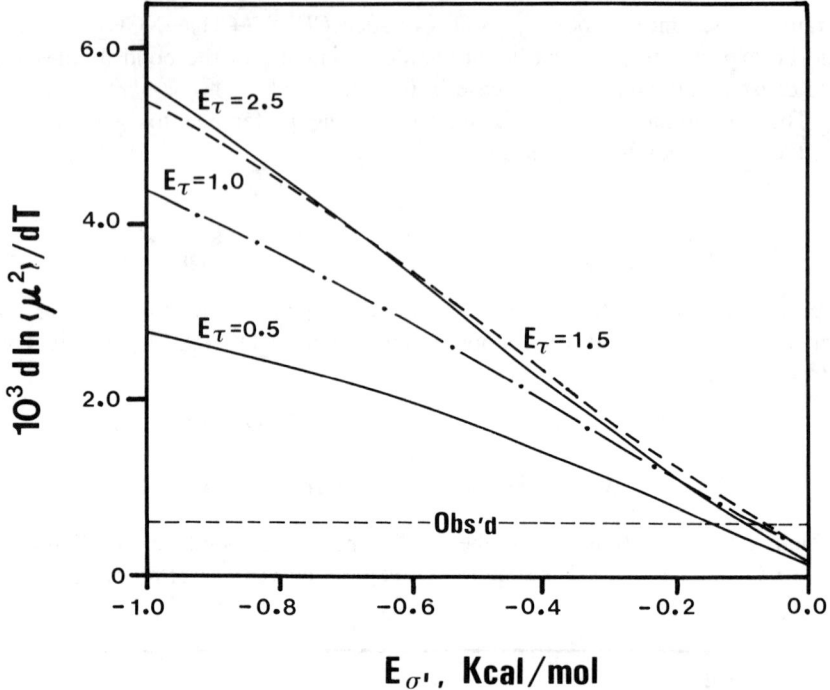

Figure 4.28 Variation of the temperature coefficient of the dipole moment ratio of poly(3,3-dimethyl thietane) with $E_{\sigma'}$ at 30°C for several values of E_τ, taking $E_\omega = 1.8$ kcal mol^{-1}. (Redrawn from reference [113] by courtesy of the American Chemical Society.)

tions are represented in Figure 4.30. The rotational states about S—C bonds are located at 0, ±120°, and the energy E_σ corresponding to the gauche states is about 0.35 kcal mol^{-1} below that of the trans states. The potential curve corresponding to O—C bonds indicates that the rotational isomeric states are located at 0, ±105°. The calculations also suggest that gauche states about these bonds have an energy of 0.5 kcal mol^{-1} above that of the corresponding trans states [114].

Experimental values of $\langle\mu^2\rangle$ and $\langle\mu^2\rangle/nm^2$, respectively, for 9-oxa-7,11-dithiaheptadecane (ODTH) and poly(3-oxa-1,5-dithiadecamethylene) (PODTD), which contain the CH_2SCH_2O—CH_2—SCH_2 sequence in their structures, are given in Table 4.8. The dependence of the mean-square dipole moment of ODTH and the dipole moment ratio of PODTD on both E_α and E_β, shown in Figures 4.31 and 4.32, respectively, suggest that the use of the values of the conformational energies obtained by semiempirical methods overestimates the polarity of these chains [114]. As can be seen in these figures, good agreement between theory and experiment can be reached by assuming that $E_\alpha = E_\beta = -0.9$ kcal mol^{-1}. Therefore, it can be

TABLE 4.8 Experimental Values of the Dipole Moment Ratio D_x and Its Temperature Coefficient (TC $= 10^3$ d ln $(\mu^2)/dT$) for Symmetric Chains with Oxygen and Sulfur Atoms in Their Structures

Polymer	Repeating Unit	D_x	TC	Reference
Poly(thiodiethylene glycol)	—S(CH₂)₂O(CH₂)₂—	0.61	1.6	[115]
Poly(1,3-dioxa-6-thiocane)	—(CH₂)₂S(CH₂)₂OCH₂O—	0.42	0.7	[116]
Poly(3-oxa-1,5-dithia-decamethylene)	—SCH₂OCH₂S(CH₂)₅—	0.32	2.2	[114]
Alternating copolymer of 1,3-dioxolane and 1,3-dithiolane	—(CH₂)₂SCH₂S(CH₂)₂OCH₂O—	0.26	4.2	[115]

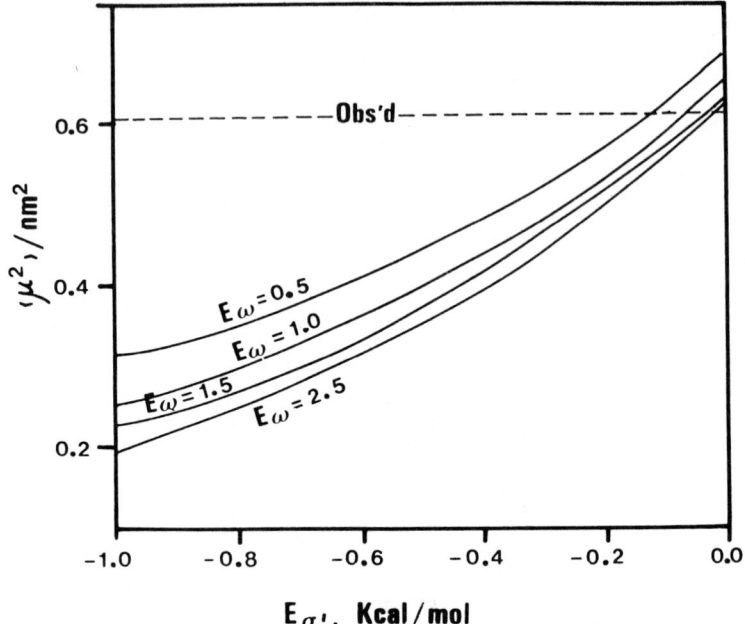

Figure 4.29 Calculated values of the dipole moment ratio of poly(3,3-dimethyl thietane) at 30°C as a function of $E_{\sigma'}$ for several values of E_{ω} (in kcal mol^{-1}). A value of $E_{\tau} = 1.8$ kcal mol^{-1} was used in the calculations. (Redrawn from reference [113] by courtesy of the American Chemical Society.)

concluded that gauche rotations about the central bonds of both CH_2O—CH_2S and CH_2O—CH_2O sequences are strongly favored with respect to the alternative trans states.

Gauche states about C—C bonds for poly(ethylene oxide) and poly(ethylene sulfide)

bring O and S atoms, respectively, into close proximity. However, whereas the energy of these states in the former sequence is about 0.4 kcal mol^{-1} below the trans states [69], the energy of gauche states is about 0.5 kcal mol^{-1} above that of the trans in the latter [108, 109]. Gauche rotations about C—C bonds in the sequence

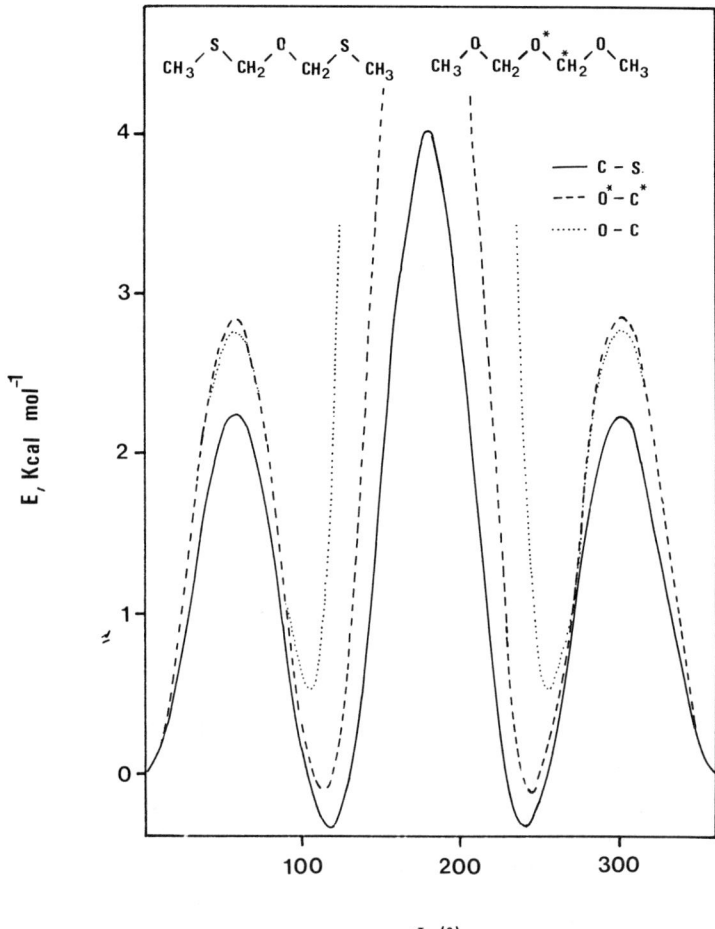

Figure 4.30 First-order interactions for rotations over the skeletal bonds shown. (Redrawn from reference [114] by courtesy of the American Chemical Society.)

cause first-order S $\cdots$ O interactions. Potential calculations predict an energy $E_{\sigma'}$ for gauche states of about 0.6 kcal mol^{-1} above that of the alternative trans states. Interactions of this kind appear in poly(thiodiethylene glycol (PTDG), an alternating copolymer of ethylene oxide and ethylene sulfide. Experimental values of the dipole moment ratio and its temperature cofficient for PTDG are shown in Table 4.8,

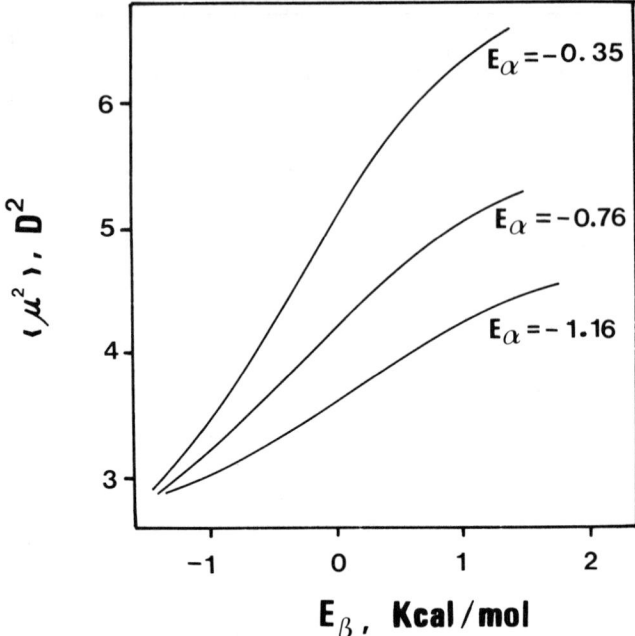

Figure 4.31   Variation of $\langle \mu^2 \rangle$ with energy E_β of gauche states about CH_2SCH_2—OCH_2 bonds for 9-oxa-7,11-dithiaheptadecane. The calculations were performed for different values of E_α (in kcal mol^{-1}) about CH_2OCH_2—SCH_2 bonds. (Redrawn from reference [114] by courtesy of the American Chemical Society.)

whereas the dependence of $\langle \mu^2 \rangle/nm^2$ on $E_{\sigma'}$ is represented in Figure 4.33. Here coincidence between theoretical and experimental results is found [115, 116] for $E_{\sigma'}$ ≈ 0.4 kcal mol^{-1}. The temperature dependence of the dipole moment ratio d ln $\langle \mu^2 \rangle/dT$ is calculated to be 1.0×10^{-3} K^{-1} with $E_{\sigma'} = 0.4$ kcal mol^{-1}, which is in fair agreement with the experimental result (1.6×10^{-3} K^{-1}).

Polyesters with Aromatic Rings in the Acid Residue

Terephthalic-acid-based linear polyesters with repeating unit —$O*C*(C_6H_4)C*O*O(CH_2)_m$— are an important and interesting kind of polymers from a practical and basic point of view. Owing to the insolubility of most of these polymers in nonpolar solvents, information on the values of the mean-square dipole moment corresponding to each member of the series is sparse. However, by schematically substituting one or more methylene groups in the structural unit for oxygen atoms, terephthalic-acid-based polyesters are obtained that are soluble in nonpolar organic solvents and, therefore, their conformational characteristics can be determined from the analysis of their experimental mean-square dipole moments.

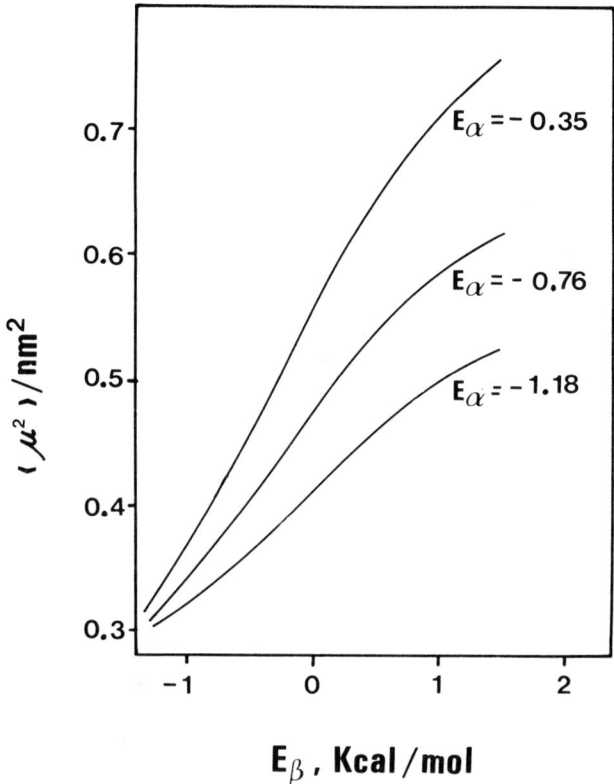

E_β , Kcal/mol

Figure 4.32 Variation of $\langle\mu^2\rangle$ with energy E_β of gauche states about CH_2SCH_2—OCH_2 bonds for poly(3-oxa-1,5-dithiadecamethylene). The calculations were performed for different values of E_α (in kcal mol^{-1}) about CH_2OCH_2—SCH_2 bonds. (Redrawn from reference [114] by courtesy of the American Chemical Society.)

Experimental values of $\langle\mu^2\rangle/nm^2$ and its temperature coefficient for poly(diethylene glycol terephthalate) —O*C*(C$_6$H$_4$)C*O*O(CH$_2$)$_2$O(CH$_2$)$_2$ (PDET) [117], poly(triethylene glycol terephthalate) —O*C*(C$_6$H$_4$)C*O*O(CH$_2$)$_2$-O(CH$_2$)$_2$O(CH$_2$)$_2$ (PTET) [118], poly(ditrimehtylene glycol) —O*C*(C$_6$H$_4$)C*O*O(CH$_2$)$_3$O(CH$_2$)$_3$ (PDTM) [119], polypropylene glycol and polydipropylene glycol therephthalate [120, 121] are shown in Table 4.9. The dipole moment associated with the ester group has a value of 1.89 D and its direction makes an angle of 123° with the C^{Ph}—C* bond [122], whereas the dipole moment of the C—O bonds of the ether group has a value [69] of 1.07 D and lies along the bond.

Infrared spectra of aliphatic esters show the ester group to be planar and the trans conformation to be preponderant over the cis [123]. In the esters of aromatic

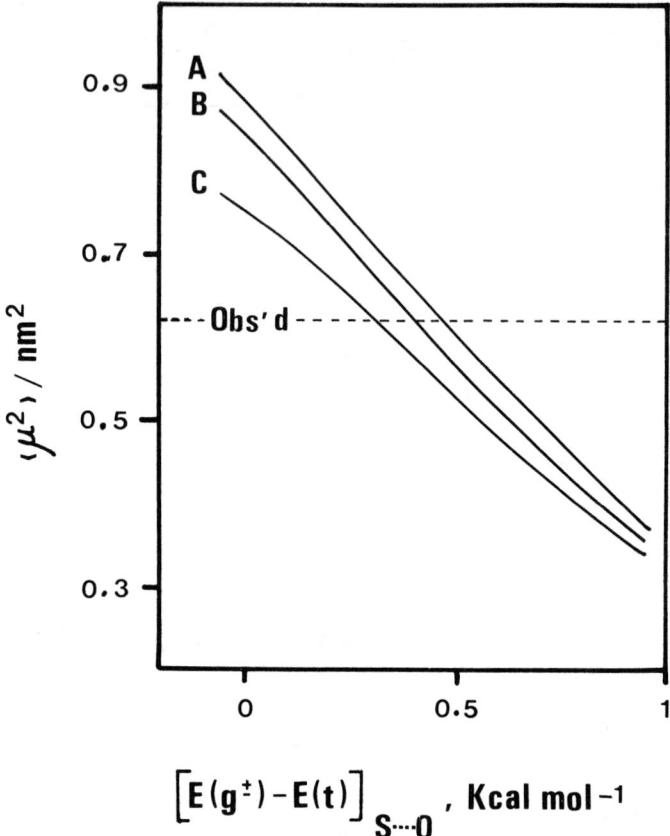

$$\left[E(g^{\pm}) - E(t) \right]_{S\cdots O} , \; \text{Kcal mol}^{-1}$$

Figure 4.33 Theoretical calculations showing the dependence of the dipole moment ratio of poly(thiodiethylene glycol) on the energy difference between gauche and trans states in which the interacting species are S and O atoms. Curves A and B were obtained assuming $\phi_g = \pm 120°$ and $\phi_g = \pm 115°$, respectively. In both cases, the values used for the skeletal bond angles were $\langle CSC = 100°$, $\langle SCC = 114°$, and $\langle OCC = \langle COC = 110°$. Curve C was obtained assuming $\phi_g = \pm 120°$ and all the skeletal bond angles equal to 110°. (Redrawn from reference [115] by courtesy of the American Chemical Society.)

acids, the cis form must be further disfavored by steric repulsion between alkyl CH_2 and aromatic CH. Furthermore, since coplanarity of the carbonyl group with the benzene ring guarantees maximum overlapping of electrons of the participating atoms, the terephthaloyl residue is restricted to cis and trans conformations [124]. The statistical weight matrices associated with bonds i through $i + 3$ in Figure 4.34 are [19, 117]

TABLE 4.9 Experimental Values (at 30°C) of the Dipole Moment Ratio D_x and Its Temperature Coefficient TC $= 10^3 \, d[\ln(D_x)]/dT$ for Several Polyesters Based on Terephthalic Acid and Its Isomers

Polymer	Structural Unit	D_x	TC	Reference
Poly(diethylene glycol terephthalate)	—OOCC$_6$H$_4$COO(CH$_2$)$_2$O(CH$_2$)$_2$—	0.66	0.1	[117]
Poly(diethylene glycol isophthalate)	—OOCC$_6$H$_4$COO(CH$_2$)$_2$O(CH$_2$)$_2$—	0.70	1.1	[129]
Poly(diethylene glycol phthalate)	—OOCC$_6$H$_4$COO(CH$_2$)$_2$O(CH$_2$)$_2$—	0.64	2.1	[130]
Poly(propylene glycol terephthalate)	—OOCC$_6$H$_4$COOCH$_2$CH(CH$_3$)—	0.58	1.5	[120]
Poly(dipropylene glycol terephthalate)	—OOCC$_6$H$_4$COOCH(CH$_3$)CH$_2$OCH$_2$CH(CH$_3$)—	0.58	1.6	[121]
Poly(ethylene terepthalate)[†]	—OOCC$_6$H$_4$COO(CH$_2$)$_2$—	0.53	1.0	[120]
Poly(triethylene glycol terepthalate)	—OOCC$_6$H$_4$COO(CH$_2$)$_2$O(CH$_2$)$_2$O(CH$_2$)$_2$—	0.68	0.1	[118]
Poly(ditrimethylene glycol terephthalate)	—OOCC$_6$H$_4$COO(CH$_2$)$_3$O(CH$_2$)$_3$—	0.81	−0.24	[119]

[†]Calculated values.

$$\mathbf{U}_i = [1] \qquad \mathbf{U}_{i+1} = [1 \quad \gamma]$$

$$\mathbf{U}_{i+2} = \begin{bmatrix} 1 \\ 1 \end{bmatrix} \qquad \mathbf{U}_{i+3} = [1 \quad \sigma_k \quad \sigma_k] \tag{4.30}$$

where γ weighs the cis ($\phi = 180°$) relative to the trans ($\phi = 0°$) conformations, and σ_k reflects the statistical weight of gauche states with respect to the trans for O—CH_2 bonds ($i + 3$ and $i + 8$) of the ester groups. In fact, the cis conformation seems to be slightly favored ($E_\gamma \approx 50$ cal mol^{-1}) over the trans and $E_{\sigma k} = 0.4$ kcal mol^{-1}. The critical analysis of the ^{1}H NMR spectrum of diethylene glycol dibenzoate (DGD) indicates [125] that the gauche population X_g about C—C bonds amounts to 0.90 ± 0.05. Values of $E_{\sigma'} = E(g^\pm) - E(t)$ lying in the range -0.8 to -1.2 kcal mol^{-1} reproduce satisfactorily the value of X_g found for this compound. This energy is about 0.5 kcal mol^{-1} lower than that found for similar conformations in PEO, suggesting that the ester group stabilizes gauche states about C—C bonds, which causes $O \cdots O$ interactions between an ether and an ester group. It should be pointed out, however, that the experimental value of the dipole moment of ethylene glycol dibenzoate, a low molecular weight analogous of poly(ethylene terephthalate), can only be reproduced [126] with $E_{\sigma'} \approx -0.6$ kcal mol^{-1}.

Stabilizing effects also appear for gauche states about C—C bonds that produce first-order $S \cdots O$ interactions in thioesters. Thus, the analysis of the ^{1}H NMR spectrum of thiodiethylene glycol dibenzoate (SDB) indicates [125] that $X_g = 0.60 \pm 0.05$, suggesting that $E_{\sigma'} \approx 0$. This value is also significantly lower than the value $E_{\sigma'} \approx 0.4$ kcal mol^{-1} found for similar states about similar bonds in poly(thiodiethylene glycol) [115]. The conformational energies for the rotational states of the remaining skeletal bonds of both DGD and SDB are similar to those used in poly(ethylene oxide) and poly(thiodiethylene glycol). These conformational energies very satisfactorily reproduce the values of the dipole moment ratio of PDET [127], PTET [118], and poly(thiodiethylene glycol terephthalate) (PSDTS) [128], as well as the dipole moments of DGD and SDB. The low value of $E_{\sigma'}$ is also responsible for the high temperature coefficient exhibited by the unperturbed dimensions of PDET chains.

The enhancing effects that the presence of carbonyl groups seem to produce in the fraction of gauche states about C—C bonds in PDET are not extended to conformations in which first-order $CH_2 \cdots O$ interactions take place, as occurs with gauche states about C—C bonds in poly(ditrimethylene glycol terephthalate) (PDTT) [119] (see Figure 4.35). The dependence of both the dipole moment ratio and the unperturbed dimensions on $E_{\sigma'} = E(g^\pm) - E(t)$, representing the first-order $CH_2 \cdots O$ interaction, is shown in Figure 4.36. The ratio $\langle \mu^2 \rangle / nm^2$ decreases with increasing $E_{\sigma'}$ because the all-trans conformation in which the dipoles are in nearly antiparallel direction increases. The unperturbed dimensions, on the contrary, follow a different trend because an increase in $E_{\sigma'}$ favors the spatial extension of the chains. Agreement between theoretical and experimental results is achieved for $E_{\sigma'} = -0.1$ to -0.2 kcal mol^{-1}, a value that is somewhat higher than that found for

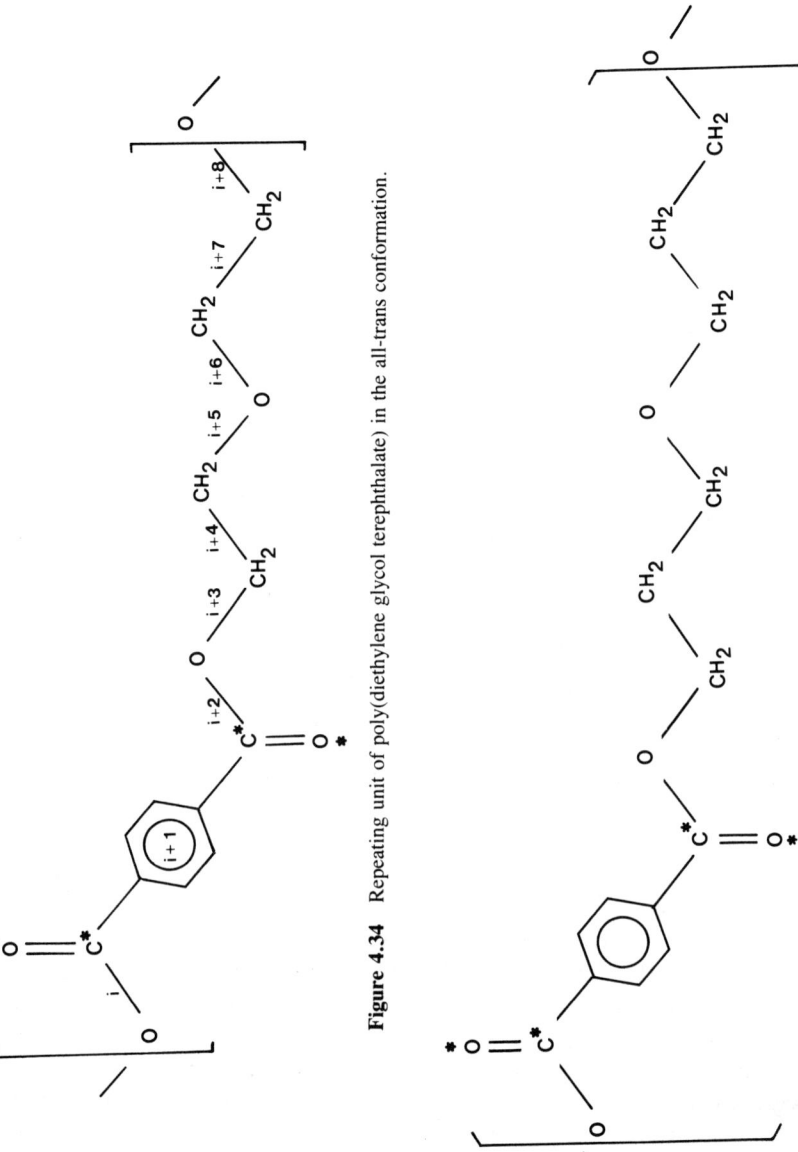

Figure 4.34 Repeating unit of poly(diethylene glycol terephthalate) in the all-trans conformation.

Figure 4.35 Repeating unit of poly(ditrimethylene glycol terephthalate) in its planar all-trans conformation.

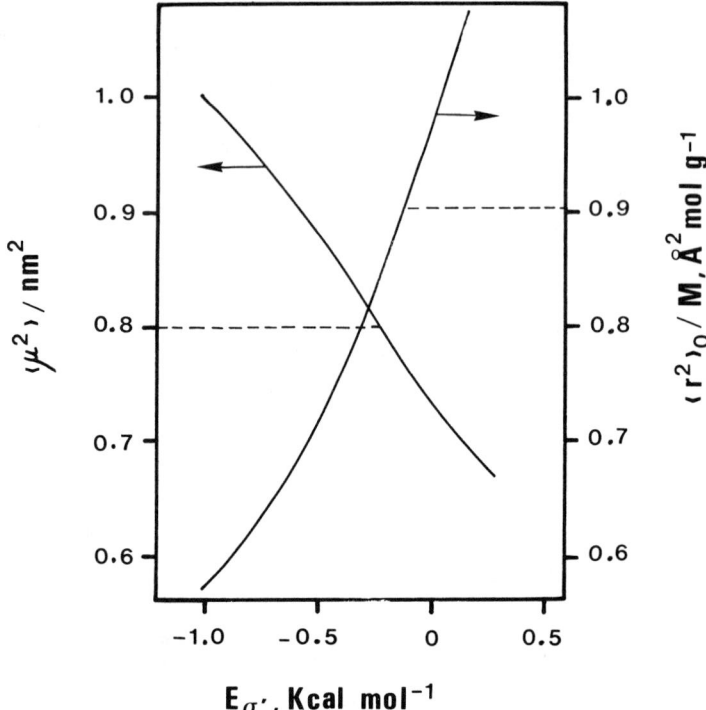

Figure 4.36 Dependence of both $\langle\mu^2\rangle/nm^2$ and $\langle r^2\rangle_0/M$ poly(ditrimethylene glycol terephthalate) on the conformational energy $E_{\sigma'}$ associated with gauche states about CC—CO bonds of the glycol residue. Horizontal lines represent the experimental values. (Redrawn from reference [119] by courtesy of the American Chemical Society.)

this quantity (-0.4 kcal mol^{-1}) in the conformational analysis of poly(trimethylene oxide) [69]. Values of $E_{\sigma'} = -0.2$ kcal mol^{-1} also reproduce satisfactorily the dipole moment of trimethylene glycol dibenzoate [126]. For values of $x > 4$ in polyesters with repeating unit [—O*C*(C$_6$H$_4$)C*O*O(CH$_2$)$_x$], correlations between the ester groups become negligible and the dipole moment ratio approaches unity.

In terephthalic- and isophthalic-based polyesters, interactions between the two ester groups attached to the same phenyl group are negligible. The carbonyl and phenyl rings are coplanar and as a result, the terepthaloyl [124] and isophthaloyl [129] residues are planar, favoring the crystallization of polyesters in which the glycol residue is symmetric. In phthalic-based polyesters (Figure 4.37), however, strong interactions among the ester groups attached to the same phenyl group occur, so that coplanarity of the carbonyl and phenyl rings is disfavored [130]. Potential-energy curves corresponding to C^{Ph}—C* bonds suggest that the rotational angles

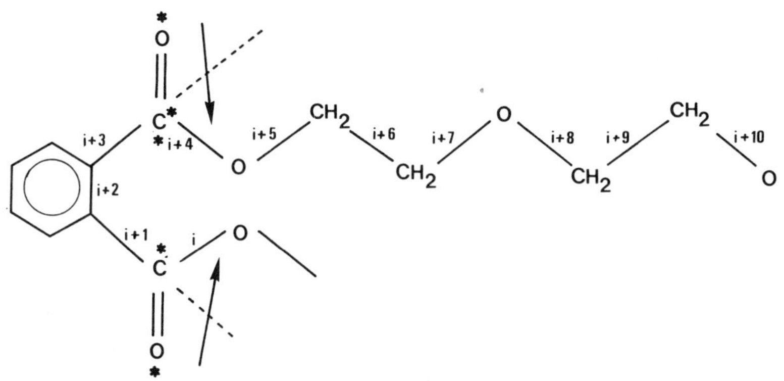

Figure 4.37 Repeating unit of poly(diethylene glycol phthalate) in an arbitrary planar conformation.

about these bonds are located at $\pm 90°$ and the conformations $g^{\pm}cg^{\mp}$ corresponding to bonds $i + 1$ through $i + 3$ in Figure 4.37 have an energy E_γ of about 0.3 kcal mol^{-1} below that of the alternative $g^{\pm}cg^{\pm}$ conformations. The theoretical evaluation of the dielectric conformation-dependent properties of poly(diethylene glycol phthalate) as a function of E_γ indicates (see Figure 4.38) that values of E_γ lying in the range of 0.35 to 0.60 kcal mol^{-1} give a good account of the experimental values of the dipole moment ratio and its temperature coefficient [130].

Polyesters with Cyclohexane Rings in the Main Chain

Polyesters can be prepared in which the cyclohexane rings are incorporated in the acid residue, in the glycol residue, or in both. As can be seen in Table 4.10, where the experimental values for the dipole moment ratio of several cycloaliphatic polyesters are given, the polarity of polyesters is very sensitive to the substitution equatorial–equatorial (trans isomers) or equatorial–axial (cis isomers) of the hydrogen atoms bonded to the carbon atoms located at the 1,4 position of the cyclohexane ring. For polyesters prepared from cyclohexane dimethanol and an aliphatic diacid, the differences in polarity arise from the different conformational characteristics of the $C_{cy}H$—CH_2 bonds for trans and cis isomers [131, 132]. In the former case, gauche states about $C_{cy}H$—CH_2 bonds, which give rise to interactions between a methylene group of the cyclohexane ring and an oxygen atom of the ester moiety, have an energy about 0.5 kcal mol^{-1} below that of the corresponding trans states, whereas in the latter, $C_{cy}H$—CH_2 bonds are restricted to $g^{\pm}$ states, with both states having the same probability.

Cycloaliphatic polyesters can also be prepared from 1,4-trans-dicarboxilic acid and a glycol. Potential-energy calculations for $C_{cy}H$—C^* bonds of the acid residue suggest that the rotational states of these bonds are located at 0° and 180°, where the carbonyl bond is cis and trans with the axial C—H bond, respectively

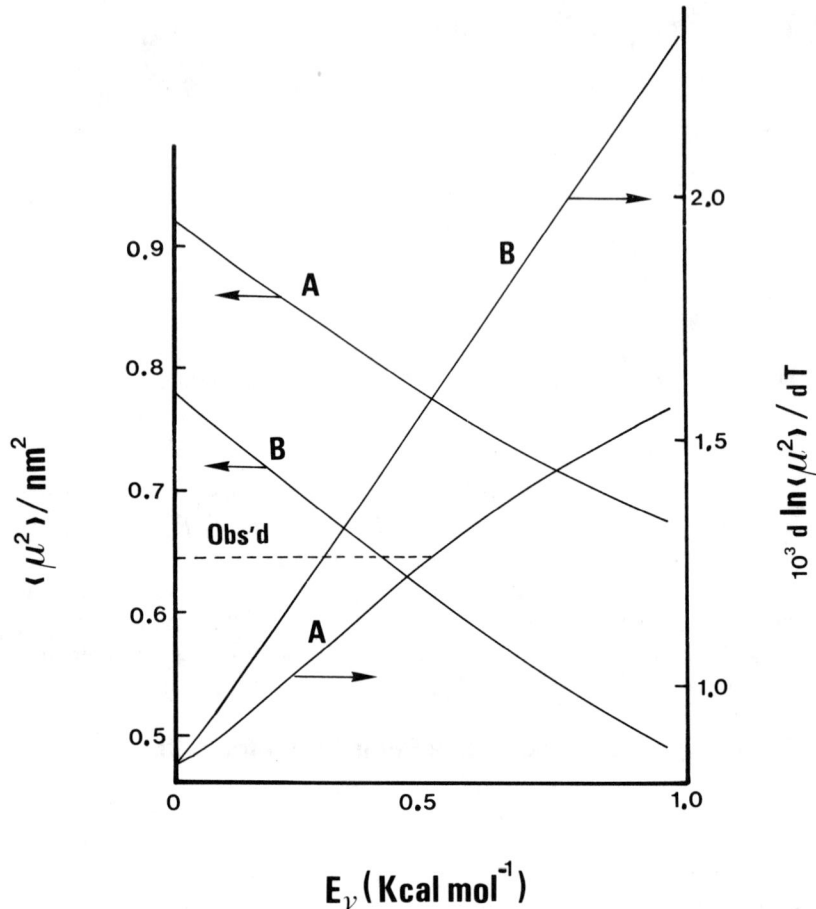

Figure 4.38 Dependence of the dipole moment ratio and its temperature coefficient of poly(diethylene glycol phthalate) on E_γ. See the text for details. Curves A and B were obtained for $\phi(Ph\text{—}C^*) = \pm 110°$ and $\pm 90°$, respectively. (Redrawn from reference [130] by courtesy of the American Chemical Society.)

[122]. The polarity of the polyesters is strongly dependent on the conformational energy $E_\gamma = E_{180} - E_0$ about $C_{cy}\text{—}C^*$ bonds [133]. This behavior is a consequence of the fact that for positive/negative values of E_γ, the favored conformations *tctct/ttttt* corresponding to $i + 1$ through $i + 5$ in Figure 4.39 have the dipoles associated with the ester groups in a nearly antiparallel direction; hence, the dipole moment ratio decreases as E_γ departs from zero. When E_γ approaches 0, the probability of occurrence of the conformation *tcttt* and its symmetric *tttct*, which places the dipoles in a nearly parallel direction, reaches a maximum, and, conse-

TABLE 4.10 Experimental Values (at 30°C) of the Dipole Moment Ratio D_x and Its Temperature Coefficient TC = $10^3 \, d[\ln(D_x)]/dT$ for Some Cycloaliphatic and Aliphatic Polyesters

Polymer	Structural Unit	D_x	TC	Reference
Poly(trans-1,4-cyclohexane dimethanol succinate)	—OOCCH$_2$C$_6$H$_{10}$CH$_2$COO(CH$_2$)$_2$—	0.45	—	[132]
Poly(trans-1,4-cyclohexane dimethanol adipate)	—OOCCH$_2$C$_6$H$_{10}$CH$_2$COO(CH$_2$)$_4$—	0.60	1.0	[132]
Poly(cis-1,4-cyclohexane dimethanol adipate)	—OOCCH$_2$C$_6$H$_{10}$CH$_2$COO(CH$_2$)$_4$—	0.93	0.4	[132]
Poly(trans-1,4-cyclohexane dimethanol sebacate)	—OOCH$_2$C$_6$H$_{10}$CH$_2$COO(CH$_2$)$_8$—	0.68	3.9	[131]
Poly(cis-1,4-cyclohexane dimethanol sebacate)	—OOCCH$_2$C$_6$H$_{10}$CH$_2$COO(CH$_2$)$_8$—	0.98	0.7	[131]
Poly(diethylene glycol-1,4-trans-cyclohexane-dicarboxylate)	—OOCC$_6$H$_{10}$COO(CH$_2$)$_2$O(CH$_2$)$_2$—	0.59	1.3	[133]
Poly(neopentyl glycol adipate)	—OCO(CH$_2$)$_4$COOCH$_2$C(CH$_3$)$_2$CH$_2$—	0.65	1.7	[138]
Poly(neopentyl glycol succinate)	—OCO(CH$_2$)$_2$COOCH$_2$C(CH$_3$)$_2$CH$_2$—	0.55	2.2	[139]

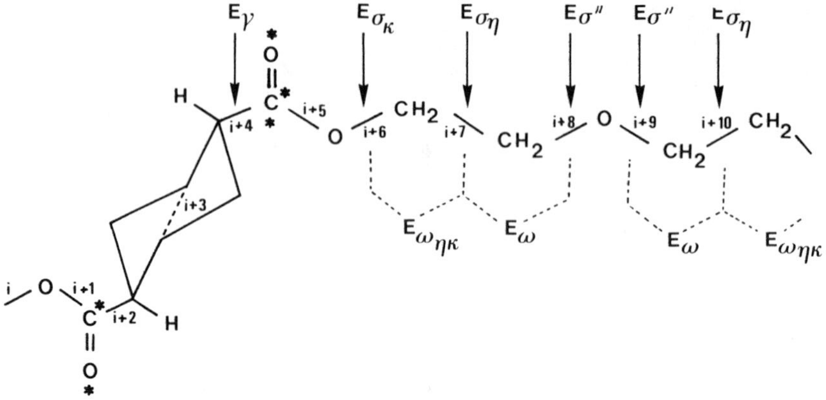

Figure 4.39 Planar segment of poly(diethylene glycol 1,4-trans-cyclohexanedicarboxilate). First- and second-order conformational energies are indicated above and below the skeleton, respectively.

quently, the value of the dipole moment is also maximum. This behavior is shown in Figure 4.40, where the dependences of the dipole moment ratio and its temperature coefficient on E_γ for poly(diethylene glycol 1,4-trans-cyclohexanedicarboxylate) are shown [133]. Agreement between theory and experiment is obtained for values of E_γ in the vicinity of zero; this result is close to 0.3 kcal mol^{-1}, the value obtained for this quantity by using semiempirical potential functions.

Aliphatic Polyesters

An important issue in the study of the conformation-dependent properties of aliphatic polyesters is to elucidate the preferred states about the CH_2CH_2—C*O*OCH$_2$ esters in aliphatic polyesters. The available information is highly controversial. From spectroscopic studies carried out on methyl propionate, some suggest that the lowest energy conformation arises when the C*=O* group eclipses the C—H bonds [134], whereas others concluded that the C*=O*/C—C eclipsed form is more stable than the C*=O*/C—H eclipsed form [135]. An intermediate value of 0.08 kcal mol^{-1} has also been reported for the energy of the gauche states about CH_2—C* bonds with respect to that of the corresponding trans state [136]. More recent work performed by Abe [137] on dialkyl esters of type $(CH_2)_n(C*O*O*Et)_2$ suggests that the dipole moment of most members of the series can be reproduced by assuming that the trans states about CH_2—C* bonds are preferred over the alternative gauche states. This problem was reexamined by studying the dipole moments of poly(neopentyl glycol adipate) [138] and poly(neopentyl glycol succinate) [139]. The structural unit of the latter polymer is shown in Figure 4.41. From the study of the statistics of oxetanes, it can be concluded that gauche

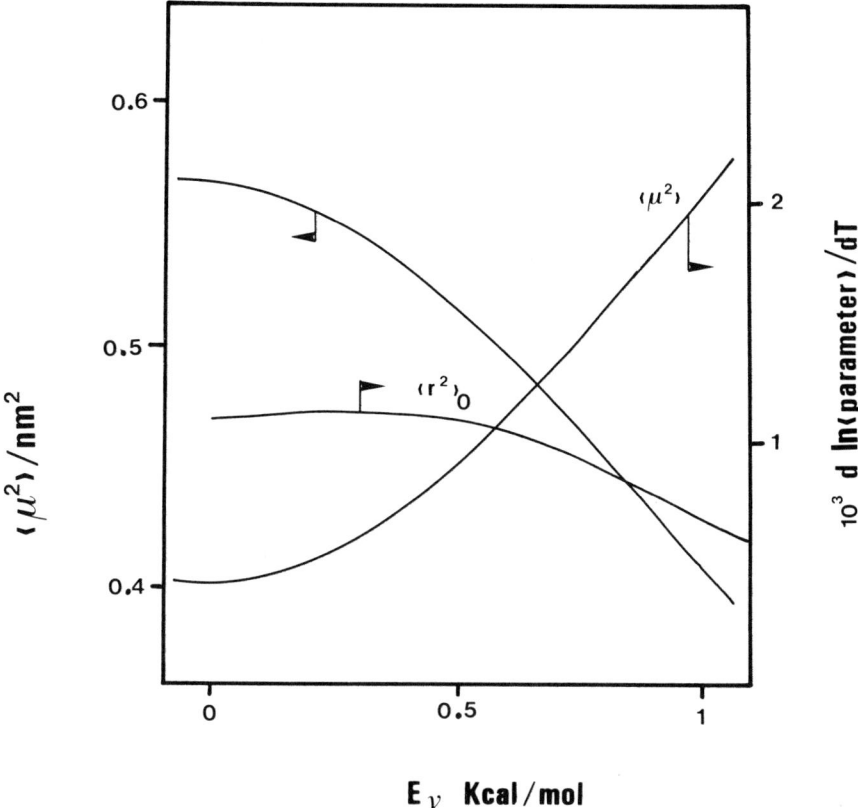

Figure 4.40 Theoretical values of the dipole moment ratio and its temperature coefficient of both the unperturbed dimensions and the dipole moment for poly(diethylene glycol 1,4-trans-cyclohexanedicarboxilate) as a function of $E_\gamma = E(\pi) - E(0)$ about C_{cy}—C^* bonds. (Redrawn from reference [133] by courtesy of the American Chemical Society.)

states about C—C bonds of the glycol residue should have an energy $E_{\sigma'}$ of about 0.6 kcal mol^{-1} below that of the trans states. The value of $\langle \mu^2 \rangle / nm^2$ increases as $E_{\sigma'}$ goes up, because the trans states about the C—C bonds of the glycol residue, which place the dipoles associated with the two ester groups in a nearly parallel direction, increase. As can be seen in Figure 4.42, agreement between theory and experiment is found for $E_{\sigma'} = -0.6$ kcal mol^{-1}. The dipole moment is sensitive to $E_{\sigma\alpha}$, the energy associated with gauche states about CH$_2$— C* bonds of the acid residue, whenever the number of skeletal bonds separating the two carbonyl groups in the acid residue is smaller than 4 [132]. This is reflected in Figure 4.43, where the

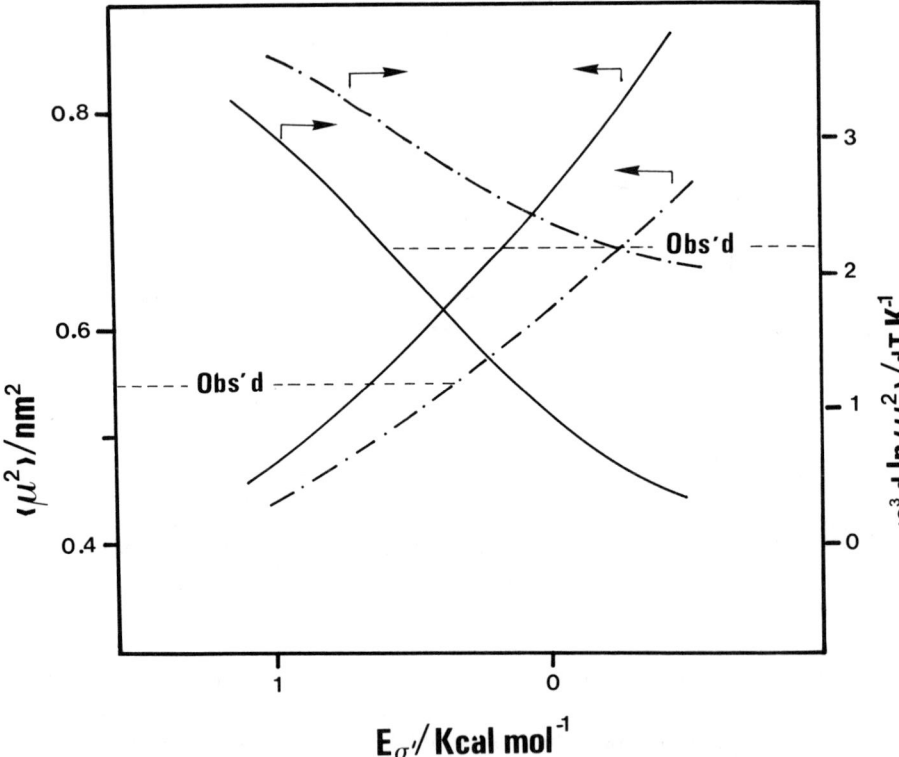

Figure 4.41 Repeating unit of poly(neopentyl glycol succinate) in its planar all-trans conformation.

Figure 4.42 Variation of the dipole moment ratio and its temperature coefficient with the energy $E_{\sigma'}$ of gauche states about C—C bonds of the glycol residue of poly(neopentyl glycol succinate). Continuous and dashed-dotted lines were calculated assuming that the conformational energy $E_{\sigma\alpha}$ of gauche states about C—C* bonds of the acid residue are 0 and 1.2 kcal mol^{-1}, respectively. (Redrawn from reference [139] by courtesy of the American Chemical Society.)

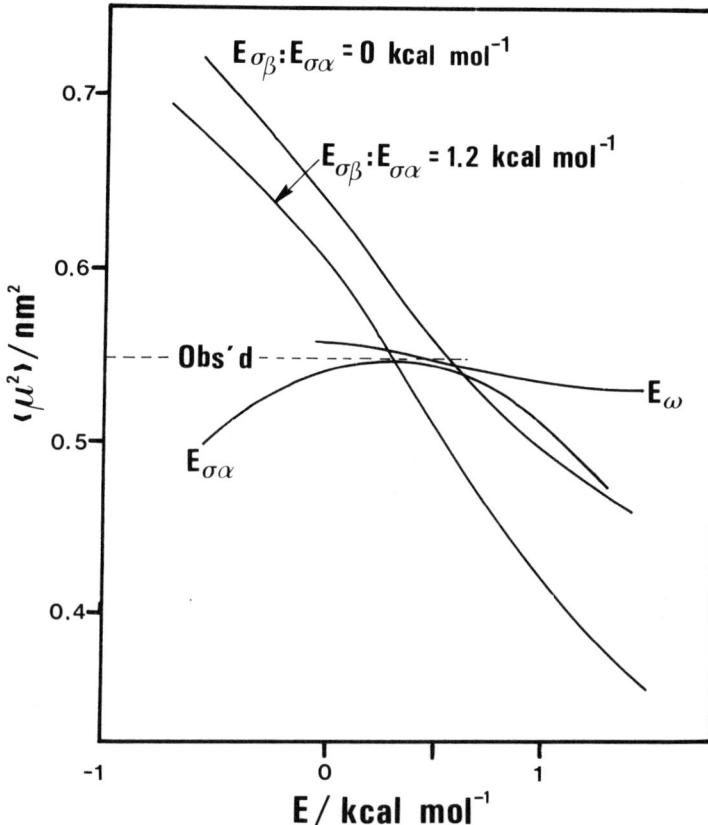

Figure 4.43 Dependence of the dipole moment ratio of poly(neopentyl glycol succi-nate) on the conformational energies indicated in Figure 4.41. (Redrawn from reference [139] by courtesy of the American Chemical Society.)

theoretical values of the dipole moment ratio and the temperature coefficient are represented for poly(neopentyl glycol succinate) [139]. The theory only gives a good account of the experimental results by assuming that $E_{\sigma\alpha}$ is equal to or lower than zero, and the value of the energy $E_{\sigma\beta}$ corresponding to gauche states about CH_2—CH_2 bonds in the acid residue, which give rise to interactions between two carbonyl groups, is 0.7 kcal mol^{-1} above that of the trans. Moreover, the positive value of the temperature coefficient of the unperturbed dimensions of poly(neopentyl glycol adipate) and poly(neopentyl glycol succinate) can only be obtained by assuming that gauche states about CH_2—C^* bonds are favored over the alternative trans. These conclusions seem to be in disagreement with recent NMR spectroscopic studies performed on molecular models of aliphatic polyesters [140].

Polyamides

The theoretical calculation of the dielectric-dependent properties of poly-amides requires the knowledge of both the dipole moment and its orientation on monosubstituted amide groups. However, some problems arise concerning the experimental measurements due to the difficulty of finding suitable solvents for the dielectric measurements [141], that is, solvents with low dielectric permittivity in which the polyamides are soluble, and to the ability of these polymers to form intermolecular associates that can be of different kind depending on the type of solvent, concentration, etc. [142, 143]. In order to overcome this difficulty, the evaluation of the modulus and the orientation of the dipole moment associated to the amide group in polyamides can be obtained from the analysis of the dielectric properties of model compounds containing one or two amide groups together with a detailed conformational analysis of these model compounds.

Figure 4.44(a) shows the amide group in its planar conformation, which is, by overwhelming difference in energy, the preferred structure, and, therefore, the OC—NH bond is considered to be restricted to the trans state [144, 145]. The positive direction is indicated by the arrow and its orientation is determined by the β angle that it makes with the direction of the R—C* group. Values of 120, 119, 110 $\pm$ 5, 110, and 109° are reported for β in the literature [33, 146–149]. Experimental values of the dipole moment for several model compound amides are given [150] in Table 4.11.

The values reported for $\langle \mu^2 \rangle^{1/2}$ of N-methylacetamide (NMA) range from 3.82 to 4.22 D, this quantity being dependent on the ability of the solvent to break associations between the molecules of solute. The molecular structures of N,N'-dimethyloxamide (NDMO), N,N'-dimethyl malonamide (NDMM), N,N'-dimethylterephthal amide (NDMT), and N,N'-dimethyl-trans-1,4-cyclohexanedicarboxamide (NDCD) are shown in planar trans conformation in Figures 4.44(b) to 4.44(e). Geometric considerations suggest that the average dipole moments corresponding to these model compounds can be written as

$$\langle \mu^2 \rangle = 2\mu_A^2 \sin^2 \beta \, (1 - \langle \cos \phi \rangle) \quad \text{for NDMO} \tag{4.31}$$

$$\langle \mu^2 \rangle = \mu_A^2 \langle F^2 \rangle \tag{4.32}$$

for NDMM, where

$$
\begin{aligned}
F^2 = & [\cos \beta \, (\cos \tau - 1) - \sin \beta \sin \tau \cos \phi_2]^2 \\
& + [\sin \beta \, (\sin \phi_1 \sin \phi_2 - \cos \tau \cos \phi_1 \cos \phi_2 - 1) \\
& - \cos \beta \sin t \cos \phi_1]^2 + \{\cos \beta \sin \tau \sin \phi_1 \\
& + \sin \beta \, [\cos \tau \sin \phi_1 \cos \phi_2 + \cos \phi_1 \sin \phi_2]^2
\end{aligned}
\tag{4.33}
$$

$$\langle \mu^2 \rangle = 2\mu_A^2 \sin^2 \beta < 1 - \cos \Phi > \quad \text{with } \Phi = \phi_1 + \phi_2$$

TABLE 4.11   Dipole Moments at 25°C of Some
Substituted Amides [150]

Compound	Solvent	$\langle \mu^2 \rangle^{1/2}$, D
NMA	1,4-dioxane	4.22 ± 0.02
NDMO		1.27 ± 0.03
NDMM		3.28 ± 0.03
NDMT	Mixture of cyclohexane and m-cresol	4.01 ± 0.03
NDCD		3.60 ± 0.05

for NDMT, and

$$\langle \mu^2 \rangle = 2\mu_A^2 \sin^2\beta \, (1 - \langle \cos \phi_1 \cos \phi_2 \rangle - \langle \sin \phi_1 \sin \phi_2 \rangle \qquad (4.34)$$

for NDCD.

Conformational-energy calculations were performed [150] for the model compounds using the van der Waals radii, polarizabilities, and the number of effective electrons given in Table 4.1. The NDMO molecule can be schematically obtained by condensation of two molecules of N-methylamide with elimination of one molecule of ethane. As indicated in Eq. (4.31), the dipole moment of each polar group μ_A can be identified with that of N-methyl acetamide. From the NMR spectrum of tetrabenzyloxamide [151], it can be concluded that the energy barrier about C*—C* bonds is 2.0 kcal mol^{-1}. By assuming that $\epsilon = 2$ for the coulombic interactions, the average $\langle \cos \phi \rangle$ in Eq. (4.31) is found to be 0.9468, which gives a value of 114° for the orientation angle β.

N-methylacetamide can also be used as the model compound for the two polar groups of NDMM because this molecule can be formally obtained by condensation of two N-methylacetamides with elimination of one molecule of methane. Quantitative information on the barrier about C—C* bonds, obtained from crystallographic analysis [152] of the unsubstituted malonamide, suggests that the energy barrier V_0 is 1.0 to 1.5 kcal mol^{-1}. Figure 4.45 shows the energy contours obtained with $\epsilon = 2$ and $V_0 = 1.0$ kcal mol^{-1}, where the presence of two minima can be seen, one located at 65° and the other at 135°, in good agreement with the preferred conformation obtained for unsubstituted malonamide, 65° and 140°. By substituting the average values of ϕ_1 and ϕ_2 obtained from the contour maps into Eq. (4.32), it is found that $\beta = 116 \pm 5°$.

The energy for a given conformation of N,N'-dimethyl terephthalamide is the sum of the energies of two N-methyl benzamide molecules (MB) and a term representing the interactions between two methyl groups:

$$E_{NDMT}(\phi_1,\phi_2) = E_{MB}(\phi_1) + E_{MB}(\phi_2) + E(\phi_1,\phi_2) \qquad (4.35)$$

Since the two amide groups of the NDMT molecule are far apart, the nonbonded and hydrogen-bonding contributions to $E(\phi_1,\phi_2)$ that depend on high powers of the

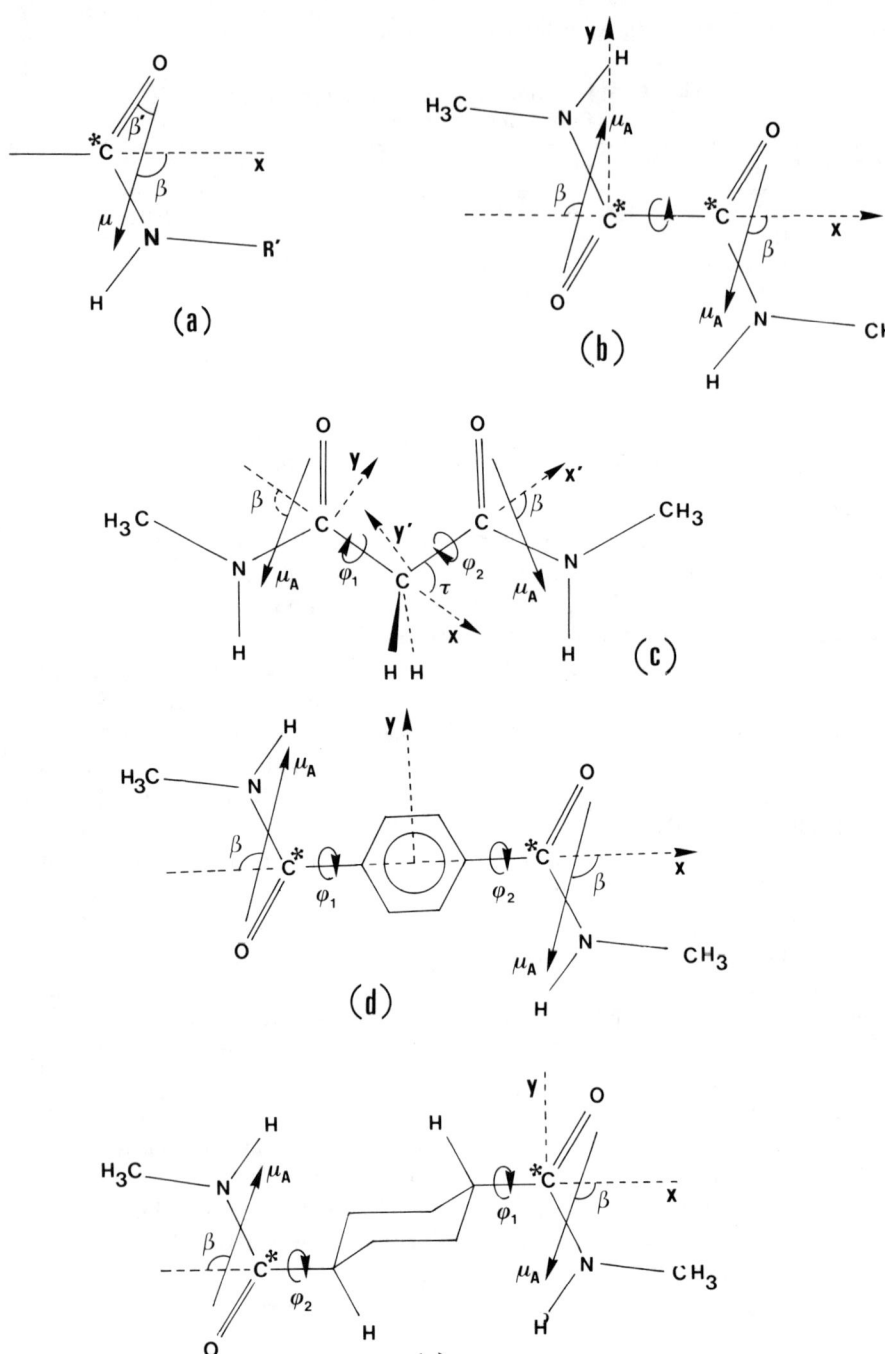

Figure 4.44 Structure of several amide compounds: (a) N-methyl acetamide, (b) N,N'-dimethyloxamide, (c) N,N'-dimethyl malonamide, (d) N,N'-dimethylterephthalamide, and (e) N,N'-dimethyl-trans-1,4-cyclohexanedicarboxamide.

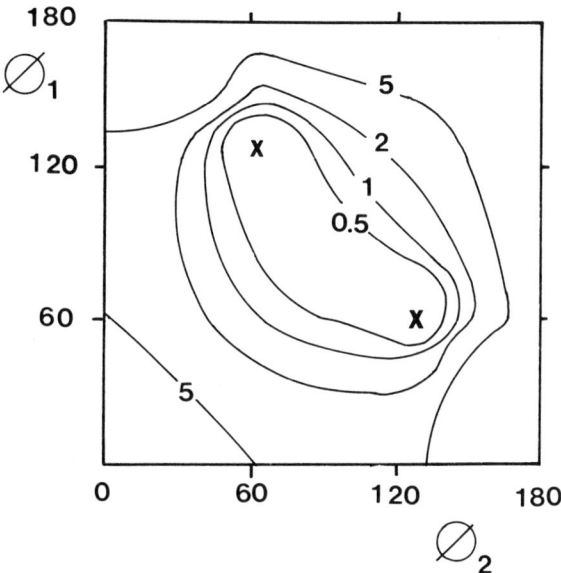

Figure 4.45 Energy contours calculated for N,N'-dimethylmalonamide. (Redrawn from reference [150] by courtesy of the American Chemical Society.)

distance are negligible, and the only significant contribution comes from coulombic interactions [150]. Hence,

$$E(\phi_1,\phi_2) = (E^0_d/2)(1 - \cos \Phi) \qquad (4.36)$$

where E^0_d is the difference in coulombic interactions between the orientations at $\Phi = 0°$ and $\Phi = 180°$. With the partial charges [150] adjusted so as to reproduce the dipole moment $\mu_A = 3.37$ D of MB, one finds $E^0_d = 0.70$ kcal mol^{-1}. The conformational analysis of MB indicates [144] that this molecule exhibits four conformations of minimum energy, all of them equivalent due to the symmetry of the molecule at $\phi = \pm\phi_m$, $180 \pm \phi_m$, with $\phi_m = 37°$; those placed at $\phi = 0°$, $180°$ are about 0.38 kcal mol^{-1} over the minimum and those at $\phi = \pm90°$ have a relative energy of about 2.75 kcal mol^{-1}. These results can be reproduced in the vicinity of the minima by a function of the form

$$E_{MB}(\phi) = E^0_{MB} \cos^2 B\phi \qquad (4.37)$$

where $E^0_{MB} = 0.38$ kcal mol^{-1} is the energy of the conformations $\phi = 0$, $180°$ relative to the minima, and B is a parameter that allows for small variations in the minima. With values of B lying in the range 3 to 4.5 kcal mol^{-1}, the minima ϕ_m lie in the range 20 to 30°. Conformational-energy calculations by using Eqs. (4.35) to (4.37) in conjunction with Eq. (4.33) give values of β ranging from 120° to 118°.

Given the high energies of the boat and skew forms of the cyclohexane ring over the chair form [153] (i.e., 5.5 and 5.0 kcal mol^{-1}, respectively), the cycle in NDCD can be considered to be in the chair form. Owing to the great separation between the two amide groups, interactions between these groups can be neglected [122]. The rotations ϕ_1 and ϕ_2 are not interdependent, so that $\langle \cos \phi_1 \cos \phi_2 \rangle = \langle \cos \phi_1 \rangle^2$ and $\langle \sin \phi_1 \sin \phi_2 \rangle = \langle \sin \phi_1 \rangle^2$. On the other hand, the symmetry of the molecule suggests that the conformations $\pm\phi_1$ have the same energy and, therefore, $\langle \sin \phi \rangle = 0$. Equation (4.34) then becomes

$$\langle \mu^2 \rangle = 2 \mu_A^2 \sin^2 \beta \, (1 - \langle \cos \phi \rangle) \tag{4.38}$$

Potential-energy calculations carried out assuming a threefold torsional function with a height barrier of 2.0 kcal mol^{-1} gave $\langle \cos \phi \rangle^2 = 0.505$. Substitution of this value in Eq. (4.38), together with the values of $\mu_A = 3.99$ D found for N-methylcyclohexane carboxyamide and the experimental result of $\mu = 3.60$ D for NDCD gives $\beta = 115°$.

The values of β obtained from the analysis of model compounds with two amide groups are in good agreement with the result obtained by Khanarian, Msek, and Moore [147] ($\beta = 110 \pm 5°$) through the analysis of substituted N-methylacetamides, with the result of the ab initio calculations of Shipman and Christoffersen [146] ($\beta = 119°$), and with the semiquantitative conclusion of Brant and Flory [154] about the dipole moment being roughly antiparallel to the C^*—O^* bond.

Vinyl Polymers with Bulky Side Groups

The dipole moments of vinyl chains show a strong dependence on the stereochemical composition of the chains [155]. The probable absence of excluded volume effects on $\langle \mu^2 \rangle$ for poly(p-chlorostyrene) chains makes this polymer suitable for investigation of the influence of intramolecular interactions on the conformations of polystyrene and related polymers. However, the most detailed experimental study, carried out by Burshtein and Stepanova [54] some time ago, seemed to suggest that the value of $\langle \mu^2 \rangle / x\mu_0^2$, where $x\mu_0^2$ is the mean-square dipole moment of a freely jointed chain of x repeating units, each having a dipole moment $\mu_0 = 1.68$ D, depends on the thermodynamic quality of the solvent. Even more, the temperature coefficient of the dipole moment reported by these authors shows an inflection at 50°C, suggesting that a change in the conformational behavior of the chain might occur at this temperature. It should be pointed out, however, that more recent studies [55] do not support excluded volume effects in the mean-square dipole moments of these chains.

Values of the dipole moment ratio of vinyl polymers with bulky planar substituent side groups, such as poly(p-chlorostyrene) (PPCS) [54, 55, 157–160], poly(p-bromostyrene) (PPBS) [160], poly(m-chlorostyrene) (PMCS) [157], and poly(vinylcarbazole) (PVK) [156] are given in Table 4.12.

The statistics of these polymers can adequately be represented by two rota-

TABLE 4.12 Experimental Values of the Dipole Moment Ratio D_x and Its Temperature Coefficient TC = $10^3\ d[\ln(D_x)]/dT$ for Some Vinyl Chains

Polymer	Structural Unit	D_x	TC	Reference
Poly(vinyl chloride)	—CH₂CHCl—	0.66–0.72	−1.1, −1.84	[177, 178]
Poly(vinyl bromide)	—CH₂CHBr—	0.45–0.53	—	[179, 180]
Poly(p-chlorostyrene)	—CH₂CH(C₆H₅Cl)—	0.53–0.80	−4.0 to +5.0	[54, 55, 157–160]
Poly(p-bromostyrene)	—CH₂CH(C₆H₅Br)—	2.20–2.44[†]	5.0	[160]
Poly(m-chlorostyrene)	—CH₂CH(C₆H₅Cl)—	1.61[†]		[157]
Poly(vinyl carbazole)	—CH₂CH(NC₁₃H₈)—	0.41	—	[156]
Poly(methyl acrylate)	—CH₂CH(COOCH₃)—	0.68	—	[164]
Poly(phenyl acrylate)	—CH₂CH(COOC₆H₅)—	0.62	> 0	[169]
Poly(p-chlorophenyl acrylate)	—CH₂CH(COOC₆H₄Cl)—	0.52	< 0	[169]
Poly(m-chlorophenyl acrylate)	—CH₂CH(COOC₆H₄Cl)—	0.57	> 0	[169]
Poly(o-chlorophenyl acrylate)	—CH₂CH(COOC₆H₄Cl)—	1.09	≈ 0	[169]
Poly(monobenzyl itaconate)	—CH₂C(COOH)(CH₂COOCH₂C₆H₅)—	3.84[†]	—	[173–176]
Poly(dibenzyl itaconate)	—CH₂C(COOCH₂C₆H₅)(CH₂COOCH₂C₆H₅)—	2.31[†]	—	[173–176]

[†]These values are given as $\langle \mu^2 \rangle / x$.

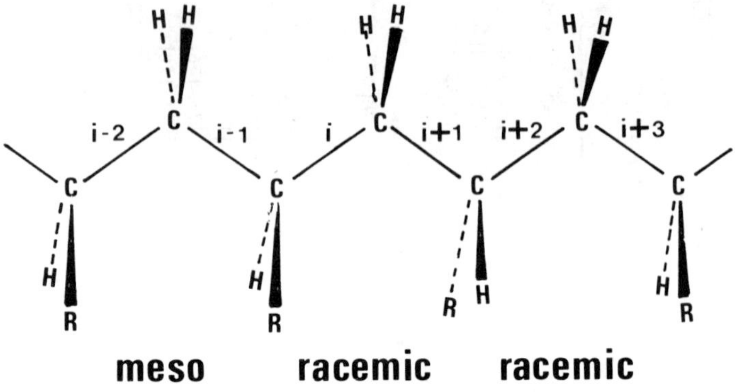

Figure 4.46 Segment of a vinyl chain in the planar all-trans conformation.

tional states, trans (t) and gauche (g), for each skeletal bond of the chain [161, 162]. The C—R bond is restricted to an orientation in which its plane is approximately perpendicular to the plane of the two skeletal bonds of the diad (e.g., bonds $i - 1$ and i in Figure 4.46) because steric repulsion by hydrogens of the adjoining methylene groups imposes this constraint. In $\bar{g}$, in which the C—R bond assumes the position of the methylene in Figure 4.46, the planar R group, orientated as stated, impinges on one of the groups, R, CH_2, or H, pendant to the neighboring substituted carbons of the diad embracing the $\bar{g}$ bond, and, as a consequence, this rotational state is suppressed. The statistical-weight matrix for a pair of bonds flanking a substituted carbon (CH_2—C^α—CH_2) is

$$U' = \begin{bmatrix} 1 & 1 \\ 1 & 0 \end{bmatrix} \tag{4.39}$$

and the statistical-weight matrices U''_m and U''_r for bond pairs (C^α—CH_2—C^α) within a meso and racemic diad are, respectively,

$$U''_m = \begin{bmatrix} \omega'' & 1/\eta \\ 1/\eta & \omega/\eta^2 \end{bmatrix} \qquad U''_r = \begin{bmatrix} 1 & \omega'/\eta \\ \omega'/\eta & 1/\eta^2 \end{bmatrix} \tag{4.40}$$

where the statistical weights are normalized to unity for racemic tt. Here the first-order parameter η measures the preference for trans over gauche. The second-order parameters ω, ω', and ω'' express the effects of repulsion between $CH_2 \cdots CH_2$, $CH_2 \cdots C_6H_5$, and $C_6H_5 \cdots C_6H_5$, respectively. From the critical interpretation of the stereochemical equilibrium in oligomers of polystyrene (PS) [162], the unperturbed dimensions and optical anisotropy of PS, the best set of statistical weight parameters were found to be $\eta = 0.8 \exp(200/T)$; $\omega = \omega' = 1.3 \exp(-100/T)$; and $\omega'' = 1.8 \exp(-1100/T)$.

Monte Carlo methods are usually utilized to generate chains of specific stereochemical composition, and the theoretical results for the dipole moment ratio and its temperature coefficient as a function of the molar fraction of meso diads are represented in Figures 4.47 and 4.48, respectively. In general, the theoretical calculations reproduce satisfactorily the experimental values of $\langle\mu^2\rangle/x\mu_0^2$ and $d[\ln(\langle\mu^2\rangle)]/dT$ for atactic chains [162] ($W_m \approx 0.30$).

The conformational analysis suggests that sequences of preferred diad conformations $|tt|$ in syndiotactic chains are preponderantly interrupted by $|gg|$ conformations, and the dipole moments of the two sequences thus joined become nearly antiparallel. As result, the ratio $\langle\mu^2\rangle/x\mu_0^2$ should vanish in the limit $x \to \infty$. However, the occurrence of tg and gt conformations is responsible for the departures from this prediction. It should be noted that the dipole moment within a $|tt|$ sequence correlates favorably, and the greater the length of y of the sequence, the larger is $\langle\mu^2\rangle/x\mu_0^2$. The average value of y was estimated to be four units [162].

The preferred conformations in isotactic chains are $|gt|$ or $|tg|$. Sequences of these conformations yield the 3_1 Natta–Corradini helix. Here a sequence $|gt|$ of y_1 units can be followed by a sequence $|tg|$ of y_2 units, but the reverse combination is forbidden. Therefore, transition $|tg| \leftrightarrow |gt|$ involves the interposition of $|tt|$ diads between the two sequences. The contribution to the dipole moment per unit in the preferred conformation of the isotactic chain is smaller than in the syndiotactic chain, and the average length of the sequence is also smaller. However, the correlations between successive sequences are favorable, and, hence, the polarity of isotactic chains is higher than that of syndiotactic.

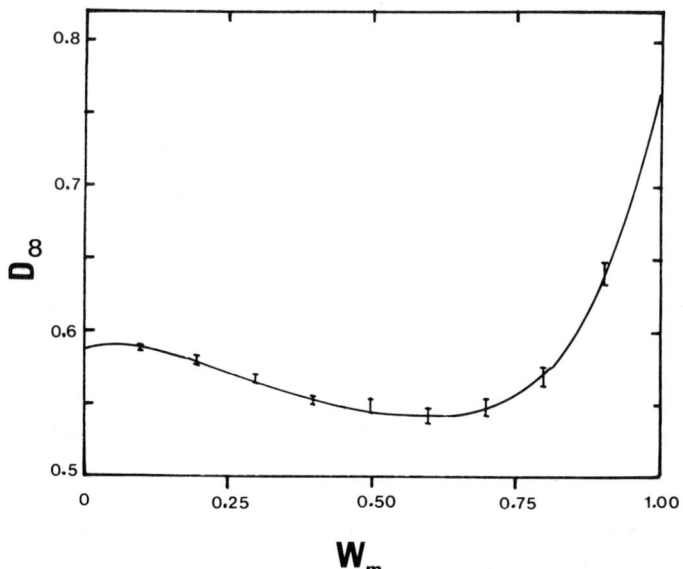

Figure 4.47 The dipole moment ratio for poly(p-chlorostyrene) calculated as a function of stereochemical composition expressed by a priori expectation of meso diad W_m. (Redrawn from reference [162] by courtesy of the American Chemical Society.)

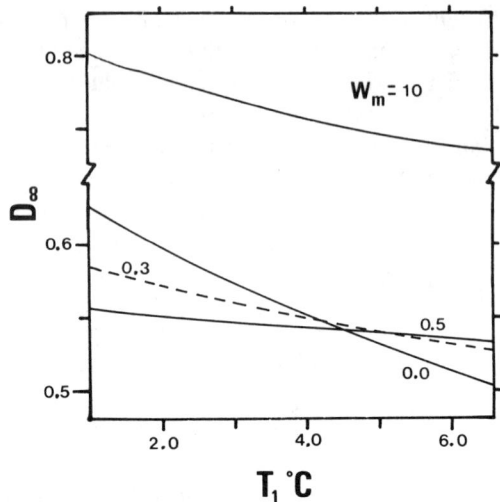

Figure 4.48 Dependence of the dipole moment ratio on temperature for poly(p-chlorostyrene) at the indicated stereochemical compositions. (Redrawn from reference [162] by courtesy of the American Chemical Society.)

Sundararajan [163] and others [156] have used a two rotational states scheme, similar to that described before, for the statistics of PVK. By symmetry, the dipole moment of the repeating unit should lie along the C^α—N bond. The component of this vector referred to the coordinate system affixed to the skeletal bonds $i - 1$ (i.e., C^α—CH_2 bonds) are $-0.3207\mu_0$, $0.4755\mu_0$, and $-0.8191\mu_0$, where μ_0 represents the modulus of the dipole moment of the repeating unit and its value is 2.93 D. As occurs with PS and its halogenated derivatives, the dipole moment ratio decreases as the isotactic fraction increases. In general, the statistics gives a good account of the experimental results.

Esters of poly(acrylic acid) with general formula $[CH_2\!\!-\!\!C^\alpha H(C^*O^*OR)\!\!-\!]_x$, where R can be methyl, phenyl halophenyl groups, etc., are another type of chains that can be described by means of a two rotational states scheme. Here the C^α—C^* bond is restricted to conformations in which the plane containing the C^α—C^*O^*—O moiety is approximately perpendicular to the skeletal bond plane of the diad in all trans conformations. Rotational angles $\chi = 0$, 180° about the C^α—C^* bond of the side groups give rise to two distinguisable conformations representing, respectively, the orientations in which the carbonyl oxygen is synperiplanar and antiperiplanar with the methine hydrogen (Figure 4.49). Rotations about C^α—C^* bonds have, in general, little influence on the rotational states about the skeletal bonds of the diads and, therefore, they are not considered in the calculation of the characteristic ratio or stereochemical equilibrium of vinyl chains. However, these rotations become essential for the evaluation of the mean-square dipole moments of

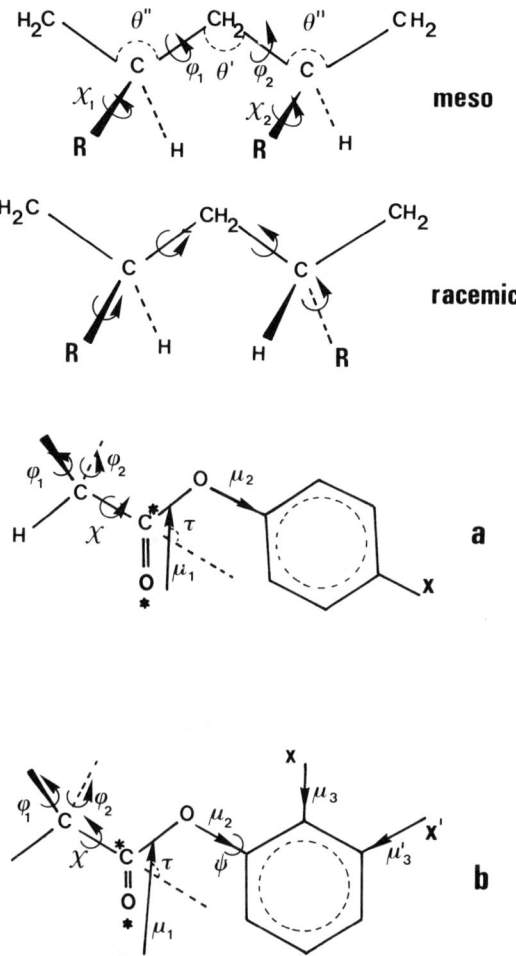

Figure 4.49 Meso and racemic diads of esters of poly(acrylic acid). (a) Side groups of phenyl (X = H) and p-chlorophenyl (X = Cl) esters of poly(acrylic acid). (b) Side groups of o-chlorophenyl (X' = H, X = Cl, $\mu_3' = 0$, and $\mu_3 = 1.6$ D) and m-chlorophenyl (X' = Cl, X = H, $\mu_3' = 1.6$ D, and $\mu_3 = 0$) esters of poly(acrylic acid). In all cases, $\mu_1 = 1.757$ D, $\mu_2 = 0.3$ D, and $\tau = 123°$.

polyacrylate chains because the polarity of the repeating unit lies in the ester side group [164–166]. Consequently, each rotational isomer of the skeleton is split into two states in order to take into account the two possible orientations of the side group. The appropriate statistical weight matrices, with the states in the order ($t,\chi = 0$), ($t,\chi = \pi$), ($g,\chi = 0$), ($g,\chi = \pi$), are [167–170]

$$U' = \begin{bmatrix} 1 & 0 & 1 & 0 \\ 0 & \rho & 0 & \rho \\ 1 & 0 & 0 & 0 \\ 0 & \rho & 0 & 0 \end{bmatrix} \tag{4.41}$$

for the CH_2—C^α—CH_2 bond pair, and

$$U'_r = \begin{bmatrix} 1 & \gamma_1 & 0 & 0 \\ \gamma_1 & \gamma_2 & 0 & 0 \\ 0 & 0 & \alpha & \alpha \\ 0 & 0 & \alpha & \alpha \end{bmatrix} \qquad U'_m = A \times \begin{bmatrix} 1 & \gamma & \beta & \beta \\ \gamma & 1 & \beta & \beta \\ \beta & \beta & 0 & 0 \\ \beta & \beta & 0 & 0 \end{bmatrix} \tag{4.42}$$

for racemic and meso configurations of the C^α—CH_2—C^α bond pair.

The statistical weight of $\chi = \pi$ relative to $\chi = 0$ is denoted by ρ, whereas the factors γ_1 and γ_2 arise from the differences in coulombic interactions of $t0,t\pi$ (or $t\pi,t0$) and $t\pi,t\pi$ relative to $t0,t0$ in the racemic diad. In the meso diad, γ comes from the difference in coulombic interactions between $t0,t\pi$ (or $t\pi,t0$), and $t0,t0$ (or $t\pi,t\pi$). A and β are combinations of statistical weights defined as $A = \omega''\delta_m/\delta_r$ and $\beta = 1/\eta\delta_m\omega''$, where ω'' represents the weight of the second-order interactions between two ester groups juxstaposed as in the meso tt state, η is the first-order statistical weight for trans versus gauche, and δ_m and δ_r represent Boltzmann factors of the coulombic interactions in $t0$, $t0$ orientations of meso and racemic diads, respectively. Finally, α governs the stability of gg versus tt in the racemic diad. A value of $\alpha = 0$ has been used in several cases, thus allowing only tt conformations for racemic diads. However, recent [1]H—NMR analysis [168] of racemic bis(o-chlorophenyl)2,4-dimethyl glutarate suggests that the fraction of gg diads may amount to $f_{gg} = 0.18$. Because

$$\alpha = Z_{\alpha\alpha}/Z \tag{4.43}$$

where $Z_{\alpha\alpha}$ and Z represent, respectively, the conformational partition function for gg conformations and the total conformational partition function in the racemic diad; the statistical weight α is related to the rest of statistical weights by the expression

$$\alpha = \frac{(1 + 2\gamma_1\rho + \gamma_2\rho^2)f_{gg}}{(1 + 2\rho + \rho^2)(1 - f_{gg})} \tag{4.44}$$

The statistics described before were used for the critical interpretation of the dipole moment ratio of poly(methyl acrylate) (PMA), poly(phenyl acrylate) (PPA), and its halogenated derivatives poly(o-chlorophenyl acrylate) (POCPA), poly(m-chlorophenyl acrylate) (PMCPA) and poly(p-chlorophenyl acrylate) (PPCPA). The experimental values of $\langle\mu^2\rangle/x\mu_0^2$ for these polymers are given in Table 4.12, where the values of μ_0^2 are considered to be those indicated in Table 4.13 for methyl propionate, phenyl acetate (PA), and the corresponding halogenated derivatives,

TABLE 4.13   Dipole Moments
(at 30°C) of Methyl Propionate (MP),
Phenyl Propionate (PP),
p-Chlorophenyl Propionate (PCPP),
m-Chlorophenyl Propionate (MCPP),
and o-Chlorophenyl Propionate
(OCPP) [171]

Compound	μ, D
MP	1.76
PP	1.65
PCPP	2.56
MCPP	2.45
OCPP	1.79

whose polar characteristics are similar to those of the side groups of the polymers under consideration [171].

For assymmetric esters such as o-chlorophenyl propionate (OCPP) and m-chlorophenyl propionate (MCPP), the value of μ_0 is strongly dependent on the rotational angle Φ about the O—Ph bond. Potential-energy calculations show that the conformational energy of OCPP is similar to that of PA with the difference that the minimum appearing at $\pm 60°$ in PA is shifted to $\pm 75°$ in OCPP due to the interactions of Cl with both C^* and O^* atoms, which in this range of values of Φ are attractive and reach their minima at $\pm 75°$ for the Cl,C^* pair ($d_{ClC^*} = 3.55$ Å) and at $\Phi \approx 100°$ for Cl,O^* pair ($d_{ClO^*} = 3.35$Å). As Φ increases, the repulsive interactions are dominant and the conformation at $\Phi = \pm 120°$ has an energy around 4 kcal mol^{-1} higher than that of $\Phi = \pm 75°$. Interactions between the Cl atom and the carbonyl group of the ester residue are negligible in MCPP, and the minima for this compound, as for PA, are located at 0, $\pm 120°$. The distribution of dipoles associated with the side groups are represented in Figure 4.49. The contribution of the dipoles, evaluated using the rotational minima about O—Ph bonds for OCPP and MCPP, and the statistics described before give a good account of the experimental dipole moments of phenyl and chlorophenyl esters of both 2,4 dimethylglutaric acid and polyacrylic acid by refining some statistical-weight parameters. It should be pointed out that in all cases, parameter ρ was found to be close to unity, suggesting that the occurrence of the two rotational angles about C^α—C^* bonds have approximately the same probability.

Particular importance also involves the studies of the conformation-dependent properties of the esters of poly(itaconic acid). The value of $(\langle\mu^2\rangle/x)$ for poly-monobenzyl and polydibenzyl itaconate are given in Table 4.12. The critical interpretation of the experimental results shows that statistics based on two rotational states, similar to that proposed by Sundararajan and Flory [172] for poly(methyl methacrylate), give a good account of both the dipole moment and the unperturbed dimensions reported for these polymers [173–176]. These studies also suggest that the dipole moment is not very sensitive to the stereoregularity of the chains.

Vinyl Polymers with Halogen Atoms Attached to the Main Chain

The experimental results on the dipole moments of poly(vinyl chloride) (PVC) [177, 178] and poly(vinyl bromide) (PVB) [179, 180] are sparse due to the possibility of aggregates formation and to the instability of the former and latter polymers, respectively. Experimental values of $\langle \mu^2 \rangle / x\mu_0^2$ for both polymers, given in Table 4.12, indicate that PVC is more polar than PVB. Here the values of μ_0^2 for PVC and PVB were assumed to be those of ethyl chloride and ethyl bromide, which amount to 2.00 and 2.08 D, respectively. The statistical-weight matrices associated with the skeletal bonds of the repeating unit can be expressed in a three rotational states scheme by [181, 182]

$$ \mathbf{U'} = \begin{bmatrix} 1 & 1 & 1 \\ 1 & 1 & \omega \\ 1 & \omega & 1 \end{bmatrix} \tag{4.45} $$

for the CH_2—C^α—CH_2 pair of bonds, and

$$ \mathbf{U''_m} = \begin{bmatrix} \omega''\eta^2 & \eta\tau\omega' & \eta \\ \eta & \tau\omega' & \omega \\ \eta\tau\omega' & \tau^2\omega\omega'' & \tau\omega' \end{bmatrix} \qquad \mathbf{U''_r} = \begin{bmatrix} \eta^2 & \eta\omega' & \eta\tau\omega' \\ \eta\omega' & 1 & \tau\omega \\ \eta\tau\omega'' & \tau\omega & \tau^2\omega'^2 \end{bmatrix} \tag{4.46} $$

for meso and racemic diads about the pair of bonds C^α—CH_2—C^α.

The statistical-weight factors appearing on those matrices have been obtained by Flory and Williams [183] by qualitative estimates based on the relative sizes of interacting atoms and by analysis of the stereochemical equilibrium composition of 2,4-dichloro-n-pentane. These values, at 25°C, were $\eta = 4.2$, $\tau = 0.45$, $\omega = \omega'' = 0.032$, and $\omega' = 0.071$. Theoretical values of the dipole moment ratio as a function of the stereochemical composition were obtained by assuming either that the dipole moment associated with the repeating unit is only due to the contribution of the C—Cl bond or that it is the result of the contribution of both the C—Cl bond and the C—H methine bond, as indicated in Figure 4.50. The results obtained as a function of the angle β, which forms the dipole moment of the repeating unit with the C—Cl bond, indicate [178] that the experimental results can only be reproduced for values of β lying in the range 10 to 15°.

Potential-energy calculations were performed on 2,4-dibromopentane, a model compound of PVB, using the parameters indicated [180] in Table 4.1. Comparison of the partition functions Z with the statistical-weight matrices for vinyl polymers allows the calculation of all statistical weights (e.g., $\eta = Z^m_{tg}/Z^r_{gg}$), which are then written as the Boltzmann exponential of the corresponding energies (e.g., η

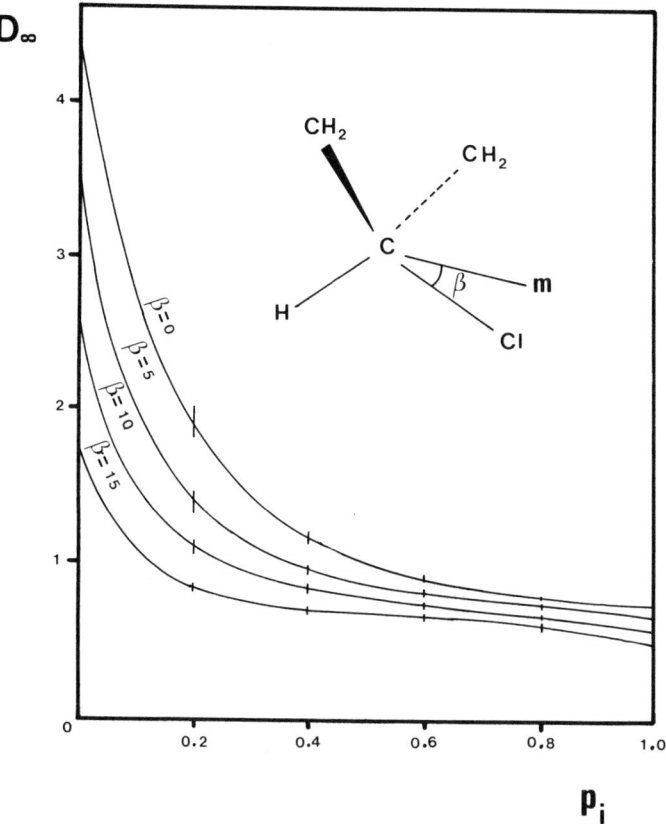

Figure 4.50 Changes on the dipole moment ratio of poly(vinyl chloride) with the probability of isotactic replacement for different values of β. (Redrawn from reference [178] by courtesy of the American Chemical Society.)

$= \exp[-E_\eta/RT])$. The values at 25°C of the statistical-weight parameters are $\eta = 1.29$, $\tau = 0.43$, $\omega = 0.015$, $\omega' = 0.034$, and $\omega'' = 0$. As in PVC, the rotational angles were found to differ from the perfectly staggered values (0, 120, and $-120°$) by less than 5°. Agreement between theoretical and experimental results for $\langle\mu^2\rangle/nm^2$ is achieved by assuming that the angle β that forms the dipole moment of the repeating unit and the C—Br bonds lies in the vicinity of 5°, which is significantly lower than that found for PVC. A lower value for PVB than for PVC is expected because of the smaller partial charge located over the tertiary carbon of the polymer due to the longer C—Br bond length compared with that of C—Cl bond.

The great difference in the values of η for PVB and PVC is due to the difference in size between Cl and Br atoms. Thus, in a racemic diad in its *tt* conformation, there is a significant interaction between the Br and the H atoms attached to the next tertiary C, and even between this tertiary C and Br, these interactions being relieved in the *gg* conformation. Although *tt* is still the preferred conformation, the difference in energy with *gg* is much smaller than in the case of PVC, where such interactions are negligible. For meso diads, the value of η is not so important, because it is still much higher than the value of ω and, as a consequence, the conformations *tg* and *gt* are preferred. The theoretical results predict that the value of the dipole moment ratio of syndiotactic chains is above that of isotactic chains, but whereas in the case of PVC the difference is large, it is small for PVB. The explanation lies in that η governs the probability of occurrence of *tt* conformations with respect to *gg* conformations in syndiotactic chains. Therefore a high value of η, as occurs in syndiotactic PVC, suggests the occurrence of long *tt . . . tt* sequences with favorable polarity. However, for values of η close to unity, as occurs in PVB, both *tt* and *gg* conformations have almost the same probability. The addition of projections of the dipoles takes place practically on isolated diads, and, hence, the values of $\langle \mu^2 \rangle$ for isotactic and syndiotactic PVB chains differ very little.

The dependence of the polarity of the chains on the stereochemical structure decreases as the separation between two consecutive side groups increases [184]. This can be seen in Table 4.14, where values of $\langle \mu^2 \rangle / nm^2$ as a function of the stereochemical structure is shown for several polymers. Thus, for poly(4,methyl-1,3-dioxolane), the distance between the side groups is much too large for much

TABLE 4.14 Influence of Tacticity on the Dipole
Moment Ratio of Several Polymers [184]

Polymer	Structural Unit	D_x
Poly(vinyl chloride)	$-CH_2CHCl-$:	
Isotactic		0.59
Atactic		1.0
Syndiotactic		4.0
Poly(*p*-chlorostyrene)	$-CH_2CH(C_6H_4Cl)-$:	
Isotactic		5.5
Atactic		0.65
Syndiotactic		0.92
Poly(propylene oxide)	$-CH(CH_3)CH_2O-$:	
Isotactic		0.50
Atactic		0.45
Syndiotactic		0.42
Poly(propylene sulfide)	$-CH(CH_3)CH_2S-$:	
Isotactic		0.41
Atactic		0.39
Poly(4-methyl-1,3-dioxolane)	$-CH_2OCH(CH_3)CH_2O-$:	
Isotactic		0.20
Atactic		0.22
Syndiotactic		0.20

correlation, thus explaining why the stereochemical structure has little effect on statistical properties such as dipole moment. The results of Table 4.14 suggest that correlations are somewhat larger, but still relatively small, for chains in which the side groups are separated by three skeletal bonds as occurs in poly(propylene oxide) (PPO) and poly(propylene sulfide) (PPS).

The situation for typical vinyl chains, where the side groups are separated by only two skeletal bonds, is very different, as illustrated in the last part of the table. As was indicated before, the large increase in the dipole moment ratio with decreasing isotacticity in PVC is because the preferred conformation of the syndiotactic modification of this polymer is planar zigzag, in which the group dipoles have components that are nearly parallel. In polymers with bulkier side groups, such as poly(p-chlorostyrene), steric interferences involving these groups make the preferred conformation of the isotactic form a helical arrangement, with each group dipole having a component pointing in the same direction along the helix axis; hence, the high polarity exhibited by perfectly isotactic chains.

Silicon-Based Polymers

Poly(dimethylsiloxane) is by far the most important silicon-based polymer. An interesting feature of this polymer is the possibility of synthesizing both linear chains and cycles $[Si(CH_3)_2O]_x$ covering a wide range of degree of polymerization [185]. Another important characteristic of poly(dimethyl siloxanes) is that the atomic polarization P_A is large and, therefore, cannot be neglected in the evaluation of the dipole moments of the chains. The value of P_A in Å^3 is related to the length of the chain by the equation [186]

$$P_A = 7.81 + 55.66x \tag{4.47}$$

where x is the degree of polymerization.

Experimental values of the dipole moment ratio are given in Figure 4.51, where the dipole moment associated with each skeletal bond was estimated to have a value of 0.6 D [186]. The results obtained in the cycles indicate that they have values of the dipole moment ratio similar to those of the corresponding linear chains [185, 187]. As a consequence, the dipole moment thus seems less generally useful than other properties that have been used to characterize differences between PDMS linear chains and cycles [188]. Moreover, the differences between the results in different solvents and in the undiluted state are most likely a manifestation of a "specific solvent effect," that is, segment–solvent interactions that presumably change conformational sequences along the chain backbone [53]. This effect has previously been demonstrated for molecular dimensions.

The critical interpretation of the unperturbed dimensions of the chains suggests that gauche states are disfavored by about 0.85 kcal mol^{-1} with respect to the trans. On the other hand, the conformations that give rise to pentane-type interferences are forbidden when the interacting species are pairs of $Si(CH_3)_2$ and

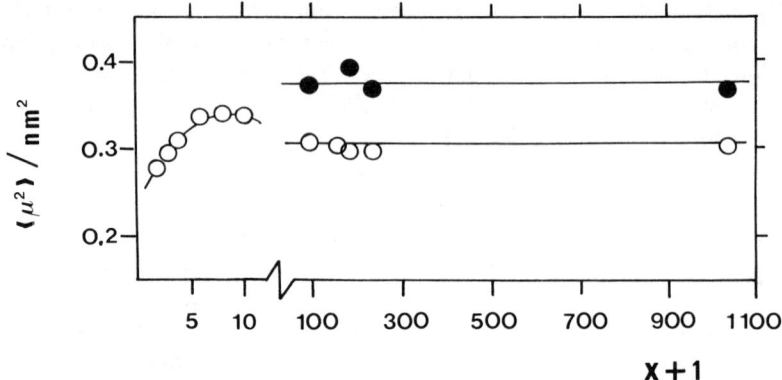

Figure 4.51 Experimental values of the ratio $\langle \mu^2 \rangle / nm^2$ for dimethyl siloxane chains shown as a function of the degree of polymerization. In the upper portion of the figure, the filled circles refer to dielectric results obtained in cyclohexane at 25°C and the open circles to results obtained on the undiluted chains at the same temperature. (Redrawn from reference [155] by courtesy of the American Chemical Society.)

partially suppressed when they are pairs of oxygen atoms [19, 189]. Fairly good agreement between theory and experiment is achieved for very short chains and long chains in which the experimental dipole moments were obtained from dielectric measurements performed in the undiluted state [53, 186]. However, for the dipole moments of long chains obtained in solution, the agreement is not as good, probably due to specific solvent interactions [155].

The high preference for trans states explains the low polarity exhibited by poly(dimethylsiloxane) chains. Displaying the chain in all trans conformations of minimum conformational energy gives a closed polygon, such as is shown in Figure 4.52, whose dipole moment is essentially zero.

Experimental studies on polysiloxane chains with assymmetric centers are scarce. A recent study [190, 191] carried out on a well-characterized poly(methylphenylsiloxane) (PMPS) and copolymers of dimethylsiloxano-methylphenylsiloxane showed that the polarity of the chains is similar to that of high molecular-weight samples of poly(dimethylsiloxane) although the dipole moment ratio of the copolymers seems to be somewhat lower than that of PMPS. The dipole moment of these polymers shows, in general, an anomalous dependence on temperature, characterized by a maximum at $T = 47.5$°C, probably caused by specific solvent effects between polymer and solvent (cyclohexane) on the one hand and by changes on the atomic polarization on the other hand. In fact, this effect seems to be smaller for the dipole moments obtained from dielectric measurements on the bulk [191].

Polytetrafluoroethylene (PTFE)

Quantitative studies on the conformation-dependent properties of PTFE are sparse due to the lack of solvents at suitable temperatures that permit convenient measurements of light scattering, viscosity, dipole moments, etc. A useful approach to obtain information on the statistics of this polymer is the critical interpretation of the dipole moment of low molecular-weight analogues of PTFE, such as α,ω-dihydroperfluoralkane compounds, $H(CF_2)_{n-1}H$, having values of n in the range of 5 to 11 (Figure 4.53). Values of $\langle\mu^2\rangle^{1/2}$ for these compounds are given in Figure 4.54. Although PTFE is structurally similar to polyethylene, X-ray measurements on the crystalline former polymer show that the most stable conformation of a perfluoroalkane chain in the crystal is a slowly twisting helix in which each C—C bond is rotated in the same sign about $17°$ from the planar trans conformation. Semiempirical calculations predict that the conformation with $\phi = 17 \pm 2°$ has an energy of 0.2 to 0.3 kcal mol^{-1} below that of $\phi = 0°$, presumably as a consequence of the relatively large van der Waals radius of the F atom and the polarity of the C—F bonds. A four states model has been used with states t^+ and $t^- \approx \pm15°$ and g^+ and $g^- \approx \pm120°$. By assigning an arbitrary statistical weight of unity to the helical conformation $t^+t^+t^+t^+$ or $t^-t^-t^-t^-$, the statistical-weight matrices for internal bonds k ($4 \le k \le n - 1$) is [192]

$$
\mathbf{U}_k =
\begin{array}{c}
 \\
15 \\
120 \\
-120 \\
-15
\end{array}
\begin{array}{cccc}
15 & 120 & -120 & -15 \\
\left[\begin{array}{cccc}
1 & \sigma & 0 & \omega \\
1 & \sigma & 0 & 0 \\
0 & 0 & \sigma & 1 \\
\omega & 0 & \sigma & 1
\end{array}\right]
\end{array}
\tag{4.48}
$$

where $t^+_{k-1}g^-_k$ (or $t^-_{k-1}g^+_k$) and $g^+_{k-1}g^-_k$ (or $g^-_{k-1}g^+_k$) conformations are suppressed as result of severe steric interactions between CF_2 groups separated by three C—C and four C—C bonds, respectively.

Semiempirical calculations indicate that there is very little hindrance to rotation about bond 2, and the corresponding statistical-weight matrix is

$$
\mathbf{U}_2 =
\begin{array}{c}
 \\
15 \\
120 \\
-120 \\
-15
\end{array}
\begin{array}{cccc}
15 & 120 & -120 & -15 \\
\left[\begin{array}{cccc}
1 & 0 & 0 & 0 \\
0 & \sigma' & 0 & 0 \\
0 & 0 & \sigma' & 0 \\
0 & 0 & 0 & \sigma'
\end{array}\right]
\end{array}
\tag{4.49}
$$

For the statistical-weight matrix associated with the third bond, U_3, $g^-_{k-1}g^+_k$ conformations are not forbidden because $CF_2 \cdots H$ nonbonded interactions do not repulse as much as the corresponding $CF_2 \cdots CF_2$ interactions. This matrix is given by

$$
\mathbf{U}_3 =
\begin{array}{c}
 \\
15 \\
120 \\
-120 \\
-15
\end{array}
\begin{array}{cccc}
15 & 120 & -120 & -15 \\
\left[\begin{array}{cccc}
1 & \sigma & 0 & \omega \\
1 & \sigma & \sigma\omega & 1 \\
1 & \sigma\beta & \sigma & 1 \\
\omega & 0 & \sigma & 1
\end{array}\right]
\end{array}
\tag{4.50}
$$

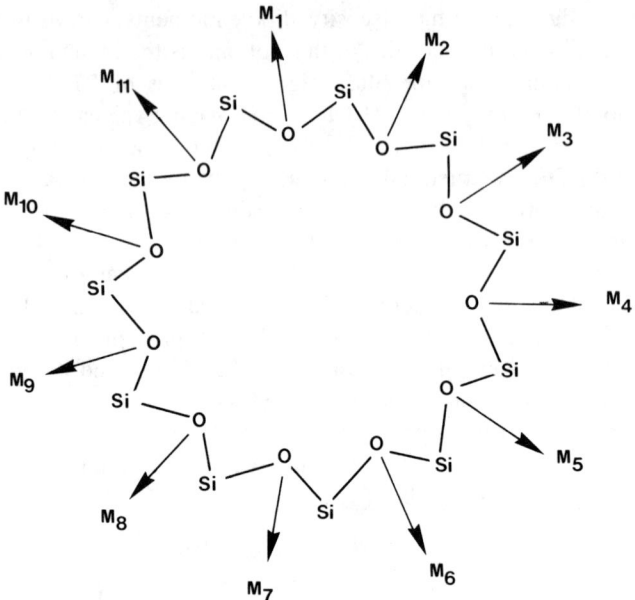

Figure 4.52 A segment of the dimethyl siloxane chain at its minimum energy all-trans planar conformation. The arrows refer to the dipoles associated with each $Si(CH_3)_2$—O—$Si(CH_3)_2$ pair of bonds.

whereas, for consistency, U_n must be written as

$$
\mathbf{U}_n = \begin{array}{c}
15 \\
120 \\
-120 \\
-15
\end{array}
\begin{array}{cccc}
15 & 120 & -120 & -15 \\
\left[\begin{array}{cccc}
1 & \sigma' & \sigma' & \omega \\
1 & \sigma' & \sigma'\beta & 0 \\
0 & \sigma'\beta & \sigma' & 1 \\
\omega & \sigma' & \sigma' & 1
\end{array}\right]
\end{array}
\tag{4.51}
$$

In the evaluation of $\langle\mu^2\rangle$ by the RIS model, it was assumed that the only contributions to the molecular dipole moment come from the terminal H—C and C—H bonds (bonds 1 and $n + 1$ in Figure 4.53). Identification of the dipole moment of the C—H ($\mu_{C—H}$) bond with the dipole moment of 1.60 D found for

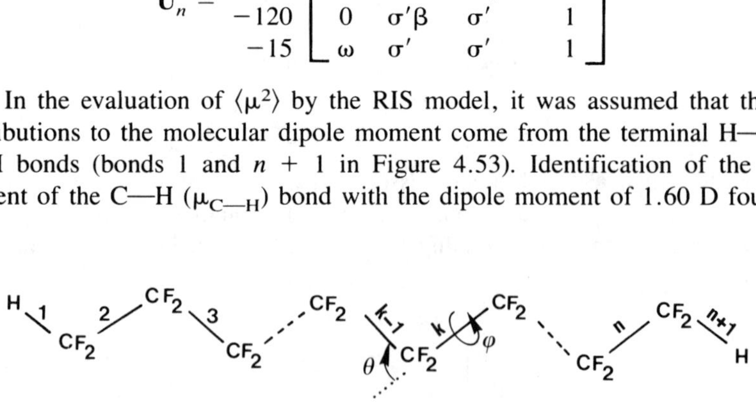

Figure 4.53 The α,ω-dihydroperfluoroalkane chain, $H(F_2)_nH$.

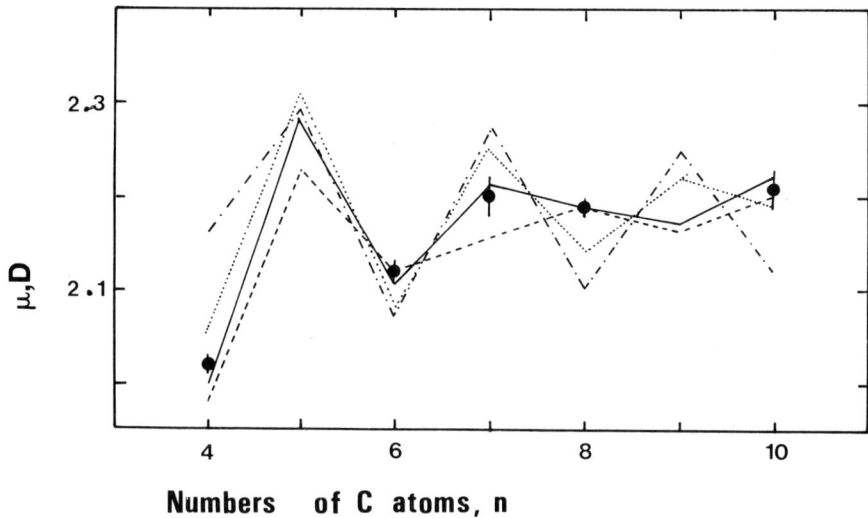

Figure 4.54 Experimental (filled circles) and calculated values of the dipole moments for $H(CF_2)_nH$. The lines connect points calculated using the four states model with $\theta = 64°$, $\phi(t) = \pm15°$, $\phi(g) = \pm120°$, $\mu_1 = 160$ D, $\sigma' = 2.0$, $\sigma\sigma'\beta = 0.20$, and with various values of the statistical weights σ and ω being as follows: dashed line—$\sigma = 0.50$ and $\omega = 0.20$; continuous line—$\sigma = 0.20$ and $\omega = 0.05$; dotted line—$\sigma = 0.05$ and $\omega = 0.05$; and dashed-dotted line—$\sigma = 0.01$ and $\omega = 0.20$.

$H(CF_2)_7F$ was assumed. Theoretical values of $\langle\mu^2\rangle$ as a function of the statistical-weight parameters are represented [192] in Figure 4.54. The results indicate that the best set of statistical weights that reproduce the experimental values are $\sigma = 0.20$, $\omega = 0.05$, $\sigma' = 2.0$, and $\sigma\sigma'\beta = 0.20$. It should be pointed out, however, that a less realistic three states scheme, which assumes that the trans conformation is planar ($\phi = 0$), also gives a good account of the experimental results. Fortunately, this ambiguity does not seriously jeopardize the assessment of the gauche trans-energy difference. Thus, the four states model gives $E_\sigma = E(t^+g^+) - E(t^+t^+) = 1.4 \pm 0.4$ kcal mol^{-1} for perfluoralkane chains, whereas the corresponding energy difference in the three states model is $E_\sigma = E(tg) - E(tt) = 1.2 \pm 0.2$ kcal mol^{-1}. These values are in satisfactory accord with estimates from infrared spectra and from semiempirical energy calculations [192].

Chemical Copolymers

The use of configuration-dependent properties is important to determine the sequence distribution in chemical copolymers [193]. Although high-resolution nuclear magnetic resonance is generally used for this purpose, stereoregularity effects may superimpose upon sequence distribution, leading to complex NMR spectra

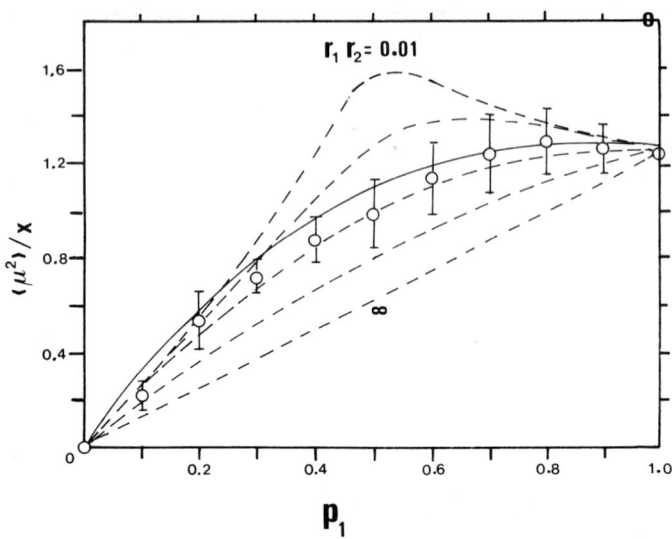

Figure 4.55 The mean-square dipole moment per monomer unit for atactic poly(p-chlorostyrene-p-methylstyrene) shown as a function of the probability of occurrence of p-chlorostyrene units. The circles and standard deviation lines represent theoretical results obtained using values of the replication probability and reactivity ratio product, $r_1 r_2$, appropriate for such chains. The continuous curve represents the experimentally observed dependence. The dashed lines represent calculated values for $r_1 r_2 = 0.01, 0.10, 1.00$, and 10.0, respectively. (Redrawn from reference [194] by courtesy of the American Chemical Society.)

with many overlapping resonances that can preclude an unambiguous analysis of monomer sequence distribution. As was indicated before, dipole moments and molar Kerr constants are more sensitive to structure than the molecular dimensions because the skeletal bonds often vary much more in polarity than they do in length. The chemical-sequence distribution of copolymers made up of units of type A and B is controlled by the reactivity ratio product $r_A r_B$. Thus, for $r_A r_B = 0.0, 1, 1000$, and ∞, the values of the sequence lengths $n_A = n_B$ are, respectively, 1.0 (perfectly alternating copolymers), 2.0 (ideal copolymers), 24 (block copolymers), and ∞ (mixtures of homopolymers) [194].

Experimental values of the mean-square dipole moment per repeating unit, $\langle \mu^2 \rangle / x$, for atactic p-chlorostyrene-p-methylstyrene are given in Figure 4.55. Here the dependence of $\langle \mu^2 \rangle / x$ on the fraction p_1 of p-chlorostyrene units shows a marked positive deviation from the simple linearity between the values of $\langle \mu^2 \rangle / x$ characterizing the parent homopolymers. Theoretical values of $\langle \mu^2 \rangle / x$ as a function of the chemical composition for different values of $r_A r_B$ are also given in Figure 4.55. Monte Carlo methods are generally used to obtain chains with a given composition, and, within each chemical composition, the same methods are utilized to obtain chains in which the stereochemical composition is atactic. Since the statistics for poly(p-chlorostyrene) and poly(p-mehylstyrene) is similar, the average value of $\langle \mu^2 \rangle / x$ for each composition can readily be calculated, and the results obtained are shown in Figure 4.55. A noteworthy feature of these results is that $\langle \mu^2 \rangle / x$ is quite sensitive to sequence distribution at constant chemical composition. Therefore, measurements of dipole moments can be used to obtain reliable values of $r_A r_B$ and, thus, values of the chemical sequence length in the copolymer. Similar sensitivity of the dipole moment to chemical composition has been reported for other chemical copolymers [195]. Thus, the dipole moments of poly(styrene-co-p-chlorostyrene), poly(styrene-co-p-methoxystyrene), poly(styrene-co-4-vinylpyridine), and poly(styrene-co-N-vinylcarbazole) were recently measured in toluene at different temperatures to investigate the effect of the nature of the polar group, composition, and temperature on the dipole moment per structural unit [196]. It was found in all the cases that the plot of this quantity versus composition shows a positive deviation from linearity, in agreement with the calculation reported earlier by Mark [194].

5
Dielectric Relaxations

In general, the term *relaxation* means the time-dependent return to equilibrium of a system that has been perturbed by a change in an applied constraint. The change in the constraint may be discontinuous or periodic, and the experimental observation is of some property that changes with time. When the constraint that undergoes a change is the electric field, the observable quantity that varies with time is the polarization, giving rise to a dielectric relaxation. Among all the contributions to the total polarization, the orientation polarization is the most important from a structural point of view, because it depends on the internal structure of the molecules of the dielectric. Therefore, the critical interpretation of the dielectric relaxation produced by the orientation polarization is important to obtain information on the molecular structure and intermolecular arrangement of the molecules of the system.

Under the effect of an applied field, the polar molecules of the system rotate toward the direction of the field, reaching an equilibrium distribution in molecular orientation with a corresponding dielectric polarization. It is obvious that the value of this quantity decreases with the molecular size, the viscosity of the medium, and the frequency of the alternating electric field, because the orientation of the molecules does not allow attainment of equilibrium with the field. Consequently, the polarization is a complex quantity with a component out of phase with the field, whereas the other component in phase is a conductance quantity that accounts for the displacement current that results in the thermal dissipation of energy.

PHENOMENOLOGICAL APPROACH

The application of a constant electric field E to a dielectric resolves into the development of an electric displacement D that increases with time until a constant value is reached if time is sufficiently long. After removal of the electric field, the electric displacement returns with time to zero. These results, somewhat idealized, are shown in Figure 5.1. At sufficiently small electric displacements, $\epsilon = D/E$, that is, the system is linear, and, consequently, it conforms to the superposition principle [58, 197, 198]. The dielectric constant is a time-dependent property, its value creasing from ϵ_∞ for $t = 0$ to ϵ_0 for $t \rightarrow \infty$. The electric displacement can obviously be written as

$$D(t) = [\epsilon_\infty + (\epsilon_0 - \epsilon_\infty)\Phi(t)]E \qquad (5.1)$$

where $\Phi(t)$ is the built-up normalized dielectric function whose extreme values are

$$\Phi(t) = 1 - \phi(t) = 0 \qquad \text{for } t = 0 \qquad (5.2)$$

$$\Phi(t) = 1 - \phi(t) = 1 \qquad \text{for } t = \infty \qquad (5.3)$$

where $\phi(t)$, the normalized decay function of the polarization when a steady macroscopic electric field is removed from the medium, acquires the values 1 and 0 at $t = 0$ and $t = \infty$, respectively. If an electric field is applied at $t = 0$, but it increases by an infinitesimal amount, dE, at $t = \theta$ $(0 < \theta < t)$, the increase in electric displacement at t will be

$$dD = \epsilon_\infty \, dE + (\epsilon_0 - \epsilon_\infty)\Phi(t - \theta) \, dE \qquad (5.4)$$

In linear systems, the total displacement at time t caused by a variable electric field $E(\theta)$ is the result of the superposition of all the increments, dD and

$$D(t) = \epsilon_\infty E(t) + (\epsilon_0 - \epsilon_\infty) \int_{-\infty}^{t} \frac{dE(\theta)}{d\theta} \Phi(t - \theta) \, d\theta \qquad (5.5)$$

By making the substitution $t - \theta = u$ and integrating by parts, one finds

$$D(t) = \epsilon_\infty E(t) + (\epsilon_0 - \epsilon_\infty) \int_{0}^{\infty} \frac{d\Phi(u)}{du} E(t - u) \, du \qquad (5.6)$$

For an alternating electric field, $E = E_0 \exp(i\omega t)$, Eq. (5.6) leads to

$$\epsilon^*(\omega) = D(t)/E(t) = \epsilon_\infty + (\epsilon_0 - \epsilon_\infty) \int_{0}^{\infty} \exp(-i\omega u) \left[-\frac{d\phi(u)}{du} \right] du \qquad (5.7)$$

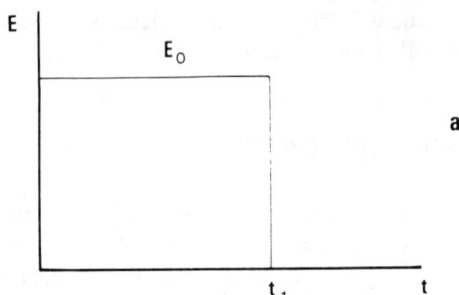

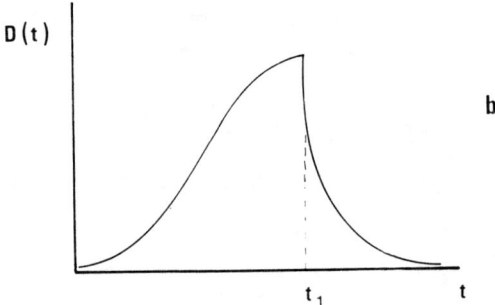

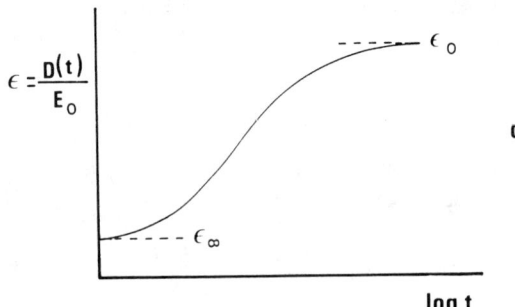

Figure 5.1 Graphs describing (b) the variation of the electric displacement $D(t)$ under the effect of a step-function experiment (a). Curve c shows the variation of the permittivity ϵ with time until steady-state conditions are attained.

where $-d\phi(u)/du = d\Phi(u)/du$, and $\epsilon^*(T\omega) = \epsilon'(\omega) - i\epsilon''(\omega)$ is the complex dielectric permitivity. Rearrangement of the terms of Eq. (5.7) gives [38, 58, 199–201]

$$\frac{\epsilon^*(\omega) - \epsilon_\infty}{\epsilon_0 - \epsilon_\infty} = \int_0^\infty (-d\phi/dt)\exp(-i\omega t)\,dt = \mathcal{L}(-d\phi/dt) \qquad (5.8)$$

indicating that the first term of Eq. (5.8) is related to $-d\phi/dt$ through the Laplace transform.

Solution of the integral requires knowing the decay function over the entire relaxation range. In an ideal relaxation range, the rate of return to the equilibrium value of an observable quantity P is dependent on the distance from equilibrium, P_{eq}, and, as a first approximation, the phenomenon may be described by a linear first-order approximation with the solution $P - P_{eq} \approx \exp(-t/\tau)$. Parameter τ is called the relaxation time, which may be defined as the time in which the observable quantity is reduced to $1/e$ times its original value. For a dielectric relaxation involving a single relaxation time, $\phi(t) = \exp(-t/\tau)$ and Eq. (5.8) becomes

$$\frac{\epsilon^*(\omega) - \epsilon_\infty}{\epsilon_0 - \epsilon_\infty} = \frac{1}{1 + i\omega\tau} \qquad (5.9)$$

from which the real $\epsilon'(\omega)$ and loss $\epsilon''(\omega)$ components of ϵ^* are given by

$$\epsilon'(\omega) = \epsilon_\infty + \frac{\epsilon_0 - \epsilon_\infty}{1 + \omega^2\tau^2} \qquad (5.10)$$

$$\epsilon''(\omega) = \frac{(\epsilon_0 - \epsilon_\infty)\omega\tau}{1 + \omega^2\tau^2} \qquad (5.11)$$

In the limit of low and high frequencies, Eqs. (5.10) and (5.11) lead to

$$\lim_{\omega \to 0} \epsilon'(\omega) = \epsilon_0 \qquad (5.12)$$

$$\lim_{\omega \to \infty} \epsilon'(\omega) = \epsilon_\infty \qquad (5.13)$$

A close examination of Eqs. (5.10) and (5.11) suggests that whereas ϵ' is a continuously decreasing function of frequency, ϵ'' approaches zero both for small and for large values of frequency, reaching a maximum at

$$\omega\tau = 1 \qquad (5.14)$$

that is, the relaxation time is equal to the reciprocal of the angular frequency at the maximum of the loss absorption (Figure 5.2). Although very simple liquids obey these equations, in most cases, and principally in polymers, the dispersion commonly occurs over a wider frequency than that predicted by Eq. (5.13). This behavior suggests that many dielectric mechanisms are involved in the relaxation, each associated with a different relaxation time τ_i. Accordingly, $\phi(t)$ should be expressed by a weighted sum of exponential decay functions, $\phi(t) = \Sigma_i \exp(-t/\tau_i)$, where $\Sigma\omega_i = 1$. Then Eq. (5.8) becomes [201]

$$\frac{\epsilon^*(\omega) - \epsilon_\infty}{\epsilon_0 - \epsilon_\infty} = \sum_i \frac{w_i}{1 + \omega^2\tau_i^2} \qquad (5.15)$$

The integral analogue of Eq. (5.15) involves a normalized continuous distribution of relaxation times $\phi'[\ln(\tau)]$, so that Eqs. (5.10) and (5.11) can be written as [58]

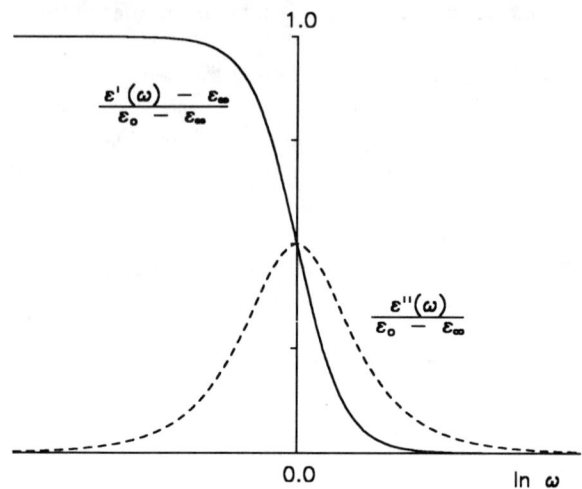

Figure 5.2 Variation of the real ϵ' and loss ϵ'' components of the complex dielectric constant ϵ^* with the frequency for an ideal dielectric with a single relaxation time.

$$\epsilon'(\omega) = \epsilon_\infty + (\epsilon_0 - \epsilon_\infty) \int_{-\infty}^{\infty} \frac{\phi'[\ln(\tau)] \, d[\ln(\tau)]}{1 + \omega^2\tau^2} \tag{5.16}$$

$$\epsilon''(\omega) = (\epsilon_0 - \epsilon_\infty) \int_{-\infty}^{\infty} \frac{\phi'[\ln(\tau)]\omega\tau \, d[\ln(\tau)]}{1 + \omega^2\tau^2} \tag{5.17}$$

Williams [201] has emphasized that if $\phi(t)$ is not a single exponential decay in linear time, the numerical fitting of $\phi(t)$ to a weighted sum of different exponential decays, as indicated in Eq. (5.15), does not necessarily mean that a distribution of relaxation times is really present. A decay function with a non-exponential form in linear time could also give a good account of the dielectric behavior of complex molecules.

The knowledge of the decay function is paramount to interpret the dielectric behavior of most polar molecules including long chains or macromolecules. Equation (5.8) states [202, 203] that the normalized complex dielectric constant is a Fourier transform of $-d\phi(t)/dt$:

$$\frac{\epsilon'(\omega) - \epsilon_\infty}{\epsilon_0 - \epsilon_\infty} = \int_0^{\infty} \left[-\frac{d\phi(t)}{dt} \right] \cos \omega t \, dt \tag{5.18}$$

$$\frac{\epsilon''(\omega)}{\epsilon_0 - \epsilon_\infty} = \int_0^{\infty} \left[-\frac{d\phi(t)}{dt} \right] \sin \omega t \, dt \tag{5.19}$$

These equations may be inverted, giving

$$\frac{d\phi(t)}{dt} = \frac{2}{\pi} \int_0^\infty \frac{\epsilon'(\omega) - \epsilon_\infty}{\epsilon_0 - \epsilon_\infty} \cos \omega t \, d\omega = \frac{2}{\pi} \int_0^\infty \frac{\epsilon''(\omega)}{\epsilon_0 - \epsilon_\infty} \sin \omega t \, d\omega \qquad (5.20)$$

It is obvious that if one of the normalized components of the complex dielectric constant is known over the entire relaxation range, then the decay function $\phi(t)$ can be determined. Actually, this function can be obtained by integration of Eq. (5.20):

$$\phi(t) = \frac{2}{\pi} \int_{-\infty}^\infty \frac{\epsilon_\infty - \epsilon'(\omega)}{\epsilon_0 - \epsilon_\infty} \sin \omega t \, d[\ln (\omega)]$$

$$= \frac{2}{\pi} \int_{-\infty}^\infty \frac{\epsilon''(\omega)}{\epsilon_0 - \epsilon_\infty} \cos \omega t \, d[\ln (\omega)] \qquad (5.21)$$

Since the decay function is unity at $t = 0$, the relaxation strength $\Delta\epsilon = \epsilon_0 - \epsilon_\infty$ is directly related to the loss permittivity by means of the expression

$$\epsilon_0 - \epsilon_\infty = \frac{2}{\pi} \int_{-\infty}^\infty \epsilon''(\omega) \, d[\ln (\omega)] \qquad (5.22)$$

THE DECAY FUNCTION

The evaluation of the decay function $\phi(t)$ requires the determination of auto and cross-correlated functions for polymer chains, but this has not been done yet in a realistic way for complex systems such as polymers in bulk. In fact, only time independent auto and cross-correlated functions can be evaluated in a realistic way for isolated polymer chains.

In the case of a system described by a single relaxation time, the decay function would be given by the relation $\phi(t) = \exp(-t/\tau)$. Thus by substituting this expression in Eq. (5.8), the single relaxation time equation for ϵ^* (Eq. 5.9) is obtained. Several decay functions have been proposed to fit the experimental results, among the most important being the Cole–Cole [204], Davidson–Cole [205], and Williams–Watts [206–209] equations. In general, the Davidson–Cole and the Williams–Watts empirical equations give a reasonable fit to the assymmetric dielectric relaxation exhibited by polymers. The derivatives $-d\phi(t)/dt$ of these empirical functions for the Cole–Cole, Davidson–Cole, and Williams–Watts empirical functions are given by Eqs. (5.23), (5.24), and (5.25), respectively,

$$-\frac{d\phi(t)}{dt} = \frac{n}{\tau\Gamma(1 + n)} \left(\frac{t}{\tau}\right)^{-(1-n)} \quad \text{for } t/\tau \ll 1$$

$$= \frac{n}{\tau\Gamma(1 - n)} \left(\frac{t}{\tau}\right)^{-(1+n)} \quad \text{for } t/\tau \gg 1 \qquad (5.23)$$

$$-\frac{d\phi(t)}{dt} = \frac{1}{\tau\Gamma(\sigma)}\left(\frac{t}{\tau}\right)^{-(1-\sigma)}\exp\left(\frac{-t}{\tau}\right) \tag{5.24}$$

$$-\frac{d\phi(t)}{dt} = \frac{1}{\tau_0}\left(\frac{t}{\tau_0}\right)^{-(1-\beta)}\left[\exp\left(\frac{-t}{\tau_0}\right)\right]^{\beta} \quad 0 < \beta < 1 \tag{5.25}$$

By comparing the three functions, one can see that the Davidson–Cole function drops much more rapidly on the left side of the α absorption than the Cole–Cole function does, due to the exponential factor. The Williams–Watts (WW) function, on the contrary, drops more slowly than do the Cole–Cole function and the Davidson–Cole function in the low-frequency region (Figure 5.3). This is because $\exp(-t/\tau_0)^{\beta}$ decays less rapidly than $\exp(-t/\tau)$ for $0 < \beta < 1$.

Most of the experiments at hand seem to indicate that independently of the nature of the relaxation experiment, near and above the glass transition temperature of amorphous polymers, the appropriate correlation or relaxation function of the local segmental motions is well described by the Williams–Watts function, also known as the Kohlrausch–Williams–Watts (KWW) equation.

The determination of $\epsilon^*(\omega)$ from Eq. (5.8) using the Williams–Watts ex-

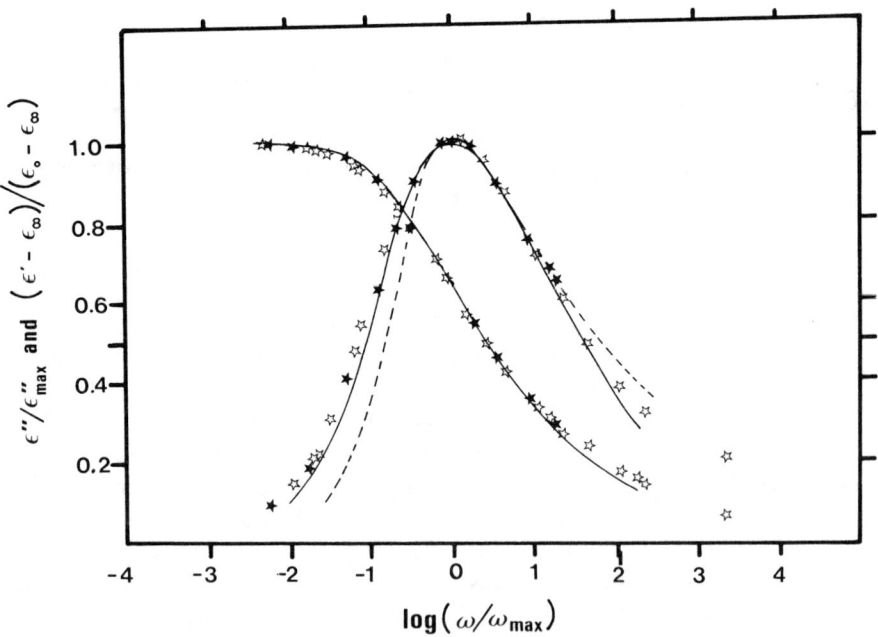

Figure 5.3 Normalized loss curves for the KWW function with $\beta = 0.38$ (solid curve) and the Davidson–Cole equation with $\sigma = 0.217$ (dashed line). Symbols represent experimental values for poly(ethyl acrylate) at different applied pressures: (open star) 1.01×10^5 N-m^{-2} (1 atm); (filled star) 4.20×10^7 N-m^{-2}; and (square) 8.27×10^7 N-m^{-2}. (Redrawn from reference [206] by courtesy of the Royal Society of Chemistry.)

pression for $-d\phi/dt$, Eq. (5.25), is rather complicated, unless $\beta = 0.5$. In this case, after some mathematical handling, Eq. (5.8) reduces to

$$\frac{\epsilon^*(\omega) - \epsilon_\infty}{\epsilon_0 - \epsilon_\infty} = \frac{1}{2} \left(\frac{\pi}{\tau_0}\right)^{1/2} \int_0^\infty \left(\frac{\pi}{t}\right)^{-1/2} \exp\left\{-\left[\left(\frac{t}{\tau_0}\right)^{1/2} + i\omega t\right]\right\} dt \qquad (5.26)$$

This is a standard form that gives

$$\frac{\epsilon^*(\omega) - \epsilon_\infty}{\epsilon_0 - \epsilon_\infty} = (\pi)^{1/2} \frac{1 - i}{\sigma} w(z) \qquad (5.27)$$

where $\sigma = (8\omega\tau_0)^{1/2}$, $z = (1 + i)/\sigma$, and $w(z) = \exp(-z^2) \operatorname{erfc}(-iz)$. In this way, $\epsilon^*(\omega)$ can be evaluated for $\beta = 0.5$. For other values of β, Williams et al. [209] outlined a procedure by which Eq. (5.8) can be obtained. Thus, writing $\delta = \tau_0^{-\beta}$ and $s = \delta_0 + i\omega$, where δ_0 can be chosen as close to zero as one desires, Eq. (5.8) becomes

$$\frac{\epsilon^*(\omega) - \epsilon_\infty}{\epsilon_0 - \epsilon_\infty} = \beta\delta \int [\exp(-st)\exp(-\delta t^\beta)]t^{\beta - 1} dt \qquad (5.28)$$

One approach to solve Eq. (5.28) consists in expressing $-\delta t^\beta$ as a series:

$$\exp(-\delta t^\beta) = \sum_{n=1}^\infty \frac{(-1)^{n-1} \delta^{n-1} t^{(n-1)\beta}}{(n-1)!} \qquad (5.29)$$

and after substituting this equation in Eq. (5.28), one obtains

$$\frac{\epsilon^*(\omega) - \epsilon_\infty}{\epsilon_0 - \epsilon_\infty} = \sum_{n=1}^\infty (-1)^{n-1} \frac{\delta^n}{n!} n\beta \int_0^\infty \exp(-st)t^{n\beta - 1} dt$$

$$= \sum_{n=1}^\infty (-1)^{n-1} \frac{\delta^n}{n!} n\beta \frac{\Gamma(n\beta)}{s^{n\beta}} \qquad (5.30)$$

$$\approx \sum_{n=1}^\infty (-1)^{n-1} \frac{1}{(\omega\tau_0)^{n\beta}} \frac{\Gamma(n\beta + 1)}{\Gamma(n + 1)} [\cos n\beta\pi/2 - i \sin n\beta\pi/2]$$

where it was taken that $\delta_0 = 0$ and the expression $i = e^{i\pi/2}$ was used in s. Hence, the first member of Eq. (5.8) can be obtained for given values of $\omega\tau_0$ and β. The reader interested in mathematical details can find them in reference [209].

COMPLEX-PLANE REPRESENTATION OF DIELECTRIC PROCESSES

Cole and Cole [210] suggested a procedure by which Eqs. (5.10) and (5.11) can be tested. The method consists in plotting the real part of the complex permittivity against the loss at each frequency. Rearrangement of those equations gives

$$\left(\epsilon' - \frac{\epsilon_0 + \epsilon_\infty}{2}\right)^2 + (\epsilon'')^2 = \left(\frac{\epsilon_0 - \epsilon_\infty}{2}\right)^2 \qquad (5.31)$$

which represents a semicircle of radius $(\epsilon_0 - \epsilon_\infty)/2$ and having a center lying on the abscissa at a distance $(\epsilon_0 + \epsilon_\infty)/2$ from the origin. It is obvious that the Cole–Cole plot is inapplicable whenever a single relaxation time is not sufficient to describe the relaxation. The shape of the dielectric process in the complex-plane plot for simple inorganic or organic compounds can take the form of either a circular arc or a skewed semicircle. Complex-plane representations of dielectric processes for two polymers and a low molecular-weight compound are given in Figure 5.4.

Cole and Cole [210] developed two empirical dispersions functions to represent these two dispersions with which the data can be represented with high accuracy. For polymer systems, the complex plot is more complicated; in most cases, the loci of dielectric dispersion are circular arcs at low frequencies and linear at high frequencies. Havriliak and Nagami [211–213] analyzed the shapes of dielectric

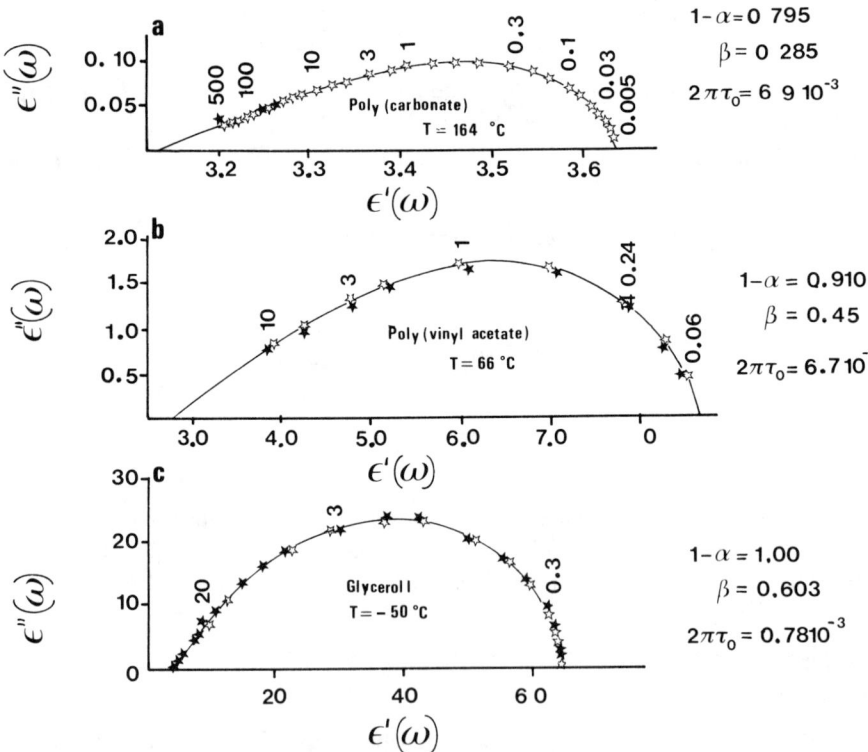

Figure 5.4 Complex-plane plot of the dielectric constant for three materials. (Redrawn from reference [213] by courtesy and permission of Butterworth-Heinemann Ltd.)

dispersions for many polymeric systems, finding that they could be described by the empirical relaxation function

$$\frac{\epsilon^*(\omega) - \epsilon_\infty}{\epsilon_0 - \epsilon_\infty} = [1 + (i\omega\tau_0)^{1-\alpha}]^{-\beta} \tag{5.32}$$

This equation is obviously a generalization of the Cole–Cole equation (because it admits symmetric broadening about the central relaxation time τ_0 for $\alpha < 1$) and the Davidson–Cole equation [205,214] (because it incorporates asymmetric high-frequency-side broadening through $\beta < 1$). The determination of the α, β, and τ parameters is straightforward. Thus, the term $1 + (i\omega\tau_0)^{1-\alpha}$ is a complex number r^* that can be written as

$$\begin{aligned} r^* &= 1 + (\omega\tau_0)^{1-\alpha} \exp[i(\pi/2)(1 - \alpha)] \\ &= [1 + (\omega\tau_0)^{1-\alpha} \sin \alpha\pi/2] - i[(\omega\tau_0)^{1-\alpha} \cos \alpha\pi/2] \end{aligned} \tag{5.33}$$

The θ angle between the real and imaginary components of r^* is

$$\theta = \arctan\left[\frac{(\omega\tau_0)^{1-\alpha} \cos \alpha\pi/2}{1 + (\omega\tau_0)^{1-\alpha} \sin \alpha\pi/2}\right] \tag{5.34}$$

Therefore, the real and loss components of ϵ^* in Eq. (5.32) are

$$\epsilon'(\omega) - \epsilon_\infty = r^{-\beta/2}(\epsilon_0 - \epsilon_\infty) \cos \beta\theta \tag{5.35}$$

$$\epsilon''(\omega) = r^{-\beta/2}(\epsilon_0 - \epsilon_\infty) \sin \beta\theta \tag{5.36}$$

where r is the modulus of r^*, and θ is given by Eq. (5.34). In the limit $\omega\tau \to 0$, $r \to 1$, $\theta \to 0$, and $\epsilon^*(\omega\tau_0 \to 0) \to \epsilon_0$ because $\sin \omega t \to 0$. For $\omega\tau_0 \to \infty$, $r^{-\beta/2} \to 0$ and $\epsilon^*(\omega\tau_0 \to \infty) \to \epsilon_\infty$. This indicates that the high- and low-frequency intercepts of the complex plot with the real axis are ϵ_∞ and ϵ_0, respectively.

Numerical values for the parameters of the Havriliak–Negami equation are shown, for different systems, in Figure 5.4. Parameters α and β can be obtained from the angle limit φ_L that the plot makes with the real axis at high frequencies. The angle φ_L can be obtained by considering that

$$\tan \phi_L = \lim_{\omega\tau_0 \to \infty} \left(\frac{\epsilon''(\omega)}{\epsilon'(\omega) - \epsilon_\infty}\right) = \tan \beta\theta \tag{5.37}$$

It follows from Eqs. (5.34) and (5.37) that when $\omega\tau_0 \to \infty$,

$$\tan \varphi_L/\beta = \cot \alpha\pi/2 = \tan (1 - \alpha)\pi/2 \tag{5.38}$$

so that $\varphi_L = (1 - \alpha)\beta\pi/2$.

Parameter τ_0 can be determined by defining the relaxation time as the condition $\omega\tau_0 = 1$. It follows from Eqs. (5.34), (5.35), and (5.36) that

$$\frac{\epsilon''_{\omega\tau=1}}{\epsilon'_{\omega\tau=1} - \epsilon_\infty} = \tan \gamma = \tan \beta \arctan\left(\frac{\cos \alpha\pi/2}{1 + \sin \alpha\pi/2}\right) \tag{5.39}$$

Because

$$1 + \sin \alpha\pi/2 = 1 + \cos (\alpha - 1)\pi/2$$
$$= 2 \cos^2[(\alpha - 1)\pi/4]$$

and

$$\cos \alpha\pi/2 = \sin (\alpha - 1)\pi/2$$
$$= 2 \sin (\alpha - 1)\pi/4 \cos (\alpha - 1)\pi/4$$

one easily obtains $\gamma = \varphi_L/2$. That is, the angle bisector of φ_L intercepts the plot at τ_0. The value of α can be obtained by taking the length R of the line from ϵ_∞ to $\epsilon^*(\omega\tau_0 = 1)$.

$$R^2 = [\epsilon''(\omega\tau_0 = 1)]^2 + [\epsilon'(\omega\tau_0 = 1) - \epsilon_\infty]^2 \tag{5.40}$$

Because $r = 2(1 + \sin \alpha\pi/2)$ for $\omega\tau_0 = 1$, it follows from Eqs. (5.35), (5.36), and (5.38) that

$$\frac{1}{\phi_L} \log \frac{R}{\epsilon_0 - \epsilon_\infty} = - \frac{1}{\pi(1 - \alpha)} \log \left(2 + 2 \sin \frac{\alpha\pi}{2}\right) \tag{5.41}$$

It should be pointed out that the Havriliak–Negami equation could be considered as a generalized way of writing the two functions of Cole–Cole [210]. Thus, when $\beta = 0$ and $0 \le \alpha \le 1$ in Eq. (5.32) the circular arc is obtained, whereas when $\alpha = 0$ and $0 \le \beta \le 1$, the skew semicircle is obtained.

CHAIN DYNAMICS

Polymers in solution and in the melt continuously change both their shape and position randomly by thermal agitation. This random motion, called Brownian motion, dominates various time-dependent phenomena such as viscoelasticity, birefringence, polarity, light scattering, etc. An important quantity to characterize the Brownian motion is the time-correlation function. Let us consider a dynamical variable A of a system of Brownian particles whose macroscopic thermodynamic properties are independent of time. Variable A can be, for example, the position r or the dipole moment μ, suitably defined, for a molecular chain in an assembly of molecules. The time-autocorrelation function $C_{AA}(t)$ is defined as the assemble-averaged quantity [215]:

$$C_{AA}(t) = \iint A(p,q;\theta) \cdot A(p,q;t + \theta)f(p,q)\, dp\, dq$$
$$= \langle A(\theta) \cdot A(t + \theta)\rangle = \langle A(0)A(t)\rangle \tag{5.42}$$

where $A(t + \theta)$ is the value of A for a molecule at time $t + \theta$, given that it was $A(\theta)$ at time θ; $f(p,q)$ is the equilibrium-phase space-distribution function; and $f(p,q)\, dp\, dq$ is the probability that a molecule has the conjugated momenta p and coordinates q in the intervals p to $p + dp$ and q to $q + dq$. For a stationary system, $\langle A(\theta) \cdot A(t + \theta) \rangle$ is only dependent on t but not on the arbitrary time θ. Owing to thermal motion, variable A describes in time t a trajectory in the phase space. Finding $C_{AA}(t)$ implies determining $A(\theta) \cdot A(\theta + t)$ for each trajectory and averaging this product over all the trajectories that may occur in interval t. The average product is then weighted by the probability $f(p,q) \, dp \, dq$ of having the initial (p,q) conditions of the molecule. The process is repeated for all allowed (p,q) starting conditions and $C_{AA}(t)$ is finally obtained using Eq. (5.42). If $t = 0$,

$$C_{AA}(0) = \int\int A^2(p,q;\theta) f(p,q) \, dp \, dq = \langle A^2(\theta) \rangle = \langle A^2(0) \rangle \qquad (5.43)$$

which is the mean-square value of variable A in the equilibrium state, and it can be calculated from time-independent statistical mechanics. For long values of t, the correlation between $A(\theta)$ and $A(t + \theta)$ vanishes and Eq. (5.43) can be written as

$$C_{AA}(t) = \int A(p,q;\theta) f(p,q) \, dp \, dq \int A(p,q;\theta + t) f(p,q) \, dp \, dq = \langle A(0) \rangle^2 \quad (5.44)$$

Function $C_{AA}(t)$ represents the time-autocorrelation function for a reference molecule. Cross-correlation functions, that is, correlation functions between different molecules, are also very important in polymers, as will be shown in what follows. The time cross-correlation function for variable A is then defined as

$$C_{A_i A_j}(t) = \langle A_i(0) A_j(t) \rangle \qquad (5.45)$$

where A_i and A_j represent the dynamical variable A for molecules i and j ($i \neq j$), respectively. Time-autocorrelation and cross-correlation functions are discussed in detail in references [215–219].

According to the Frölich theory, the static dielectric constant and the constituent dipole moments of a polymer system are related by the expression [39, 220]:

$$\epsilon_0 - \epsilon_\infty = \left[\frac{4\pi}{3k_BT} \frac{3\epsilon_0(2\epsilon_0 + \epsilon_\infty)}{(2\epsilon_0 + 1)^2} \right] \frac{\langle \mathbf{M}(0) \cdot \mathbf{M}(0) \rangle}{V} \qquad (5.46)$$

where $\mathbf{M}(0)$ is the instantaneous dipole moment of the macroscopic sphere of volume V in the system; and $\langle \mathbf{M}(0) \cdot \mathbf{M}(0) \rangle$ is the mean-square dipole moment, the average being taken in absence of an external electric field, over all the configurations of the ensemble contained in the sphere. The instantaneous dipole moment associated with a chain i, contained in volume V, is $\Sigma \boldsymbol{\mu}_{ij}(0)$, where $\boldsymbol{\mu}_{ij}$ represents the elementary dipole moment in the chain. If volume V contains N chains, each of them with n_{si} elemental dipoles, it follows that [220]

$$\langle \mathbf{M}(0) \cdot \mathbf{M}(0) \rangle = \sum_{i=1}^{N} \left[\sum_{j=1}^{n_{si}} \boldsymbol{\mu}_{ji}(0) \cdot \sum_{k=1}^{n_{si}} \boldsymbol{\mu}_{ki}(0) \right]$$

$$= \sum_{i=1}^{N} \left[n_{si}\mu^2 + 2 \sum_{j=2}^{n_{si}} \sum_{k=1}^{j-1} \langle \boldsymbol{\mu}_{ji}(0) \cdot \boldsymbol{\mu}_{ki}(0) \rangle \right] \qquad (5.47)$$

where reference dipole j and dipole k are attached to the same chain. Moreover, if the concentration of dipoles is c_r and intermolecular dipole–dipole interactions are neglected, the average of Eq. (5.46) for a high molecular-weight polymer becomes

$$\langle \mathbf{M}(0) \cdot \mathbf{M}(0) \rangle / V = c_r \left[\mu^2 + \sum_{j \neq k} \langle \mathbf{\mu}_j(0) \cdot \mathbf{\mu}_k(0) \rangle \right] \tag{5.48}$$

Here, for simplicity, it has been assumed that $\mu_{ij} = \mu_{ik} = \mu$. The cross-correlation term that appears in this equation represents the contribution of the connectivity between dipoles in the polymer chain to the dielectric constant of the medium. For a nonassociated small-molecule medium, the cross-correlation term is nil.

By following the same approach, the time-dependent quantity $\langle \mathbf{M}(0) \cdot \mathbf{M}(t) \rangle / V$ can be expressed by

$$\langle \mathbf{M}(0) \cdot \mathbf{M}(t) \rangle / V = c_r [\langle \mathbf{\mu}_j(0) \cdot \mathbf{\mu}_j(t) \rangle] + \sum_{j \neq k} \langle \mathbf{\mu}_j(0) \cdot \mathbf{\mu}_k(t) \rangle \tag{5.49}$$

where dipoles j and k are linked to the same chain. The correlation between $\mathbf{\mu}_j(0)$ and $\mathbf{\mu}_j(t)$ decreases as t increases, so that at long times [201],

$$\langle \mathbf{\mu}_j(0) \cdot \mathbf{\mu}_j(t) \rangle = \mu^2 \langle \cos \theta(t) \rangle \tag{5.50}$$

because above T_g and as time develops, the average projection of the vector on the initial direction decreases and eventually reaches zero. Consequently, the first term in Eq. (5.50) can be written as

$$\langle \mathbf{\mu}_j(0) \cdot \mathbf{\mu}_j(t) \rangle = \mu^2 \Gamma_{jj}(t) \tag{5.51}$$

where $\Gamma_{jj}(t)$ is a decay function that has the limiting values $\Gamma_{jj}(0) = 1$ and $\Gamma_{jj}(t \to \infty) = 0$. The cross-correlation term $\langle \mathbf{\mu}_j(0) \cdot \mathbf{\mu}_k(t) \rangle$ represents the time-dependent correlations between dipoles j and k. It is evaluated as the weighted sum of the decay terms obtained for given initial directions j and k, the weighting factors being the probabilities of obtaining the given relative orientations. The decay-correlation function $\Gamma(t)$ can then be defined as [220]

$$\Gamma(t) = \frac{\langle \mathbf{M}(0) \cdot \mathbf{M}(t) \rangle}{\langle \mathbf{M}(0) \cdot \mathbf{M}(0) \rangle} = \frac{\langle \mathbf{\mu}_j(0) \cdot \mathbf{\mu}_j(t) \rangle + \sum_{j \neq k} \langle \mathbf{\mu}_j(0) \cdot \mathbf{\mu}_k(t) \rangle}{\mu^2 + \sum_{j \neq k} \langle \mathbf{\mu}_j(0) \cdot \mathbf{\mu}_k(0) \rangle} \tag{5.52}$$

For nonassociated liquids, the complex dielectric constant $\epsilon^*(\omega)$ is related to the decay function by an equation of the form

$$\frac{\epsilon^*(\omega) - \epsilon_\infty}{\epsilon_0 - \epsilon_\infty} p(\omega) = \mathcal{L}\left(-\frac{d\Gamma_0}{dt} \right) \tag{5.53}$$

where $\Gamma_0(t) = \langle \mathbf{\mu}(0) \cdot \mathbf{\mu}(t) \rangle / \langle \mathbf{\mu}(0) \cdot \mathbf{\mu}(0) \rangle$, and $p(\omega)$ depends upon internal field considerations. Several expressions have been given for $p(\omega)$, but, in any case, its value seems to be close to unity. For polymers, one can write [220]:

$$\frac{\epsilon^*(\omega) - \epsilon_\infty}{\epsilon_0 - \epsilon_\infty} \, p(\omega) = \mathscr{L}\left(-\frac{d\Gamma}{dt}\right) \tag{5.54}$$

Comparing Eqs. (5.54) and (5.8) suggests that $\Gamma(t) = \phi(t)$. It is obvious that the differences in the dielectric behavior of polymers and simple substances arise from the cross-correlation terms that appear in Eq. (5.52).

DYNAMICS OF DILUTE SOLUTIONS

The dielectric response of polymer solutions in nonpolar solvents to an electric field depends on both the flexibility of the chains and the type of dipoles (A, B, or C) attached to the chains. In flexible polymers, polarization involves local mode motions together with rotatory diffusion, whereas dipole orientation in rigid chains in solution requires rotatory diffusion of the macromolecule as a whole so that local modes are not involved. Relaxation modes in dielectric processes of dilute solutions of polymers with dipole components parallel to the chain contour (type A) have been discussed by Stockmayer [30], taking the normal mode analysis of Rouse and Zimm [24] as a framework. In order to avoid complications in the internal field by effect of the solvent, it is advisable to perform relaxation studies using nonpolar solvents. The first few modes involved in the polarization of flexible polymers are indicated in Figure 5.5. The translational mode, $k = 0$, is electrically active only in

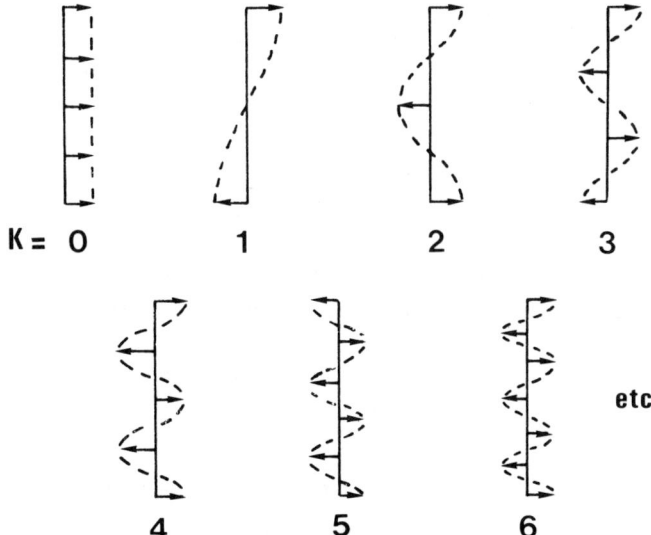

Figure 5.5 Schematic representation of the normal modes for a linear polymer chain. The solid vertical lines represent the chain contour, and the dashed lines indicate relative magnitudes of the displacements. Arrows are placed at positions of maximum amplitude.

ions. Dielectric relaxation involves odd-numbered modes, principally the first mode, in which the electric field drives the two chain ends in opposite directions as consequence of the unreversed sequence of type A dipoles. This relaxation caused by fluctuations of the end-to-end vector of the chain is called the normal mode process. It is obvious that the time-autocorrelation function for such dipole assembly average as the chain bond vectors and, consequently, the spring-bead model developed by Rouse and Zimm is useful to describe the dielectric relaxation of these chains. Furthermore, the decay function in Eq. (5.8) is strongly dependent on molecular weight.

Unlike type A dipoles, the vector sum of a sequence of dipoles perpendicular to the chain contour (type B dipoles) does not correlate with the displacement length. The relevant modes in chain diffusion are short-range or local modes and, therefore, are independent of molecular weight. In this case, the dielectric behavior cannot be described in terms of the Rouse–Zimm model since the relaxation processes will occur at very high frequency at which the eigenfunctions and the eigenvalues of the diffusion operator are not known.

Dielectric relaxation experiments performed on dilute solutions of polar polymers in which the repeat-unit electric dipole moment has a component perpendicular $\mu_{\perp}$ and another parallel $\mu_{\parallel}$ to the chain contour exhibit a low-frequency loss peak related to the parallel component, with a frequency of maximum loss that decreases sharply with increasing molecular weight. A second high-frequency loss peak appears that is independent of molecular weight and is attributed to the perpendicular component of the repeat-unit dipole; therefore, the relaxation involves local segmental motions of the backbone.

The dielectric relaxation of parallel dipoles can be described in terms of the familiar spring-bead model. By neglecting cross-correlation terms, the Rouse–Zimm theory suggests that the decay function is related to the dipole-correlation function by [221, 222]

$$\phi(t) = \langle \mathbf{\mu}(t) \cdot \mathbf{\mu}(0) \rangle / \langle \mathbf{\mu}^2(0) \rangle = 8\pi^{-2} \sum_{p \text{ odd}} p^{-2} \exp(-t/\tau_p) \qquad (5.55)$$

where $\mathbf{\mu}(t)$ is the vector sum of the parallel dipoles of the chain at time t, and τ_p is the dielectric relaxation time for the pth normal mode. Substitution of Eq. (5.55) in Eq. (5.8) gives

$$\frac{\epsilon^*(\omega) - \epsilon_{\infty}}{\epsilon_0 - \epsilon_{\infty}} = 8\pi^{-2} \sum_{p \text{ odd}} p^{-2} (1 + i\omega\tau_p)^{-1} \qquad (5.46)$$

The relaxation time for the normal mode process can be written as [24]

$$\tau_1 = K_1 M[\eta]\eta_0/RT \qquad (5.47)$$

where η_0 is the viscosity of the solvent, $[\eta]$ is the intrinsic viscosity of the solution, M is the molecular weight of the polymer, R is the gas constant, and T is the

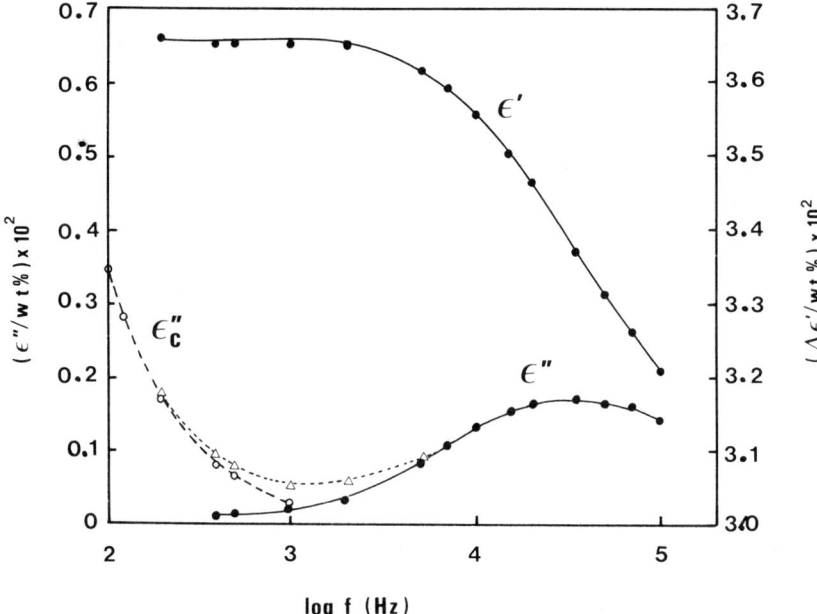

Figure 5.6 Dielectric constant and loss factor for a solution of poly(ϵ-caprolactone) in dioxane at a concentration of 3.44% by weight at 30°C. The dotted curve is the observed loss factor; the solid curve has been corrected by subtraction of the hyperbolic dashed curve due to ionic conductance. (Redrawn from reference [221] by courtesy of and permission of John Wiley & Sons, Inc.)

absolute temperature. The numerical constant K_1 has the values 1.22 and 0.85 for free-draining and nondraining conditions, respectively [24].

The theory was applied to dilute solutions of poly(ϵ-caprolactone) in dioxane [221]. The dipole moment associated with the ester group has components perpendicular and parallel to the chain contour of 1.62 and 0.83 D, respectively. The solutions exhibit two distinct frequency regions, one of which is in the audiofrequency range, shown in Figure 5.6, and is caused by parallel dipoles. This relaxation also shows strong molecular-weight dependence. The other relaxation, caused by the perpendicular components, appears in the high-frequency region. The dielectric constant ϵ' and the loss factor ϵ'' were fitted as functions of frequency to the Havriliak and Negami equation [213]:

$$\frac{\epsilon^*(\omega) - \epsilon_u}{\epsilon_u - \epsilon_\infty} = [1 + (i\omega\tau_0)^{1-\alpha}]^{-\beta} \tag{5.58}$$

where $\epsilon^*(\omega) = \Delta\epsilon' - i\epsilon''$, and the increment in dielectric constant over the pure solvent is represented by $\Delta\epsilon'$. The dispersion corresponding to the relaxation of the dipole moment parallel to the chain direction was successfully characterized in terms of the spring-bead model, and the dependence of both the relaxation time and the reduced viscosity on molecular weight was quite similar. The total change in the dielectric strength is the difference between the value of the dielectric constant at a very low audio frequency $\Delta\epsilon'_0$ and that derived from the refractive-index increment n_1 for visible light. The change in strength $\Delta\epsilon'_0 - \Delta\epsilon'_u$ is the contribution associated with the parallel dipole component, whereas $\Delta\epsilon'_u - n_1^2$ is associated with the perpendicular component. From these results and by using the equation of Guggenheim and Smith [21, 22], values of 0.85 D and 1.61 D were obtained for the parallel and perpendicular components, respectively, of the dipole associated with the ester group of low molecular-weight fractions, in very close agreement with the value reported for these quantities in methyl formate.

Dielectric spectroscopy has proved to be a powerful tool for studying dynamic and static properties of polymer molecules with parallel dipoles in the chain direction in semidilute and concentrated solutions. From the study of the normal mode processes, not only the relaxation times, but also the mean-square end-to-end distance $\langle r^2 \rangle$ can be determined. Studies of this kind were carried out on linear and star-shaped polyisoprene by Adachi and co-workers [223–229]. The polarization $\mathbf{M}(t)$ of a solution is the vector sum of all dipoles [229]:

$$\mathbf{M}(t) = \Sigma\mathbf{P}_i(t) + \Sigma\ \Sigma\mathbf{p}_{ij} + \Sigma\mathbf{p}_s \tag{5.59}$$

where $\mathbf{P}_i$ is the permanent dipole moment of the individual chains, $\mathbf{P}_{ij}$ and $\mathbf{p}_s$ are, respectively, the induced dipole moments of the repeat units and solvent molecules. By neglecting the cross-correlation terms between the ith and jth chains and ignoring the time-independent terms in Eq. (5.59), the decay function can be expressed as [220]

$$\phi(t) = \langle\mathbf{P}_i(0)\cdot\mathbf{P}_i(t)\rangle/\langle[P_i(0)]^2\rangle \tag{5.60}$$

where

$$\mathbf{P}_i(t) = \Sigma[\boldsymbol{\mu}_{ij}\|(t) + \boldsymbol{\mu}_{ij}\perp(t)] \tag{5.61}$$

and $\boldsymbol{\mu}_{ij}\|$ and $\boldsymbol{\mu}_{ij}\perp$ are the parallel and perpendicular components, respectively, to the chain contour of the dipole moment $\boldsymbol{\mu}_j$ associated with the repeat unit of the polymer chains. The average $\langle\mathbf{P}_i(0)\cdot\mathbf{P}_i(t)\rangle$ involves four summations:

$$\begin{aligned}\langle\mathbf{P}_i(0)\ \mathbf{P}_i(t)\rangle = {} & \Sigma\Sigma\langle\boldsymbol{\mu}_{ij}\|(0)\cdot\boldsymbol{\mu}_{ik}\|(t)\rangle \\ & + \Sigma\Sigma\langle\boldsymbol{\mu}_{ij}\|(t)\cdot\boldsymbol{\mu}_{ik}\perp(t) \\ & + \Sigma\Sigma\langle\boldsymbol{\mu}_{ij}\perp(0)\cdot\boldsymbol{\mu}_{ik}\|(t)\rangle \\ & + \Sigma\Sigma\langle\boldsymbol{\mu}_{ij}\perp(0)\cdot\boldsymbol{\mu}_{ik}\perp(t)\rangle\end{aligned} \tag{5.62}$$

The parallel component is related to the $\mathbf{b}_{ij}(t)$ bond vector of the jth repeat unit by

$$\boldsymbol{\mu}_{ij}\|(t) = \mu\mathbf{b}_{ij}(t) \tag{5.63}$$

where μ is the parallel dipole moment per unit of contour length. Accordingly, the time-autocorrelation function of the parallel component can be written as

$$\Sigma\Sigma\langle\boldsymbol{\mu}_{ij}\|(0)\cdot\boldsymbol{\mu}_{ik}\|(t)\rangle = \mu^2\Sigma\Sigma\langle\mathbf{b}_{ij}(0)\cdot\mathbf{b}_{ik}(t)\rangle \tag{5.64}$$

Since the end-to-end vector $\mathbf{R}_i(t) = \Sigma\mathbf{b}_{ij}(t)$, Eq. (5.64) becomes

$$\Sigma\Sigma\langle\boldsymbol{\mu}_{ij}\|(0)\cdot\boldsymbol{\mu}_{ik}\|(t)\rangle = \mu^2\langle\mathbf{R}_i(0)\ \mathbf{R}_i(t)\rangle \tag{5.65}$$

If cross-correlation terms between parallel and perpendicular components in Eq. (5.62) are further ignored, the decay function can be expressed by [229]

$$\phi(t) = \frac{\mu^2\langle\mathbf{R}_i(0)\ \mathbf{R}_i(t)\rangle + \Sigma\Sigma\langle\boldsymbol{\mu}_{ij}\perp(0)\cdot\boldsymbol{\mu}_{ik}\perp(t)\rangle}{\mu^2\langle\mathbf{R}_i(0)\ \mathbf{R}_i(0)\rangle + \Sigma\Sigma\langle\boldsymbol{\mu}_{ij}\perp(0)\cdot\boldsymbol{\mu}_{ik}\perp(0)\rangle} \tag{5.66}$$

As was indicated before, the relaxation time involved in the normal mode process is connected with molecular motions of the whole chain and, consequently, is much larger than that corresponding to the perpendicular components, which reflect conformational changes of a short number of segments. Both relaxations are well separated in the time domain, and the decay function $\phi'(t)$ that accounts only for low relaxation modes can be approximated by

$$\phi'(t) = \langle\mathbf{R}_i(0)\cdot\mathbf{R}_i(t)\rangle/\langle R^2\rangle \tag{5.67}$$

According to Eq. (5.8), the complex dielectric constant in the low-frequency region can be expressed by

$$\frac{\epsilon^*(\omega) - \epsilon_u}{\epsilon_0 - \epsilon_u} = \int_0^\infty \left[-\frac{d\phi'(t)}{dt} \right] \exp(-i\omega t)\ dt$$

$$\tag{5.68}$$

$$= \int_0^\infty -\frac{d}{dt} \left[\frac{\langle\mathbf{R}_i(0)\mathbf{R}_i(t)\rangle}{\langle R^2\rangle} \right] \exp(-i\omega t)\ dt$$

The relaxation strength $\Delta\epsilon = \epsilon_0 - \epsilon_u$ *associated with the low-frequency relaxation can be related to* $\langle\mu_\|^2\rangle$ by means of the equation of Guggenheim [21]:

$$\langle\mu_\|^2\rangle = (3k_BTM/4\pi N_A)[9/(\epsilon_1 + 2)^2]\Delta\epsilon/C \tag{5.69}$$

Since $\langle\mu_\|^2\rangle = \mu^2\langle R^2\rangle$

$$\Delta\epsilon = \epsilon_0 - \epsilon_u = 4\pi N_A\mu^2\langle R^2\rangle CF/3k_BT \tag{5.70}$$

where C is the solution concentration, and $1/F = 9/(\epsilon_1 + 2)^2$. From Eqs. (5.68) and (5.70), the value of ϵ^* for the normal mode process is given by [229]

$$\frac{\epsilon^*(\omega) - \epsilon_u}{C} = \frac{4\pi N_A F\mu^2\langle R^2\rangle}{3k_BTM} \int_0^\infty -\frac{d}{dt} \left[\frac{\langle\mathbf{R}(0)\cdot\mathbf{R}(t)\rangle}{\langle R^2\rangle} \right] \exp(-i\omega t)\ dt \tag{5.71}$$

The dielectric relaxation associated with the normal mode process was used to determine the mean-square end-to-end distance of polymer chains as a function of concentration. In the calculation, Eq. (5.70) was used. For example, the exponent in the relationship $\langle R^2\rangle \approx C^n$ for solutions of cis-polyisoprene in benzene was found to be $-\frac{1}{5}$ in the semidilute region of $0.01 < C < 0.25$ and $-\frac{1}{3}$ in the region $0.25 < C$ $<$ about 0.5 where C is given in g/cm^3. In the θ solvent, dioxane, $\langle R^2 \rangle$ was found to be independent of concentration. Consequently, dielectric relaxations can be used to study excluded volume effects in molecular chains with dipole components in the chain direction.

Results on the dielectric relaxation of dilute solutions of polymers with only dipoles perpendicular to the chain contour (type B) and/or dipoles in the side chain (type C) are sparse. The reason is that the loss peaks usually appear in the frequency range from 1 MHz to 1 GHz, and their experimental measurement presents some difficulties. These relaxations are probably related to glass–rubber and subglass absorptions, labeled α and β peaks, respectively, that appear in the dielectric relaxation spectra of solid amorphous polymers in the audio-frequency region. When the temperature is raised or the system is diluted, the characteristic relaxation times of both processes approach each other and a single relaxation often merges, which could be called $\alpha\beta$ dispersion. It has been pointed out [230] that the relative strength of both dispersions is not a simple function of molecular geometry; for example, for poly(methyl acrylate), $\alpha > \beta$, but under some conditions, the reverse is true in poly(methyl methacrylate).

In polymers with flexibly polar side groups, dispersions are also related to motions involving the side groups. In dilute solutions of poly(methyl vinyl ketone) in dioxane, two relaxation peaks, one centered at 50 MHz and the other at 1 GHz, can be detected (Figure 5.7). The lower-frequency dispersion is assigned to local conformational transitions in the chain backbone, and the higher-frequency absorption is probably connected to internal rotations of the side acetyl group. The two dispersions seem to be the sum of two Havriliak–Negami contributions, Eq. (5.32), in which $\alpha = 0.00$ for both peaks, and $\beta = 0.82$ and 1.00 for the low- and high-frequency dispersions, respectively. From the dielectric strength of the peaks, the contributions of the motions involved in both relaxations to the orientation polarization can be obtained. The relaxation times of dipoles perpendicular to the chain contour were interpreted in terms of the Kramer theory [231]:

$$\tau = A_2 \eta_0 \exp(H_a/RT) \tag{5.72}$$

where η_0 is the viscosity of the solvent, and H_a is the potential barrier height for bond rotation. The value of H_a obtained for the higher-frequency dispersion of dilute solutions of poly(propylene oxide) was 2.7 kcal mol^{-1} [232]. This value is in accordance with that of the activation energy corresponding to the barrier height for backbone motion. Kramer's rate constant between isomeric transition states such as trans–gauche is

$$k_s = \frac{3E^*}{4\pi I} \exp\left(-\frac{E^*}{RT}\right) \tag{5.73}$$

where I is the moment of inertia of the rotating unit around the chain bond, and the threefold potential energy is assumed to have the form $U = \frac{1}{2}E^*(1 - \cos 3\phi)$. Rate

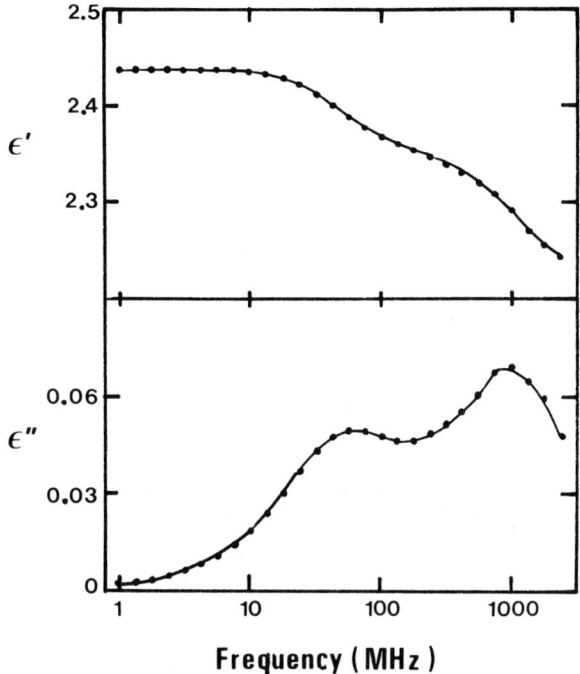

Figure 5.7 Complex permittivity plot as a function of frequency for a 4.21% (w/w) solution of poly(methyl vinyl ketone) in dioxane at 20°C. (Redrawn from reference [230] by courtesy of the American Chemical Society.)

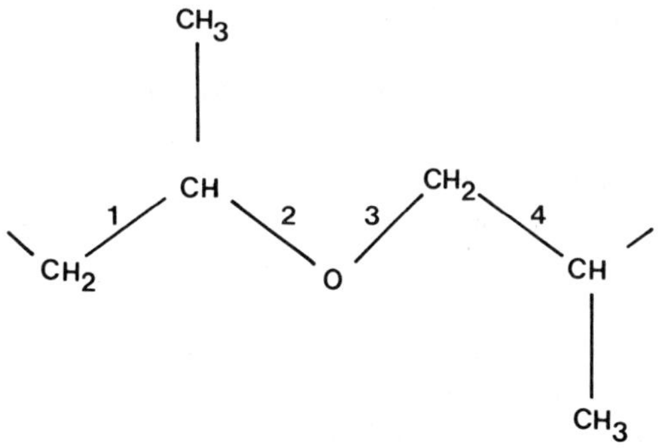

Scheme 5.1

transitions about bonds 1, 2, 3, and 4 in Scheme 5.1 may be different and a distribution of relaxation times might arise from this difference. Therefore, the values of τ_s measured are averaged values of those corresponding to correlated motions around these bonds. The values of 365, 96, and 26 ps found for the τ_s of poly(styrene oxide), poly(propylene oxide), and poly(ethylene oxide), respectively, reflect the flexibility of the chains that increases from poly(styrene oxide) to poly(ethylene oxide).

Multiplicity of peaks can appear in dilute solutions of type C polymers, that is, in polymers carrying flexible side-chain dipoles. Coupling between side-chain and backbone motions could lead to a single high-frequency process. This seems to be the behavior detected in the dielectric relaxation of dilute solutions of polymers in which the dipoles are separated for one or two bonds of the main chain, as occurs in polyacrylates, poly(vinyl isobutyl ether), and poly(methyl acrylate) [42]. In this situation, dipole orientation would require coupling side-chain–main-chain motions. However, it is expected that increasing the number of bonds of the side chain could give rise to the appearance of separate modes.

ANALYSIS OF LOCAL MOTIONS BY THE DYNAMIC ROTATIONAL ISOMERIC STATE (DRIS) MODEL

The Rouse–Zimm model is unsuitable to interpret rapid relaxation processes, such as those occurring in the dielectric relaxation of dilute solutions of type B polymers, which are attributed to local conformational transitions of the backbone. Whereas conformational details can be ignored for the description of long-wavelength modes (they are only dependent on the chain length), local modes are strongly dependent on the conformational characteristics of the chain. All the experimental evidence at hand suggests departure from Debye behavior, that is, the local backbone rearrangements cannot be described by a single exponential relaxation process. Two models have been used to explain local motions. One of the models postulates a type of crankshaft motion [233–237] that does not perturb the chain tails surrounding the mobile segment. The other model assumes conformational transition of a single bond, and, in this case, distortion and deformation of neighboring units are permitted in order to accommodate the newly created isomeric state [238–243]. In all cases, local dynamics is described by the orientational relaxation of a vector affixed to a bond of the chain through correlated rotational transitions of the neighboring bonds. The process must proceed in a way that does not require gross motions of the chain tails [244]. It should be pointed out that time-correlation functions of interest in segmental relaxation experiments have also been exactly derived in the optimized Rouse–Zimm approximation by Perico and Guenza [245]. The resulting local dynamics are given as a sensitive function of the details of the intramolecular potential and of the specific position along the chain of the relaxing segment [246–249].

Rigorous mathematical treatment of neighboring correlations presents formidable mathematical difficulties, one of the most important being associated with

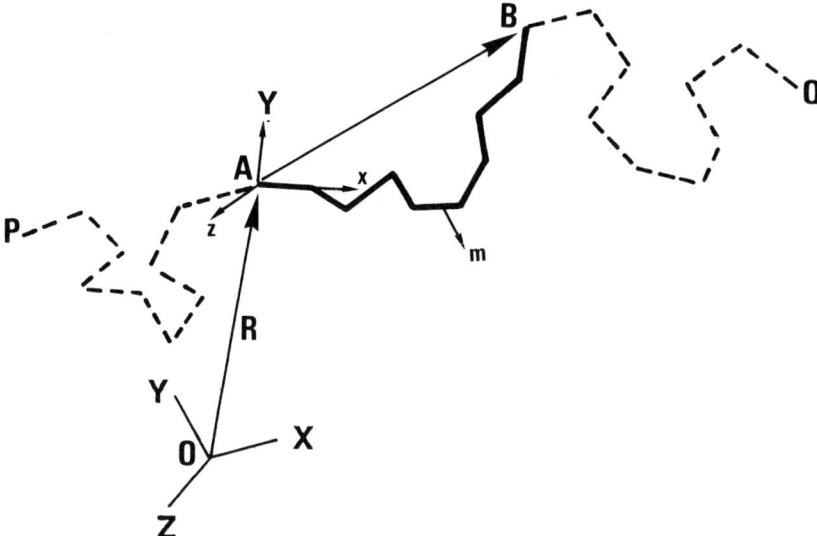

Figure 5.8 Schematic representation of a portion of a polymeric chain. A unit vector **m** is affixed to the central bond of the nine-bond sequence *AB* whose end-to-end vector is denoted by **r**.

chain connectivity and the resistance of the surroundings. Recently, a mathematical scheme was developed by Bahar and Erman [250, 251] to deal with these problems for a short sequence situated in a long molecular chain. The scheme based on the mathematical model of Jernigan [252] applies the rotational isomeric state model to chain dynamics and, consequently, the internal time-correlation functions are computed by using matrix-multiplication methods previously used in calculations of equilibrium conformation-dependent properties.

Figure 5.8 shows an instantaneous conformation of a long-chain molecule in which a short sequence that includes a dipole moment **m** is represented by a heavy line, whereas the two portions of the chain flanking the sequence are represented by dashed lines. The coordinate system *Axyz* is embedded in the chain and its instantaneous position with respect to the laboratory frame *OXYZ* is denoted by **R**. The orientational relaxation of **m** involves (a) an internal relaxation due to conformational transitions in the sequences formulated with respect to the reference frame *Axyz*, and (b) relaxation of the coordinate system *Axyz* with respect to the laboratory frame referred to as the external relaxation. Accordingly, the time-correlation function $\phi(t)$ can be expressed as the product $\phi(t) = \phi_{ext}(t)\phi_{int}(t)$, where $\phi_{ext}(t)$ and $\phi_{int}(t)$ are the external and internal correlation functions, respectively, because it is assumed that both functions are uncorrelated. The internal autocorrelation function is defined as

$$\phi_{int}(t) = \langle \mathbf{m}(0) \cdot \mathbf{m}(t) \rangle \tag{5.74}$$

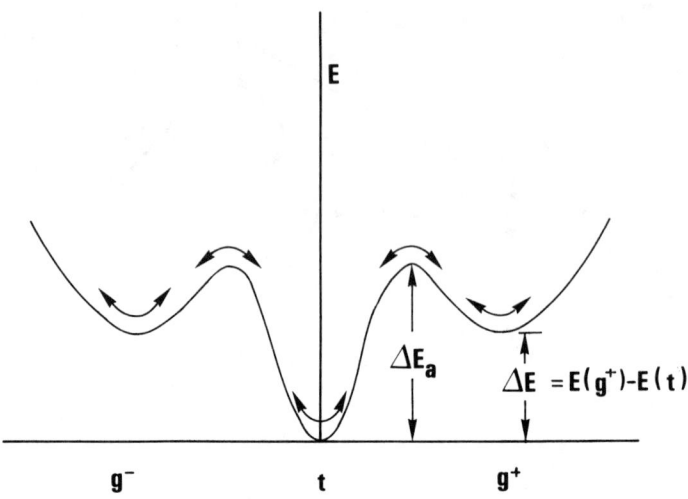

Figure 5.9   Transitions over rotational barriers.

The conformational transition rate can be expressed as the product of two terms, one dependent on the height E_a of the saddle point to be crossed with respect to the initial isomeric state, and the other accounts for the drag friction of the medium (see Figure 5.9). This leads to Kramer's expression for the rate of passage through the barrier in a high-friction medium [253, 254]:

$$r = (ff^*)^{1/2}(2\pi\xi)^{-1} \exp(-\Delta E_a/RT) \tag{5.75}$$

where f and f^* are, respectively, the force constants for oscillations in the well and in the activated state. ξ is the friction coefficient, which is related to the diffusion coefficient by

$$D = k_B T/\xi \tag{5.76}$$

For illustrative purposes, let us consider that the mobile bond undergoes an independent conformational transition [252]:

$$g^- \underset{r_1}{\overset{r_2}{\rightleftarrows}} t \underset{r_2}{\overset{r_1}{\rightleftarrows}} g^+$$

If $\mathbf{P}(t)$ is a vector of the time-dependent probabilities of all possible conformational states, that is, $\mathbf{P}(t) = \{p_t(t), p_{g^+}(t), p_{g^-}(t)\}$, the motion of the bond may be described as [250]

$$\begin{bmatrix} dp_t(t)/dt \\ dp_{g^+}(t)/dt \\ dp_{g^-}(t)/dt \end{bmatrix} = \begin{bmatrix} -2r_1 & r_2 & r_2 \\ r_1 & -r_2 & 0 \\ r_1 & 0 & -r_2 \end{bmatrix} \begin{bmatrix} p_t(t) \\ p_{g^+}(t) \\ p_{g^-}(t) \end{bmatrix} \tag{5.77}$$

or

$$dP(t)/dt = AP(t) \qquad (5.78)$$

where $\Sigma A_{ij} = A_j$ for $i \neq j$. It is obvious that each element in matrix $\mathbf{A}$ describes the momentary rate passage from state i to j. Solution of Eq. (5.78) gives

$$P(t) = \exp(\mathbf{A}t)P(0) = \mathbf{B}\Lambda\mathbf{B}^{-1}P(0) \qquad (5.79)$$

where Λ is the diagonal matrix of eigenvalues of $\mathbf{A}$:

$$\Lambda = \begin{bmatrix} \exp[\lambda_1(t)] & & \\ & \exp[\lambda_2(t)] & \\ & & \exp[\lambda_3(t)] \end{bmatrix} \qquad (5.80)$$

and $\mathbf{B}$ is the matrix formed by eigenvectors of $\mathbf{A}$. Equation (5.79) can then be written as

$$P(t) = C(t,0)P(0) \qquad (5.81)$$

where $\mathbf{C}$ is the time-dependent conditional probability matrix. Actually, the element C_{ij} in this matrix denotes the probability of the bond to be in state i at time t, assuming that the conformation was j at time $t = 0$. The elements of each column include all possible transitions starting from a fixed initial conformation, and, consequently, their sum must be unity. The components of the vector of rotational states in equilibrium $P(0)$ can be obtained in a realistic way by standard matrix-multiplication methods described in detail elsewhere [19]. The joint probability matrix $\mathbf{P}^\wedge$ can thus be obtained as

$$\mathbf{P}^\wedge = \mathbf{C} \ \text{diag} \ \mathbf{P}(0) = \begin{bmatrix} C_{tt} & C_{tg^+} & C_{tg^-} \\ C_{g^+t} & C_{g^+g^+} & C_{g^+g^-} \\ C_{g^-t} & C_{g^-g^+} & C_{g^-g^-} \end{bmatrix} \begin{bmatrix} p_t & & \\ & p_{g^+} & \\ & & p_{g^-} \end{bmatrix} \qquad (5.82)$$

The element $P_{ij}^\wedge$ denotes the joint probability for a bond having state i at time t and state j at time 0. If the transition is pairwise-dependent, the rate transition scheme becomes obviously much more complex. Assuming that each bond has accessibility to only three states, t, g^+, and g^-, the scheme is

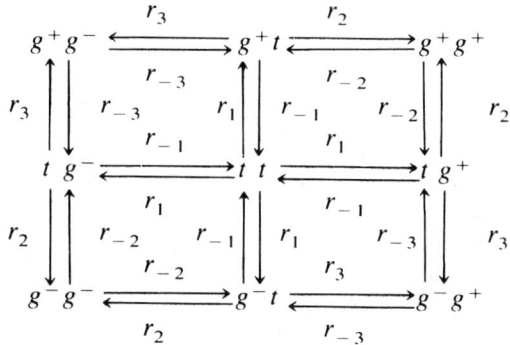

Following the same procedure outlined above, the master equation becomes

$$d\mathbf{p}^*(t)/dt = \mathbf{A}^*\mathbf{P}^*(t) \tag{5.83}$$

or

$$
\begin{bmatrix}
\dot{p}_{tt} \\
\dot{p}_{tg^+} \\
\dot{p}_{tg^-} \\
\dot{p}_{g^+t} \\
\dot{p}_{g^+g^+} \\
\dot{p}_{g^+g^-} \\
\dot{p}_{g^-t} \\
\dot{p}_{g^-g^+} \\
\dot{p}_{g^-g^-}
\end{bmatrix}
=
\begin{array}{c}
tt \\ tg^+ \\ tg^- \\ g^+t \\ g^+g^+ \\ g^+g^- \\ g^-t \\ g^-g^+ \\ g^-g^-
\end{array}
\begin{bmatrix}
-4r_1 & r_{-1} & r_{-1} & r_{-1} & 0 & 0 & r_{-1} & 0 & 0 \\
r_1 & R & 0 & 0 & r_{-2} & 0 & 0 & r_{-3} & 0 \\
r_1 & 0 & R & 0 & 0 & r_{-3} & 0 & 0 & r_{-2} \\
r_1 & 0 & 0 & R & r_{-2} & r_{-3} & 0 & 0 & 0 \\
0 & r_2 & 0 & r_2 & -2r_{-2} & 0 & 0 & 0 & 0 \\
0 & 0 & r_3 & r_3 & 0 & -2r_{-3} & 0 & 0 & 0 \\
r_1 & 0 & 0 & 0 & 0 & 0 & R & r_{-3} & r_{-2} \\
0 & r_3 & 0 & 0 & 0 & 0 & r_3 & -2r_{-3} & 0 \\
0 & 0 & r_2 & 0 & 0 & 0 & r_2 & 0 & -2r_{-2}
\end{bmatrix}
\begin{bmatrix}
p_{tt}(0) \\
p_{tg^+}(0) \\
p_{tg^-}(0) \\
p_{g^+t}(0) \\
p_{g^+g^+}(0) \\
p_{g^+g^-}(0) \\
p_{g^-t}(0) \\
p_{g^-g^+}(0) \\
p_{g^-g^-}(0)
\end{bmatrix}
$$

$$\tag{5.84}$$

where $R = -r_2 - r_3 - r_{-1}$, and the dots on the components of the left-column vector represent derivatives over time. Equation (5.83) can be solved to give [250]

$$\mathbf{P}^*(t) = \mathbf{B}e^{\mathbf{\Lambda}(t)}\mathbf{B}^{-1}\mathbf{P}^*(0) \tag{5.85}$$

where $\mathbf{\Lambda}(t)$ is the diagonal matrix of eigenvalues of $\mathbf{A}$; $\mathbf{B}$ is the matrix formed of eigenvectors of $\mathbf{A}$; and $\mathbf{B}^{-1}$ is the inverse of $\mathbf{B}$. By writing $\mathbf{P}^*(0)$ as a diagonal matrix, the time-dependent joint probability matrix can be written as

$$\mathbf{P}^{*\wedge} = \mathbf{C}^* \text{ diag } \mathbf{P}^*(0) \tag{5.76}$$

The elements $p(\alpha\beta,t;\alpha'\beta',0)$ of $\mathbf{P}^{*\wedge}$ give the joint probability of state $\alpha\beta$ at time t and $\alpha'\beta'$ at $t = 0$ for a given pair of bonds. The analysis can be extended to a sequence of N bonds in motion. Assuming that the motions of the bonds are independent, that is, each bond undergoes an independent conformational transition as in Scheme 5.1, vector $\mathbf{P}^n(t)$ may be written [250]

$$\mathbf{P}^n(t) = \mathbf{P}_1(t)\otimes\mathbf{P}_2(t)\otimes\mathbf{P}_3(t)\otimes \cdots \otimes\mathbf{P}_{n-1}(t)\otimes\mathbf{P}_n(t) \tag{5.87}$$

where $\otimes$ denotes the direct product. By differentiating Eq. (5.87) with respect to time and using Eq. (5.78), the master equation becomes

$$d\mathbf{P}^n(t)/dt = \{\Sigma\mathbf{I}_3\otimes\mathbf{I}_3\otimes \cdots \otimes\mathbf{A}_j\otimes \cdots \mathbf{I}_3\otimes\mathbf{I}_3\}\mathbf{P}^n(t) = \mathbf{Q}\mathbf{P}^n(t) \tag{5.88}$$

where $\mathbf{Q} = \Sigma(\mathbf{I}_3\otimes\mathbf{I}_3\otimes \cdots \otimes\mathbf{A}_j\otimes \cdots \mathbf{I}_3\otimes\mathbf{I}_3)$ is a matrix of $3^n \times 3^n$ elements, and $\mathbf{I}_3$ is a third-order identity matrix. The joint probability matrix $\mathbf{P}^{n\wedge}$ for the conformations of n sequences of bonds can be obtained following the procedure outlined in Eqs. (5.69) to (5.72). Each component of $\mathbf{P}^{n\wedge}$ gives the joint probability that the sequence of n bonds has the conformation $\{\phi\}_i$ at time t and $\{\phi\}_j$ at time 0. In the same way, if simultaneous transitions of the jth and $(j + 1)$th bonds are considered, the master equations take the form

$$d\mathbf{P}^{*\mathbf{n}}(t)/dt = \{\Sigma\mathbf{I}_3\otimes\mathbf{I}_3\otimes\cdots\otimes\mathbf{A}_{j,j+1}\otimes\cdots\mathbf{I}_3\otimes\mathbf{I}_3\}\,\mathbf{P}^{*\mathbf{n}}(t)$$
$$= \mathbf{Q}^*\mathbf{P}^{*\mathbf{n}}(t) \tag{5.89}$$

The joint probability matrix $\mathbf{P}^{*n\wedge}$ can then be obtained from this equation by the methods outlined before the chains in which an unique bond undergoes a transition.

Since fast motions in polymer chains are independent of molecular weight, it is expected that conformational transitions will take place in such a way that the fluctuations in $\mathbf{r}$ in Figure 5.8, $\Delta r = |\mathbf{r}(t) - \mathbf{r}(0)|$, will be confined within a spherical domain defined by $\Delta r < \delta_0$, where δ_0 denotes the radius of the sphere in which the end B can travel between successive transitions. Confinement of Δr into a sphere excludes some conformational transitions that otherwise would occur. Thus, if the transition from conformation $\{\phi\}_i$ at time $t = 0$ to $\{\phi\}_j$ at time t renders $\Delta r > \delta_0$, the element P_{ij} of the joint probability matrix $\mathbf{P}^{n\wedge}$ has to be replaced by zero. Accordingly, the elements of $\mathbf{P}^{n\wedge}$ have to be renormalized in order that $\Sigma\Sigma P_{ij} = 1$.

The internal correlation function can be written as

$$\phi_{int}(t) = \langle\mathbf{m}(0)\cdot\mathbf{m}(t)\rangle = \mathbf{M}^T[\mathbf{P}^{n\wedge}(t)\otimes\mathbf{I}_3]\mathbf{M} \tag{5.90}$$

where $\mathbf{M} = \text{column}(\mathbf{m}_1, \mathbf{m}_2, \ldots \mathbf{m}_j, \ldots \mathbf{m}_{3n})$, and $\mathbf{M}^T$ is the transpose of $\mathbf{M}$. Here $\mathbf{m}_j$ is the vector $\mathbf{m}$ corresponding to the conformation $\{\phi\}_j$. This equation can also be expressed in a more convenient form by [250]

$$\phi_{int}(t) = \Sigma\Sigma\,\mathbf{P}^{n\wedge}_{ij}\,\mathbf{m}_i\cdot\mathbf{m}_j \tag{5.91}$$

where the summations are over all the conformations of the chains. This method is important because it offers a convenient and realistic way to calculate the time-delayed joint probability matrix $\mathbf{P}^{n\wedge}$, which fully describes the conformational transitions of the investigated segment at which the dipole is associated. Therefore, the method can be used to calculate the time dependence of the internal orientational autocorrelation function. A setback of this procedure is that it requires knowing the conformational-energy path involved in the transition. The procedure has been described in some detail because it is one of the few attempts to incorporate realistic conformational characteristics into the treatment of the local dynamics associated with the internal orientational rearrangements of finite sequences within the chain.

The analysis of the dynamics of the nine-bond sequences performed on chains with conformational characteristics similar to polyethylene suggests that the average relaxation time for internal units decreases with increasing sequence length. This behavior can be interpreted on the basis that longer sequences have accessibility to a larger number of conformations to relaxation. Therefore, a more rapid decay function is attainable by allowing a longer sequence to undergo conformational transitions. Another important effect is the environmental resistance due to friction and the size of the relaxing units, which tend to slow down the motion of the unit under consideration. Finally, the application of the present analysis to the study of chain dynamics where Rouselike behavior dominates is not meaningful.

POLYELECTROLYTE SOLUTIONS

The study of the dielectric properties of polyelectrolytes has focused the attention of many researchers for more than 30 years without providing a full understanding of these properties. This is not surprising in view of the rather poor knowledge of the conformational characteristics of polyelectrolytes and the technical difficulties involved in the experimental determination of their dielectric behavior. Polyelectrolytes are macromolecules that in solution carry a relatively large number of charged groups. Interactions between the fixed charges on the macromolecules and the counterions can give rise to molecular associations that will condition the dielectric behavior of polyelectrolytes. The conformational characteristics of dilute solutions of polyelectrolytes in the presence of low molecular-weight salt added resembles that of a wormlike chain. However, the conformational aspects are much more difficult to understand in solutions in absence of salt, where intermolecular interactions can be very strong. In this case, there are even doubts about whether a concentration can be reached where intermolecular interactions are absent. Another difficulty is that the capacitance determinations are performed with blocking electrodes between which electric current circulates. The blocking electrodes give rise to a frequency-dependent polarization impedance, which is also called the *electrode effect*. All these difficulties render very difficult the measurement and interpretation of the dielectric behavior of polyelectrolytes. The reader is referred to a review [255] on the subject, where experimental results as described in the literature are summarized, and the theoretical aspects of the dielectric behavior of polyelectrolytes are discussed.

DIELECTRIC RELAXATIONS OF AMORPHOUS POLYMERS
IN THE BULK

An important parameter that has a marked influence in relaxation phenomena is the temperature [23, 58]. Polymer chains in the bulk present a glass transition temperature below which the convolutions of the skeleton bonds are largely immobilized. However, above T_g, the conformations of the chains are continuously changing so that the chains can be considered as Brownian particles. Dipolar responses in the glassy state will presumably involve configurational rearrangements of side-chain dipoles, short-range conformational orientations of dipoles rigidly attached to the main chain, coupled motions of the side chains and a short number of skeletal bonds in order to favor orientation of side-chain dipoles, etc.

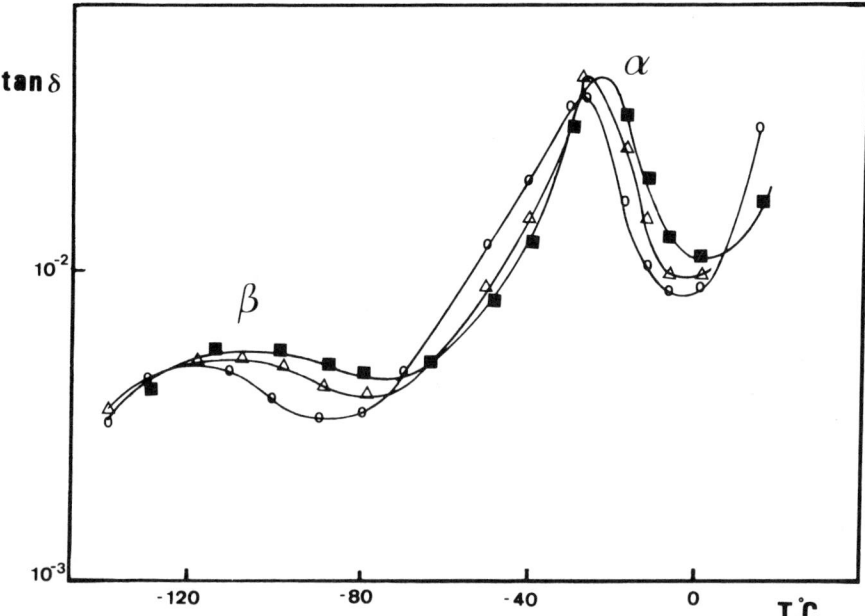

Figure 5.10 Temperature dependence of the dielectric loss tangent for poly(oxy-methylene-1,4-cis-cyclohexylenmethylenoxysebacoil) at frequencies of 0.2 (circles), 0.5 (triangles), and 1 (squares) kHz. (Redrawn from reference [257] by courtesy of the American Chemical Society.)

Dielectric spectra of amorphous polymers in the bulk exhibit a principal absorption, labeled α, and one or more secondary relaxations [42, 58, 256, 257] called β, γ, etc. (Figure 5.10). The dielectric dispersions in the frequency domain are obtained at constant temperature ($T > T_g$), changing the frequency. In this case, the α absorption appears at a lower frequency than the β, and both dispersions tend to coalesce into a single $\alpha\beta$ at high temperatures. In order to avoid experimental difficulties involved in the determination of dielectric relaxations at high frequency, the α processes are usually studied above the glass-transition temperature in the frequency range of 10^{-2} to 10^6 Hz. The β relaxations are studied in the same frequency range below T_g. The dielectric results can also be obtained at a single frequency changing the temperature. In this case, the β absorption appears at a lower temperature (below T_g) than the α dispersion, which, in turn, is centered a few degrees (depending on frequency) above T_g.

For type B polymers, the locations of both the α and β absorptions in the frequency and temperature domains are independent of chain length. The same occurs for polymers in which one of the components of the dipole of the repeat unit is parallel to the chain contour. However, in this case, another absorption is predicted that will be located at higher temperature than the α relaxation in the tem-

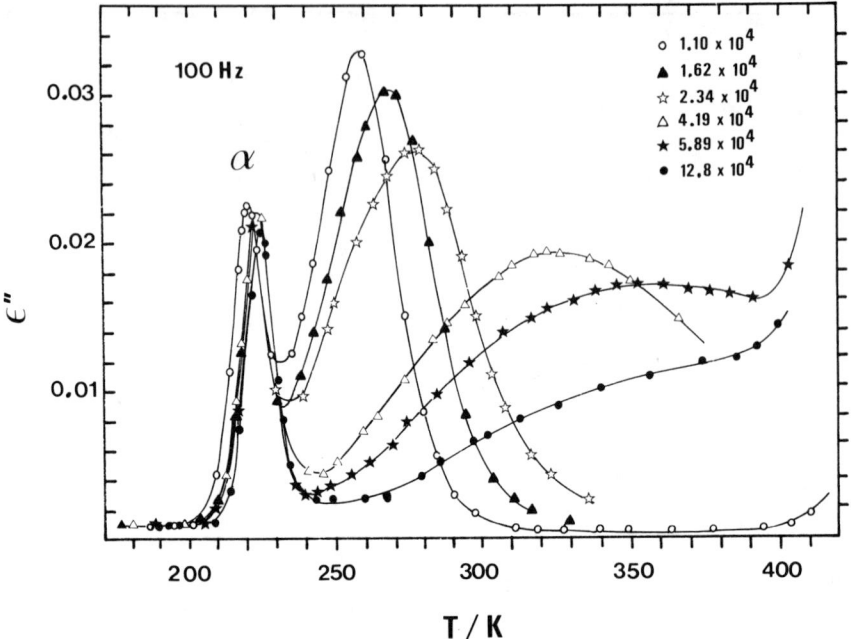

Figure 5.11 Variation of ϵ'' at 100 Hz with temperature for undiluted samples of cis-polyisoprene with various molecular weights indicated. (Redrawn from reference [223] by courtesy of the American Chemical Society.)

perature domain (Figure 5.11) or at lower frequency (Figure 5.12) in the frequency domain. The reason is that above a critical molecular weight M_c, polymer chains become highly entangled and the molecular motions of lower modes become restricted so that the relaxation times associated with long-range motions become strongly dependent on molecular weight; this behavior is depicted in Figure 5.13. Long-range motions will reflect the entanglement effects giving rise to a dispersion whose location will be shifted to a higher temperature in the temperature domain or a lower frequency in the frequency domain as the molecular weight increases [29, 42, 58, 258, 259].

Nonpolar low molecular-weight diluents will decrease the glass-transition temperature of the amorphous polymers and, consequently, the α relaxation will be shifted to a higher frequency and lower temperature in the frequency and temperature domains, respectively. It is believed that the α absorption is associated with micro-Brownian motions of the chains and, hence, it is also known as the dynamic glass transition of the polymer. The mechanisms responsible for the β processes are produced either by motions of dipoles associated with the side chain coupled with local motions of the main chain or simply by local motions of the backbone if the dipoles are rigidly linked to the main chain. In order to establish the molecular

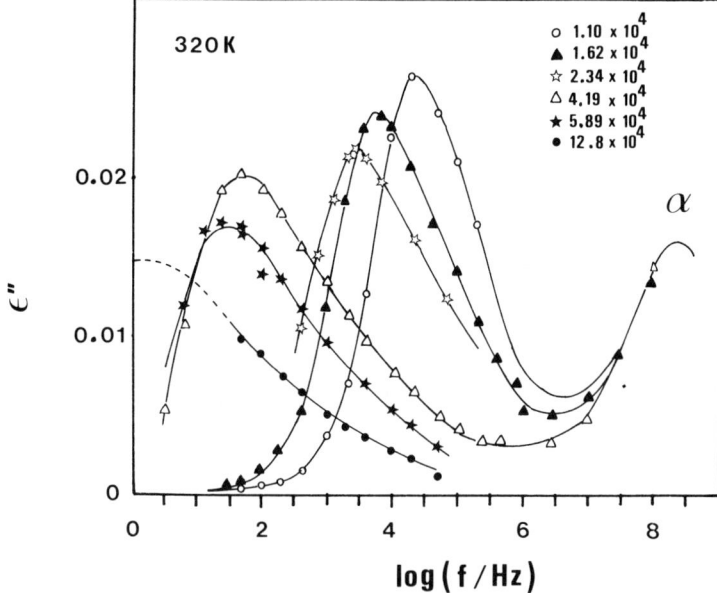

Figure 5.12 Dependence of ϵ' and ϵ'' on frequency at 320 K for undiluted cis-poly-isoprene samples. (Redrawn from reference [223] by courtesy of the American Chemical Society.)

origin of the β process, it is often necessary to combine different dynamic measurements.

CONSIDERATIONS AT THE MICROSCOPIC LEVEL ON THE α AND β PROCESSES OF AMORPHOUS POLYMERS

For most polymers, the strength of the dielectric α relaxation is larger than that of the β ($\alpha > \beta$); only for a few chains, such as atactic poly(methyl methacrylate) and poly(ethyl methacrylate), the opposite occurs [58,260–262]. The total dielectric strength is the sum of the strengths corresponding to the different absorptions. If only the α and β relaxations are present, the dielectric strength is partitioned between the α and β processes. Thus, if $\Delta\epsilon \approx C_r\langle\mu^2\rangle$, where C_r and $\langle\mu^2\rangle$ represent, respectively, the concentration and apparent mean-square dipole moment of a relaxing group, a part of $\langle\mu^2\rangle$ will relax in the β process, the remainder being relaxed in the α process [220]. The ratio of the relaxing parts depends on the relative strength of the α and β absorptions.

The contribution of the α and β relaxations to the decay function has been rationalized in an elegant way by Williams and co-workers [263–267], without

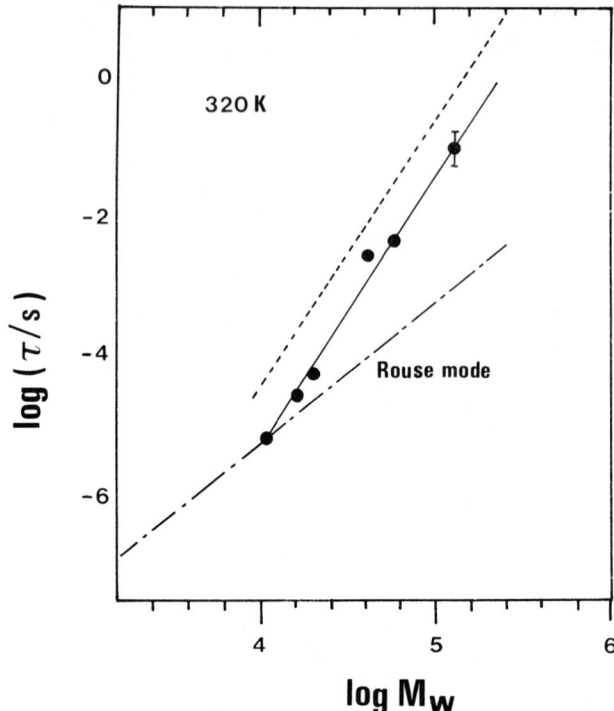

Figure 5.13   Comparison of the molecular-weight dependencies of the dielectric relaxation time at 320 K for the normal mode process (filled circle), the mechanical maximum relaxation time (dashed line), and the dielectric relaxation time predicted by the Rouse theory (dotted–dashed line). (Redrawn from reference [278] by courtesy of the American Chemical Society.)

specifying detailed models for motion, and their approach follows. It is assumed that a dipole k may find itself in a variety of temporary local environments at the arbitrary time $t = 0$. The relaxation of the dipole with time via a particular environment $r(k)$ is characterized by the relaxation function $\Phi_{\beta r(k)}(t)$. The collapse of the local environment of the dipole k causes its total relaxation via an α process, described by the relaxation function $\Phi_{\alpha(kk)}(t)$. The time-autocorrelation function $\langle \boldsymbol{\mu}_k(0) \cdot \boldsymbol{\mu}_k(t) \rangle$ is related to both relaxation functions by

$$\langle \boldsymbol{\mu}_k(0) \cdot \boldsymbol{\mu}_k(t) \rangle = \mu^2 \Phi_{\alpha(kk)}(t) \left[\sum_{r(k)} p_{r(k)} q_{\alpha r(k)} + \sum_{r(k)} p_{r(k)} q_{\beta r(k)}(t) \Phi_{\beta r(k)}(t) \right]$$

$$\equiv \mu^2 \Phi_{\alpha(kk)}(t)(A_{kk} + B_{kk} \Phi_{\beta(kk)}(t)) \tag{5.92}$$

In this expression, $q_{\alpha r(k)} = [\langle \boldsymbol{\mu}_k \rangle_{r(k)}]^2 / \mu^2$ is the fraction of μ^2 that is not relaxed by the $\beta_{r(k)}$ process and, consequently, $q_{\alpha r(k)} + q_{\beta r(k)} = 1$; $p_{r(k)}$ is the equilibrium probability of obtaining the environment $r(k)$ for the dipole k. The effective strengths for the α and β processes are A_{kk} and B_{kk}, respectively, whose sum is unity. In the same way, the cross-correlation function $\langle \boldsymbol{\mu}_k(0) \cdot \boldsymbol{\mu}_{k'}(t) \rangle$ can be written as

$$\langle \boldsymbol{\mu}_k(0) \cdot \boldsymbol{\mu}_{k'}(t) \rangle = \langle \boldsymbol{\mu}_k(0) \cdot \boldsymbol{\mu}_{k'}(0) \rangle \Phi_{\alpha(kk')}(t)[A_{kk'} + B_{kk'} \Phi_{\beta(kk')}(t)] \quad (5.93)$$

where $A_{kk'}$ and $B_{kk'}$ are, respectively, the fractions of $\langle \boldsymbol{\mu}_k(0) \cdot \boldsymbol{\mu}_{k'}(0) \rangle$ that are relaxed by long-range and local processes, and $A_{kk'} + B_{kk'} = 1$; $\Phi_{\beta(kk')}(t)$ represents a function that describes the relaxation of the k and k' dipole via local motions in the environments $r(k)$ and $r(k')$. The subsequent relaxation described by the function $\Phi_{\alpha(kk')}(t)$ occurs after collapsing the environments around k and k'. Therefore, it can be expected that $\Phi_{(kk')}(t) \approx \Phi_{\alpha(kk)}(t)$, for all k and k'. Accordingly, the total decay function can be expressed by

$$
\begin{aligned}
\phi(t) = {} & \left[\frac{\mu^2 A_{kk} \Phi_{\alpha(kk)}(t) + \displaystyle\sum_{k \neq k'} \langle \boldsymbol{\mu}_k(0) \cdot \boldsymbol{\mu}_{k'}(0) \rangle A_{kk'} \Phi_{\alpha(kk')}(t)}{\mu^2 + \displaystyle\sum_{k \neq k'} \langle \boldsymbol{\mu}_k(0) \cdot \boldsymbol{\mu}_{k'}(0) \rangle} \right] \\[2ex]
& + \left[\frac{\mu^2 B_{kk} \Phi_{\alpha(kk)}(t) \Phi_{\beta(kk)}(t)}{\mu^2 + \displaystyle\sum_{k \neq k'} \langle \boldsymbol{\mu}_k(0) \cdot \boldsymbol{\mu}_{k'}(0) \rangle} \right] \\[2ex]
& + \left[\frac{\displaystyle\sum_{k \neq k'} \langle \boldsymbol{\mu}_k(0) \cdot \boldsymbol{\mu}_{k'}(0) \rangle B_{kk'} \Phi_{\alpha(kk')}(t) \Phi_{\beta(kk')}(t)}{\mu^2 + \displaystyle\sum_{k \neq k'} \langle \boldsymbol{\mu}_k(0) \cdot \boldsymbol{\mu}_{k'}(0) \rangle} \right]
\end{aligned}
\quad (5.94)
$$

Williams [220] considers three situations. If the local motions occur at a much faster rate than the micro-Brownian motions, the relaxation functions of the β absorption control the relaxation process, so that $\Phi_{\beta(kk)}(t) \Phi_{\alpha(kk)}(t) \approx \Phi_{\beta(kk)}(t)$ and $\Phi_{\beta(kk')}(t) \Phi_{\alpha(kk')}(t) \approx \Phi_{\beta(kk')}(t)$. The decay function $\phi(t)$ decays in two stages, producing the α and β processes whose relative strength is

$$\frac{\Delta \epsilon_\beta}{\Delta \epsilon_\alpha} = \frac{\mu^2 B_{kk} + \displaystyle\sum_{k'} \langle \boldsymbol{\mu}_k(0) \cdot \boldsymbol{\mu}_{k'}(0) \rangle B_{kk'}}{\mu^2 A_{kk} + \displaystyle\sum_{k'} \langle \boldsymbol{\mu}_k(0) \cdot \boldsymbol{\mu}_{k'}(0) \rangle A_{kk'}} \quad (5.95)$$

The model predicts that both the autocorrelation and cross-correlation terms contribute to the α and β processes. However, whereas the model assumes that the α process occurs after collapse of the local environment, the β process is assumed to be the result of the weighted sum of elemental processes occurring in a variety of local environments. Hence, the basic assumptions made in the development of Eq. (5.94) suggest that the β dispersion should be broader than the α process in the frequency domain. Moreover, although the mechanisms responsible for both processes are different, they are closely interrelated because the same dipoles contribute to both.

At a temperature below T_g, the system will be in a non-equilibrium state and $\Phi_{\alpha(kk)}(t)$ and $\Phi_{\alpha(kk')}(t)$ will not decay. Equation (5.8) becomes

$$\left[\frac{\epsilon^*(\omega) - \epsilon_{\infty\beta}}{\epsilon_0 - \epsilon_\infty}\right] p(\omega) = 1 - i\omega \int_0^\infty \phi(t) \exp(-i\omega t)\, dt \qquad (5.96)$$

where $p(\omega)$ is a field factor, and $\phi(t)$ is given by

$$\phi(t) = \frac{B_{kk}\Phi_{\beta(kk)}(t) + \sum_{k'} B_{kk'}\Phi_{\beta(kk')}(t)}{B_{kk} + \sum_{k'} B_{kk'}} \qquad (5.97)$$

indicating that at $T < T_g$ only a portion of $\langle \mathbf{M}(0)\cdot\mathbf{M}(0)\rangle$ relaxes. Moreover Eqs. (5.94) and (5.96) suggest that the β process is continuous from above to below T_g.

Finally, since the temperature dependence of the α process is usually greater than that of the β, a temperature $(T > T_g)$ can be reached at which the relaxation functions $\phi_{\alpha(kk)}(t)$ and $\phi_{\alpha(kk')}(t)$ that govern the α absorption in Eq. (5.94) decay faster than their relaxation-function counterparts $\phi_{\beta(kk)}(t)$ and $\phi_{\beta(kk')}(t)$ in the β process, and from Eq. (5.94), one obtains

$$\phi(t) = \frac{\mu^2\Phi_{\alpha(kk)}(t) + \sum_{k'} \langle \boldsymbol{\mu}_k(0) \cdot \boldsymbol{\mu}_{k'}(0)\rangle\Phi_{\alpha(kk')}(t)}{\mu^2 + \sum_{k'} \langle \boldsymbol{\mu}_k(0) \cdot \boldsymbol{\mu}_{k'}(0)\rangle} \qquad (5.98)$$

Thus, the α and β processes coalesce at a high temperature, giving rise to the $\alpha\beta$ process, which relaxes via the α process extrapolated to higher temperature.

The model described gives a reasonable explanation of the relaxation processes of polymers [220], For example, it explains the reason why the β processes are broader than the α dispersions, the appearance of the $\alpha\beta$ process in the frequency domain at $T \gg T_g$, and the conservation rule $\Delta\epsilon \approx \Delta\epsilon_\alpha + \Delta\epsilon_\beta$, which seems to hold, at least semiquantitatively, in amorphous polymers. Finally, a decrease in temperature should reduce the local motions of dipoles in their environments and, as a consequence, the strength of the β relaxation in the frequency domain should decrease with decreasing temperature. This conclusion is also in agreement with experimental results.

THE COUPLING SCHEME

The physical basis of the empirical Kohlrauch–Williams–Watts (KWW) decay function was recently examined by Ngai and co-workers for entangled polymer systems [268–271]. They assume that the primary relaxation mechanism is caused by interaction of the molecular species, which are called primary species, with the heat bath. In this case, the relaxation of the measured quantity is described by a single linear exponential with a constant relaxation time. The coupling among the primary species provides a nonnegligible time-dependent effect on the relaxation that is already in progress due to the primary-species–heat-bath interaction. The theoretical context for a time-dependent relaxation rate is based on classic Liouville dynamics [272, 273] supplemented by an extension of the Dirac constraint theory [273–275] to time-dependent constraints. The treatment leads to the conclusion that at times longer than time t_c, characteristic of the strength of the intermolecular interaction, the initially constant relaxation rate W_0 slows down. For thermorheologically simple systems, the effect of the coupling interactions in the relaxation rate $W(t)$ is given by [23]

$$W(t) = W_0 f(t) \qquad (5.99)$$

where [270]

$$f(t) = \begin{cases} 1 & w_c t < 1 \\ (w_c t)^{-n} & w_c t > 1 \end{cases} \quad \text{and} \quad 0 < n < 1 \qquad (5.100)$$

In these expressions, $(w_c)^{-1}$ is a time characteristic of the complexity of the system and n is a parameter characterizing the coupling strength between the primary species.

If the time–temperature correspondence holds, that is, if the system is thermorheologically simple, the evolution of the normalized relaxation function $\phi(t)$ for a single relaxation mechanism can be written as

$$d\phi/dt = -W_0 f(t)\phi \qquad (5.101)$$

For polymers around the α relaxation and above, intermolecular coupling is important so that [270]

$$d\phi(t)/dt = -W_0 (w_c t)^{-n} \phi \qquad (5.102)$$

From Eq. (5.102), the KWW equation is obtained:

$$\phi(t) = \exp(-t/\tau^*)^{1-n} \qquad (5.103)$$

with

$$\tau^* = [(1 - n)(w_c)^n \tau_0]^{1/(1-n)} \qquad (5.104)$$

where τ^* is the effective relaxation time of the complex system, and $\tau_0 = (W_0)^{-1}$ is the primitive relaxation time. The appearance of τ^* in the KWW equation has important implications. The dynamics of polymer chains in the melt and in concentrated solutions follow Rouse's behavior for molecular weights below a critical value M_c. For $M < M_c$, the chains are not entangled and the relaxation time associated with the first relaxation mode scales as

$$\tau_{1R} < M^2 \qquad (5.105)$$

For $M > M_c$, the chains become entangled, and, according to Eq. (5.104), the primitive relaxation time of the Rouse first mode is shifted to a longer time:

$$\tau^* \approx (\tau_{1R})^{1/(1-n)} \approx M^{2/(1-n)} \tag{5.106}$$

The coupling parameter for entangled linear chains has been determined from experimental data of several polymers [268, 276, 277] and it was found to have a value that falls into the narrow range $0.40 < n < 0.46$. The model thus predicts that τ^* has the molecular-weight dependence of $M^{3.33}$ to $M^{3.70}$, in agreement with the results obtained for the terminal τ_1 relaxation time from mechanical experiments. Similar dependence is also found for the normal mode of the dielectric relaxation of entangled polymer chains with dipole components parallel to the chain contour [278]. In this case, the constraints in the lower-frequency dielectric dispersion are similar to those in the terminal region of the mechanical measurements (see Figure 5.13). However, the constraints imposed on relaxation of a dipole perpendicular to the chain contour need not be the same as those imposed on shear deformation or those imposed on bulk deformation. Therefore, it is expected that in this case, the values of n in the KWW equation will be dependent on the type of dynamic variable. In fact, an entropic reformulation of the coupling model leads to the inequalities [271]

$$n_d < n_J < n_b \tag{5.107}$$

where n_d, n_J, and n_b refer to n in the KWW function for dipole relaxation, shear compliance, and density fluctuation.

INFLUENCE OF TEMPERATURE IN DIELECTRIC RELAXATION

Dielectric measurements can be performed at constant temperature, changing the frequency (results in the frequency domain), or at constant frequency, changing the temperature (results in the temperature domain) [23, 58]. In the case of subglass relaxations, the plots of $\ln(\omega_{max})$ versus T, in the former case, or $\ln(\omega)$ versus T_{max}, in the latter, usually give straight lines. In most cases, the relaxation time associated with the maximum of the absorption $\tau = 1/\omega_{max}$ obeys Arrhenius behavior:

$$\tau \approx (\Delta H/RT) \tag{5.108}$$

where ΔH is the activation energy of the process.

On the contrary, the relaxation times associated with the glass–rubber relaxation in both mechanical and dielectric experiments are, in most systems, dependent on free volume, so that the time τ_j corresponding to the dielectric or mechanical mechanism j in the relaxation process is related to the free volume fraction v_f by the Doolittle equation [279]:

$$\tau_j = \tau_0 \exp(B/v_f) \tag{5.109}$$

Parameter B is given by the Cohen–Turnbull expression [280]:

$$B = \xi v^*/v_m \tag{5.110}$$

where v^* is the minimum required volume for a relaxation process to take place, v_m is the mean volume of the relaxing polymer segment, and ξ is a constant whose value lies in the range 0.5 to 1. It is obvious that B should depend on the relaxing variable. By assuming that v_f is a linear function of temperature,

$$v_f = \alpha(T - T_\infty) \tag{5.111}$$

Eq. (5.109) becomes the Vogel equation [281]:

$$\tau = \tau_0 \exp\left[\frac{B}{\alpha(T - T_\infty)}\right] \tag{5.112}$$

where α is the expansion coefficient, and T_∞ is the temperature at which the free volume would be zero were it not for the formation of the glassy state. Here the subindex j was withdrawn because for systems thermorheologically simple, all the dielectric mechanisms in the relaxation process show the same temperature dependence. Of course, the same occurs in the α mechanical process. The value of T_∞ is taken to be that one for which the plot $\ln(\tau)$ versus $1/(T - T_\infty)$ gives a straight line (Figure 5.14); usually, it is 50°C below T_g. It is worthy to note that the relative free volume corresponding to T_g, v_g/B is related to $m = Bv/\alpha$ of the Vogel equation by the expression

$$v_g/B = (T_g - T_\infty)/m \tag{5.113}$$

The values of v_g/B obtained from the analysis of dielectric and mechanical results lie in the range 0.025 ± 0.005 for most systems, that is, v_g is constant for most amorphous polymers. The B parameter is related to the minimum volume necessary for a relaxation process to take place and its value is believed to be close to unity.

The predictions of the coupling model for the relaxation times of dielectric (d), mechanical (j), and density fluctuation (b) variables are given by [270]

$$\tau_i^* = [(1 - n_i)(W_c)^{n_i} (W_{0i})^{-1}]^{1/(1 - n_i)} \tag{5.114}$$

where i can be d, j, and b. On the other hand, the friction coefficient near T_g varies rapidly with temperature so that the reciprocals of W_{0d}, W_{0J}, and W_{0b} are proportional to the friction coefficient χ_0. This quantity is given by

$$\chi_0 \approx \exp[B_0/(T - T_\infty)] \tag{5.115}$$

so that Eq. (5.114) becomes

$$\tau_i^* \approx \exp\{[B_0/(1 - n_i)]/(T - T_\infty)\} \tag{5.116}$$

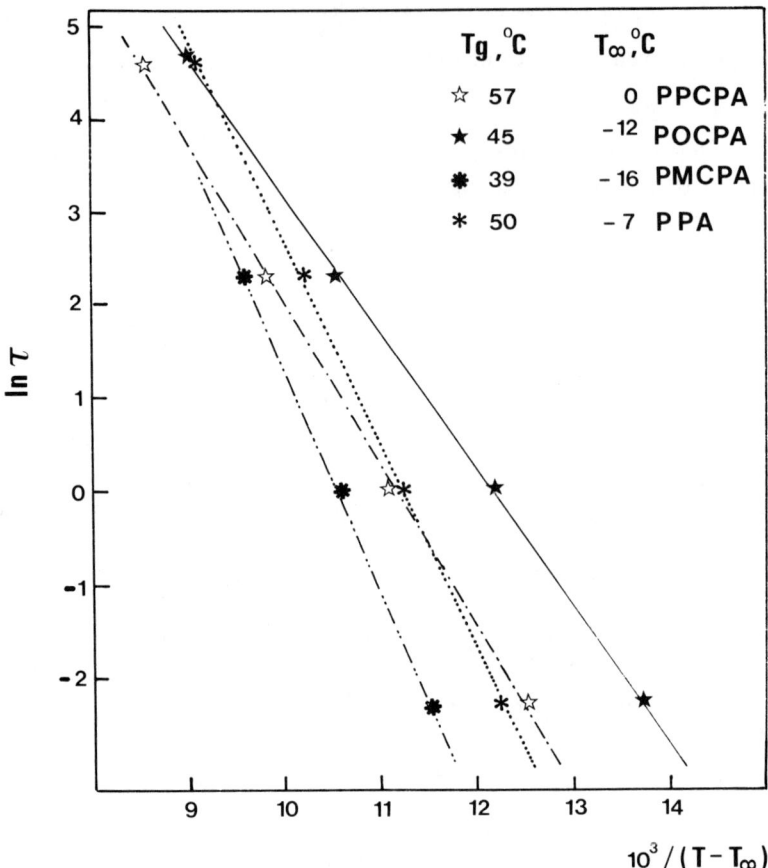

Figure 5.14 Plot showing the temperature dependence of the α process for poly(phenyl acrylate) (PPA), poly(p-chlorophenyl acrylate) (PPCPA), poly(o-chlorophenyl acrylate) (POCPA), and poly(m-chlorophenyl acrylate) (PMCPA). Values of T_g and T_∞ for the polymer chains are indicated in the figure. (Redrawn from reference [281] by courtesy of the American Chemical Society.)

where i represents d, j, and b for dielectric, mechanical, and density fluctuation relaxations, respectively. Since τ_i^* is related to the mean relaxation time τ_i of Eq. (5.112) by the expression [271]:

$$\tau_i = \int_0^\infty \exp[-(t/\tau_i^*)^{1-n_i}] \, dt = \frac{\tau_i^*}{1-n_i} \, \Gamma\left(\frac{1}{1-n_i}\right) \qquad (5.117)$$

where Γ is the gamma function, and τ_i is given by Eq. (5.112).

From Eqs. (5.116) and (5.117), one obtains

$$B_d = B_0/(1 - n_d) \tag{5.118}$$

$$B_j = B_0/(1 - n_j) \tag{5.119}$$

$$B_f = B_0/(1 - n_b) \tag{5.120}$$

Therefore, these three results can be combined into an invariance relation:

$$(1 - n_d)B_d = (1 - n_j)B_j = (1 - n_b)B_b = B_0 \tag{5.121}$$

The invariance of the product has been verified, within the error bounds, for several polymer systems [271].

RELAXATION PHENOMENA IN CRYSTALLINE POLYMERS

It is well known that polymers with regular structures can crystallize, the degree of crystallinity being strongly dependent on the thermal history [283]. In some polymers, such as poly(ethylene terephthalate) (PET), the degree of crystallinity can vary from zero for the polymer quenched from the melt to 50–60% if the crystallization is carried out at low undercoolings, $\Delta T = T_m - T_c$, where T_m is the thermodynamic temperature of melting, and T_c is the crystallization temperature. Other polymers, such as polyethylene, can never be obtained in the amorphous state whatever the thermal history used in the crystallization process is. Thus, it is convenient to distinguish between different categories of polymers according to their crystallization rate. Polymers with very slow crystallization rates, such as PET, aromatic polycarbonates, etc., are difficult to crystallize to extents beyond 50% and they will be called low-crystallinity polymers. There are polymers that are difficult to crystallize to extents higher than 60% and that are not ordinarily quenchable to the complete amorphous state; in Boyd's terminology, these are "medium-crystallinity" polymers. Examples are aliphatic polyamides and polyesters. A third category has polymers such as polyethylene and polyethers, which are rapidly crystallizing polymers, so that their crystallinity after quenching from the melt is hardly below 50%. The crystallinity of these polymers is ordinarily in the range 60–80%. Therefore, in most crystalline polymers, amorphous and crystalline regions coexist.

In general, the α and β processes observed in amorphous polymers can also appear in crystalline polymers as consequence of microbrownian motions and local motions that take place in the amorphous phase. However, a new process may also appear in crystalline polymers caused by motions in the crystalline phase. The influence of the crystallinity on ϵ' and ϵ'' and on the shape of the Cole–Cole plot is shown in Figures 5.15 and 5.16. By comparing the α process for amorphous and semicrystalline polymers, one can conclude [284–286]:

(a) The relaxation strength of the α process for a crystalline polymer is smaller than that for the same polymer in the amorphous state.

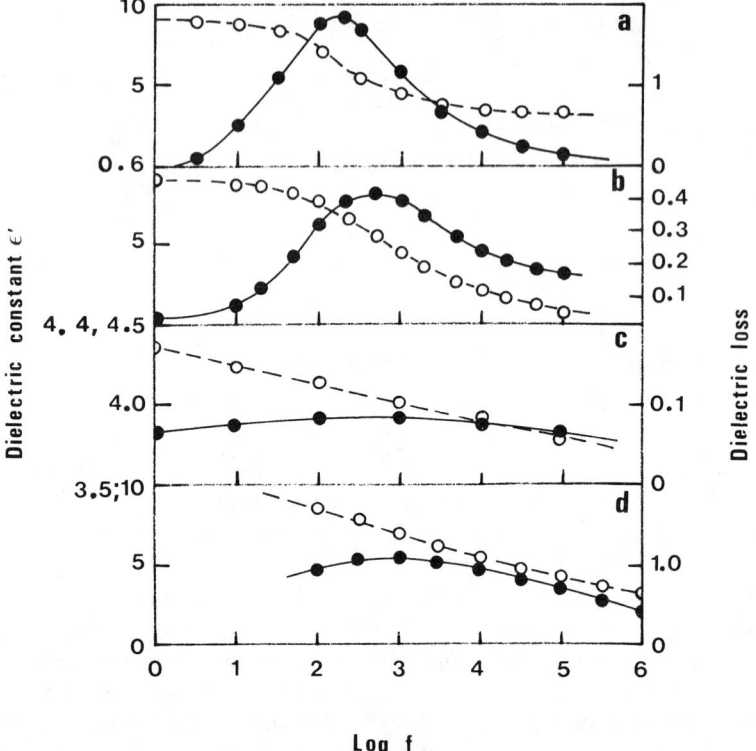

Figure 5.15 Dielectric constant ϵ' (open circles) and loss ϵ'' (filled circles) versus log frequency for completely amorphous and semicrystalline polymers in the glass–rubber relaxation. (a) Poly(vinyl acetate) (PVAC) is a completely amorphous polymer; (b) quenched poly(ethylene terephthalate) (PET) is a completely amorphous polymer and behaves similarly to PVAc; (c) a 50% crystalline PET; (d) poly(vinyl chloride) is a low-crystallinity polymer; its loss behavior, however, is seen to be more typical of a semi-crystalline polymer than a wholly amorphous one. (Redrawn from reference [256] by courtesy and permission of Butterworth-Heinemann Ltd.)

(b) The width of the α peak increases as the degree of crystallinity increases.

(c) The α process for crystalline polymers appears at a lower frequency than a similar process in the amorphous polymer.

As far as the β relaxation is concerned, the strength of this absorption decreases as the degree of crystallinity increases. However, contrary to what occurs with α relaxation, the location of the β process in the frequency domain, at a given temperature, seems to be independent of both the molelcular weight of the chains and the degree of crystallization. The decrease in the intensity of the α and β absorptions with increasing crystallinity is because fewer dipoles contribute to these

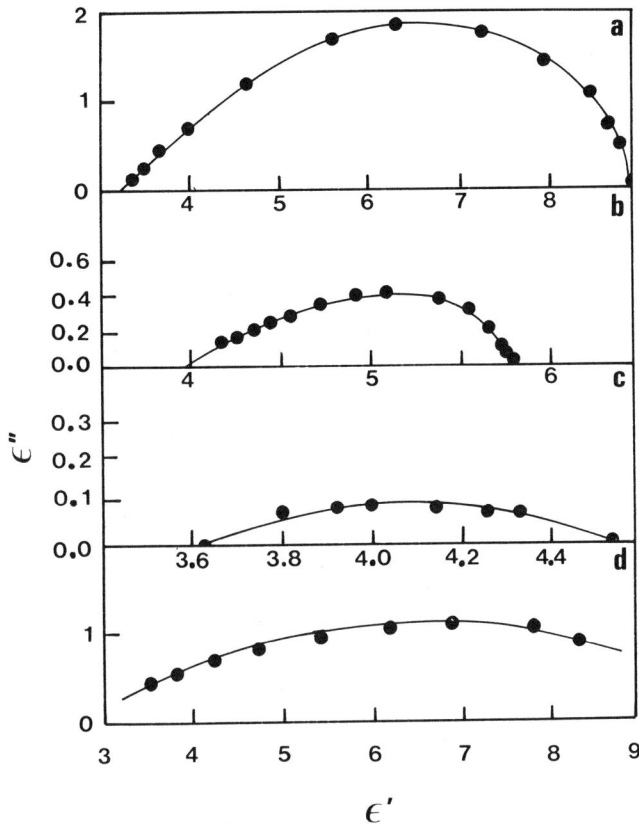

Figure 5.16 Cole–Cole plots for the polymers indicated in Figure 5.15. (Redrawn from reference [256] by courtesy and permission of Butterworth-Heinemann Ltd.)

relaxations since a large fraction of them are immobilized in the crystalline phase. The results of relating the amorphous-phase relaxed dielectric constants with the apparent dipole-correlation factors g are indicated [284] in Figure 5.17. It can be seen that for $\alpha + \beta$ processes, there is a significant reduction in g with crystallinity that is most significant with the onset of crystals. However, for the β process, the correlation factors are seen to be quite insensitive to the presence and the degree of crystallinity. For medium- and high-crystallinity polymers, relaxation processes appear that have the same interpretation as for low-crystallinity polymers (i.e., the glass–rubber relaxation of the amorphous phase). However, comparison of the relaxation behavior of these crystalline polymers with that of the completely amorphous material is not possible because of the unattainability of the latter. The reader will find an exhaustive study on the experimental behavior and molecular interpretation of the relaxation processes of crystalline polymers in references [256] and [257].

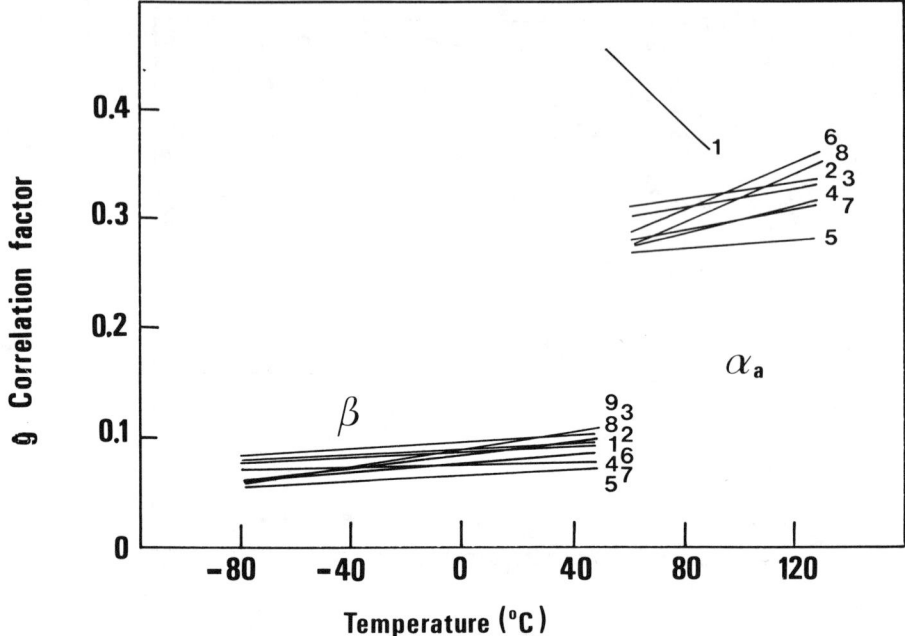

Figure 5.17 Correlation factor g for the amorphous fraction versus temperature in poly(ethylene terephthalate). Degree of crystallinity: (1) 0%, (2) 28%, (3) 31%, (4) 34%, (5) 39%, (6) 49%, (7) 53%, (8) 56%, and (9) 63%. (Redrawn from reference [256] by courtesy and permission of Butterworth-Heinemann Ltd.)

DIELECTRIC RELAXATIONS OF RIGID-BACKBONE POLYMERS

If the mean-square end-to-end distance of a polymer chain is proportional to the number of skeletal bonds ($\langle r^2 \rangle \sim n$), the chain is called flexible. If not, the chain is termed stiff and the probability-distribution end-to-end vector, unlike with what occurs in flexible chains, is not a simple Gaussian. The persistence length for a freely rotating chain can be obtained from the average projection of the end-to-end vector $\mathbf{R}$ along the chain's initial bond direction [287]:

$$\langle \mathbf{R} \cdot \mathbf{R}_1 / l \rangle = l^{-1} \sum \langle \mathbf{R}_j \cdot \mathbf{R}_1 \rangle = l^{-1} \sum l^2 (-\cos \theta)^{j-1} \qquad (5.122)$$

where $l = |\mathbf{R}_1|$. The sum of the geometric series in Eq. (5.122) gives

$$\langle \mathbf{R} \cdot \mathbf{R}_1 / l \rangle = l \, \frac{1 - (-\cos \theta)^n}{1 + \cos \theta} \qquad (5.123)$$

which leads to $l/(1 + \cos \theta) \equiv \frac{1}{2}\Gamma$ when $n \to \infty$. The term $\frac{1}{2}\Gamma$ is called the *persistence length*. The wormlike chain is the continuous chain model generated by taking the limit $l \to 0$, $n \to \infty$, $nl \to N$, $\theta \to \pi$, and $l/(1 + \cos \theta) \to \frac{1}{2}\Gamma$ of the discrete freely rotating chain. The statistics of the wormlike or stiff chains lead to the following expressions [287–291]:

$$\lim_{\Gamma N \to \infty} \langle R^2 \rangle = N/\Gamma \tag{5.124}$$

$$\lim_{\Gamma N \to 0} \langle R^2 \rangle = N^2 \tag{5.125}$$

These relations indicate that when the chain length N is much greater than the persistence length, the chain becomes Gaussian and, consequently, the effects of stiffness are negligible. However, if both quantities are comparable, the chain becomes a rigid rod. Natural polymers such as DNA, helical polypeptides [292–294], and derivatives of cellulose and other polysaccharides [295–301] are examples of rigid chains. The high equilibrium rigidity of many of these polymers favors the formation of lyotropic mesophases in their concentrated solutions. This type of behavior also appears in several aromatic polyamides and polyesters and other rigid-chain polymers [302–306]. By combining rigid-chain fragments with flexible-chain sequences, thermotropic mesophases are formed where the mesogenic chores are played by the rigid fragments [307]. The flexible components of the chains serve as "internal plasticizers" that facilitate the orientationally ordered mutual arrangement of mesogenic parts. Rigid mesogenic parts can be located both in the main chain or in the side chains.

Nematic liquid–crystal phases can be obtained from concentrated solutions of polydisperse rodlike polymers such as poly(alkyl isocyanates) [308, 309], α helical polypeptides [292–294], and certain aromatic polyamides [302–306]. At low concentrations, the solutions of these rigid chains are isotropic, until a concentration is reached at which the material is biphasic, being an equilibrium mixture of isotropic and anisotropic phases in a certain concentration range; for larger concentrations, a single anisotropic phase exists. It is obvious that the dielectric relaxation of stiff-chain solutions is strongly dependent on concentration.

The dielectric static permittivity for an isotropic solution of rigid chains can be expressed by [39]

$$\epsilon_0 - \epsilon_\infty = \frac{4\pi}{3k_B T} \frac{3\epsilon_0(2\epsilon_0 + \epsilon_\infty)}{(2\epsilon_0 + 1)^2} \frac{\langle \mathbf{M}(0) \cdot \mathbf{M}(0) \rangle}{V} \tag{5.126}$$

where $\mathbf{M}(0)$ is the instantaneous dipole moment of a macroscopic sphere of volume V. For the sake of simplicity, it is assumed that the solvent is only weakly polar and, consequently, its contribution to ϵ_0 is negligible. This condition occurs, for example, in toluene solutions of poly(alkyl isocyanates). For a polydisperse system in which the molecular-weight distribution is $W(M)$, and the instantaneous dipole moment of a chain i of molecular weight M is $\mathbf{P}_{iM}(0)$, the mean-square dipole moment $\langle \mathbf{M}(0) \cdot \mathbf{M}(0) \rangle$ in Eq. (5.120) is the sum of three contributions [310], that is,

$$\langle \mathbf{M}(0) \cdot \mathbf{M}(0) \rangle = \sum_M W(M) \langle \mathbf{P}_M^2(M) \rangle + \sum_M W(M) \sum_j \langle \mathbf{P}_{iM}(0) \cdot \mathbf{P}_{jM}(0) \rangle$$

$$+ \sum_M W(M) \sum_{M'} \sum_k \langle \mathbf{P}_{iM}(0) \cdot \mathbf{P}_{kM'}(0) \rangle \tag{5.127}$$

The first term on the right side of Eq. (5.127) represents the mean-square dipole moment averaged over the distribution. The second term indicates the correlation between a reference molecule i of molecular weight M with all the other molecules j of the same molecular weight. Finally, the last contributions arise from the correlation between the ith molecule of molecular weight M with all molecules k of molecular weight M'.

By defining

$$\langle \bar{P}^2(M) \rangle = \frac{\displaystyle\sum_M W(M) \langle P^2(M) \rangle}{\displaystyle\sum_M W(M)} \tag{5.128}$$

Eq. (5.127) can be written as

$$\langle \mathbf{M}(0) \cdot \mathbf{M}(0) \rangle = N \langle \bar{P}^2(M) \rangle \bar{g}_1 \tag{5.129}$$

where N is the number of chains in the sphere of volume V, and

$$\bar{g}_1 = 1 + \frac{\displaystyle\sum_M W(M) \sum_j \langle \mathbf{P}_{iM}(0) \cdot \mathbf{P}_{jM}(0) \rangle}{\langle \bar{P}^2(M) \rangle} + \frac{\displaystyle\sum_M W(M) \sum_{M'} \sum_j \langle \mathbf{P}_{iM}(0) \cdot \mathbf{P}_{jM'}(0) \rangle}{\langle \bar{P}^2(M) \rangle} \tag{5.130}$$

Liquid $\mathbf{P}_i$ and vacuum ${}^0\mathbf{P}_i$ dipole moments are related by

$$\mathbf{P}_i = {}^0\mathbf{P}_i \left[\frac{(2\epsilon_0 + 1)(\epsilon_\infty + 2)}{3(2\epsilon_0 + \epsilon_\infty)} \right]^2 \tag{5.131}$$

and, hence, for a solution of concentration $C_p = N/V$, Eq. (5.126) becomes

$$\epsilon_0 - \epsilon_\infty = \frac{4\pi}{3k_B T} \left(\frac{3\epsilon_0}{2\epsilon_0 + \epsilon_\infty} \right) \left(\frac{\epsilon_\infty + 2}{3} \right)^2 C_p \langle {}^0\bar{P}^2(M) \rangle \bar{g}_1 \tag{5.132}$$

Because $C_p = N_A w_p / V M_n$, where N_A is Avogadro's number, w_p is the mass of the polymer in volume V, and M_n is the number average molecular weight of the distribution, Eq. (5.132) can be written as

$$\epsilon_0 - \epsilon_\infty = \frac{4\pi N_A}{3k_B T M_n} f(\epsilon_0, \epsilon_\infty) F(\rho_p, \rho_s, C) C \langle {}^0\bar{P}^2(M) \rangle \bar{g}_1 \tag{5.133}$$

where $f(\epsilon_0, \epsilon_\infty) = (\epsilon_\infty + 2)^2 \epsilon_0 / 3(2\epsilon_0 + \epsilon_\infty)$. A simple mixing law for volumes was assumed in Eq. (5.133), so that

$$w_p / V = w_p / (w_p \rho_p^{-1} + w_s \rho_s^{-1}) = c F(\rho_p, \rho_s, C) \tag{5.134}$$

where w_s is the mass of solvent in volume V, and ρ_p and ρ_s represent the densities of polymer and solvent, respectively. Thus, Eq. (5.133) shows that once ϵ_0 and ϵ_∞ are measured for different concentrations, the values of $\langle ^0\bar{P}^2(M)\rangle \bar{g}_1$ as a function of concentration can also be obtained.

The isotropic and anisotropic phases coexist in the biphasic region, and the average $\langle \mathbf{M}(0)\cdot\mathbf{M}(0)\rangle_p$ is given by

$$\langle \mathbf{M}(0) \cdot \mathbf{M}(0)\rangle_p = {}_IN \sum_M {}_IW(M)\langle \bar{P}^2(M)\rangle({}_I\bar{g}_1)$$
$$+ {}_AN \sum_M {}_AW(M)\langle \bar{P}^2(M)\rangle({}_A\bar{g}_1) \qquad (5.135)$$

where subscripts I and A refer to the isotropic and anisotropic phases, respectively. It should be pointed out that the molecular-weight distributions ${}_IW(M)$ and ${}_AW(M)$ are different from the distribution for the unpartitioned system [311–316]. Moreover, since ${}_IN = \sum_M {}_IN(M)$ and ${}_AN = \sum_M {}_AN(M)$, the relaxation strength in the biphasic region is

$$\epsilon_0 - \epsilon_\infty = v({}_I\Omega) + (1 - v)({}_A\Omega) \qquad (5.136)$$

where v is the volume fraction of the isotropic phase. ${}_I\Omega$ and ${}_A\Omega$ correspond to the second member of Eq. (5.133) and they can be expressed as

$$_I\Omega = \frac{4\pi N_A}{3k_BT} \frac{1}{_IM_n} {}_If(\epsilon_0,\epsilon_\infty)_I F(\rho_p,\rho_s,{}_IC)_I C\langle ^0\bar{P}^2(M)\rangle({}_I\bar{g}_1) \qquad (5.137)$$

By substituting subscript I with A, the corresponding expression for ${}_A\Omega$ is obtained.

The dielectric behavior of isotropic solutions of poly(n-alkyl isocyanates) with the pendant group ranging from n-butyl to n-octyl has been studied [317–323].

Williams and co-workers [310] studied the dielectric behavior of toluene solutions of poly(n-hexyl isocyanate) as a function of concentration. A summary of their results follows. In the range $0 < C < 15\%$, ϵ_0 is a linear function of C; for the range $15\% < C < 21\%$, ϵ_0 increases with concentration, but it falls below the linear relation. For $22\% < C < 35\%$, there is a sharp decrease in the value of ϵ_0, and for $C > 35\%$, the solution is essentially anisotropic. The linear increase of ϵ_0 with concentration in Figure 5.18 arises because in this region the system is isotropic in the sense that all molecules may, through the Brownian motion, randomize their dipole vectors into a 4π solid angle. As the concentration increases, the orientation of the rods is severely hindered by the increase of viscosity that takes place in the isotropic phase. Here, the permittivity-versus-concentration plot falls below the straight line corresponding to the isotropic phase. For $C > C^*$, where C^* is the critical concentration at which the biphasic region appears, $\Delta\epsilon$ is a weighted sum of terms proportional to $(1 - \Phi_A)v_2\langle ^0\bar{P}^2(M)\rangle({}_I\bar{g}_1)$ and $\Phi_A v_2\langle ^0\bar{P}^2(M)\rangle({}_A\bar{g}_1)$. In the range $22\% < C < 35\%$, $\Delta\epsilon$ falls dramatically as a consequence of the decrease in $(1 - \Phi_A)v_2\langle ^0\bar{P}^2(M)\rangle({}_I\bar{g}_1)$. It should be pointed out that the changes in both ϵ_m'' and $\tau = 1/\omega_m$ with concentration follow the same trend as ϵ_0.

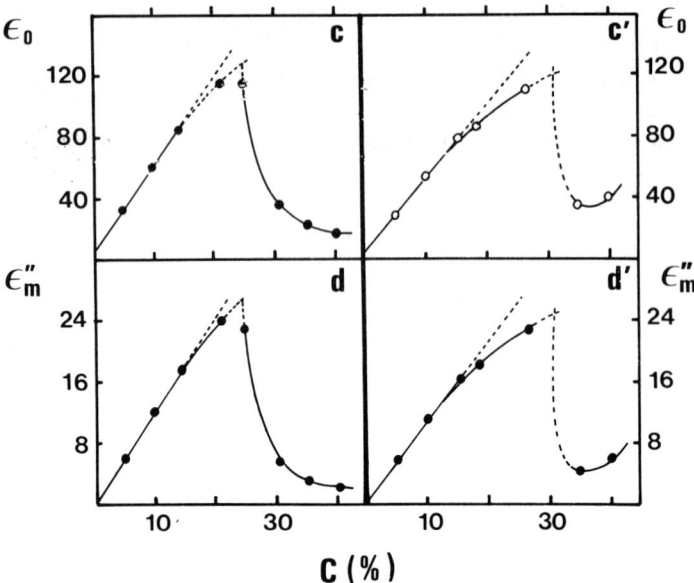

Figure 5.18   (c) ϵ_0 and (d) ϵ''_m as a function of polymer concentration [% (w/w)] for toluene solutions of poly(hexamethylene isocyanate) at 292.2K. (c′) and (d′) are the corresponding plots for a copolymer of *n*-butyl isocyanate and *n*-nonyl isocyanate in toluene solutions. (Redrawn from reference [310] by courtesy of the American Chemical Society.)

The rotational motion of solutions of rodlike polymer chains was investigated by Doi [324]. The dynamics of these chains is defined in terms of [325, 326]

$$1/L^3 \ll C \ll 1/dL^2 \tag{5.138}$$

where C is the concentration of molecular chains per unit of volume, L is the length of the chains, and d is its diameter. According to the theory, crystalline order appears at a critical concentration, $C^* \approx 1/dL^2$. The inequality $C \gg 1/dL^2$ indicates that the orientation is severely restricted by the surrounding polymers, whereas $C \ll 1/dL^2$ means that the orientation of the chains is independent because C is below the critical concentration.

Owing to the restrictions imposed by the surrounding polymer, each chain can be considered to be confined in a tubelike region of radius a and length L. On a short time scale, the unit vector **u** along a given chain fluctuates within a small angle $\theta \approx a/L$ inside the tube. This process has a characteristic relaxation time $\tau_f = \theta^2/D_0$, where D_0 is the rotational diffusion coefficient for the free reorientation of the rod in a solvent of viscosity η. On the long time scale, the chain moves out of the old tube and enters a new one, which forms with the old tube an angle of order θ. Although θ is small, repetition of this process changes **u** in large scale. This process, termed the reptation–rotation mode, has a relaxation time $\tau_D \approx (2\theta^2 D_0)^{-1}$. Doi [324] introduces a unit vector $\mathbf{n}$, which represents the direction of the tube; the condition that $\mathbf{u}$ is fluctuating around $\mathbf{n}$ is represented by the coupling potential

$$V_c(\mathbf{u} - \mathbf{n}) = (k_B T/2\theta^2)(\mathbf{u} - \mathbf{n})^2 = (k_B T/\theta^2)(1 - \mathbf{u}\cdot\mathbf{n}) \qquad (5.139)$$

At equilibrium, in the absence of an external field,

$$\langle(\mathbf{u} - \mathbf{n})^2\rangle = \frac{\int(\mathbf{u} - \mathbf{n})^2 \exp\left[-\dfrac{V_c(\mathbf{u} - \mathbf{n})}{k_B T}\right] d^2\mathbf{u}}{\int \exp\left[-\dfrac{V_c(\mathbf{u} - \mathbf{n})}{k_B T}\right] d^2\mathbf{u}} \qquad (5.140)$$

where $d^2\mathbf{u}$ represents the fact that the integral is performed over the entire surface of the sphere $|\mathbf{u}| = 1$. The result obtained is

$$\langle(\mathbf{u} - \mathbf{n})^2\rangle = 2\theta^2 \qquad (5.141)$$

Thus, the correlation between the directions of the tube and rod axes is given in terms of a small angle θ.

Expressions were obtained for the orientational distribution function $\Phi(\mathbf{u},\mathbf{n},t)$ for the case in which the system was under the effects of an applied electric field. The relations obtained from the model for the time-autocorrelation function $\langle\mathbf{u}(0)\cdot\mathbf{u}(t)\rangle$, which represents the reorientation of the dipole moment $\boldsymbol{\mu} = \mu\mathbf{n}$, and for the complex dielectric polarizability $\alpha^*(\omega)$ were [324]

$$\langle\mathbf{u}(0) \cdot \mathbf{u}(t)\rangle = (1 - 2\theta^2) \exp(-t/\tau_D) + 2\theta^2 \exp(-t/\tau_H) \qquad (5.142)$$

$$\alpha^*(\omega) = \frac{C\mu^2}{3k_B T}\left(\frac{1 - 2\theta^2}{1 + i\omega\tau_D} + \frac{2\theta^2}{1 + i\omega\tau_H}\right) \qquad (5.143)$$

where $\tau_H = \tau_F$, and τ_D are the relaxation characteristic times of the fluctuation and reptation modes, respectively, indicated before. Equations (5.142) and (5.143) suggest that the process proceeds by a fluctuation relaxation characterized by a strength factor $2\theta^2$ followed by a reptation–rotation process characterized by a strength factor $1 - 2\theta^2$. Similar expressions were obtained by Williams [327] using a model in which, at short times, the rods are permitted to diffuse freely within a cone, while the cone is allowed to undergo a small step diffusion at long times. The model used for the restricted diffusion was previously described by Warchol and Vaughan [328] and by Wang and Pecora [329] and it is called the WVWP model. The analysis of the results does not permit reaching any definitive conclusion about which of these models is more reliable. The critical interpretation of the experimental results requires knowing the molecular-weight distribution function. In the biphasic region, the complexity increases because, as was indicated before, fractionation occurs in the sense that the rods of higher molecular weight tend to pass to the anisotropic phase. Simulation of experimental dielectric relaxation data using the WVWP model was carried out by Moscicki and Williams [330]. They obtained values of

$\Delta\epsilon$, the loss maximum ϵ''_m, and the frequency of maximum loss, which vary with polymer concentration in a manner consistent with the experimental data.

The detailed analysis of the dynamics of polymer rod solutions requires relating the complex dielectric permittivity $\epsilon^*(\omega)$ to decay function $\phi(t)$, as indicated in Eq. (5.8). For solutions in which the solvent is only weakly polar, the contribution to the dielectric relaxation comes only from the chains. Consequently [310],

$$\phi_i(t) = [\langle_i\bar{P}^2(M)\rangle\langle_i\bar{g}_1\rangle]^{-1}\left\{ \sum_M [\langle_iW(M)\rangle\langle\mathbf{P}_{jM}(0) \cdot \mathbf{P}_{kM}(t)\rangle_i]\right.$$

$$+ \sum_M \left[\langle_iW(M)\rangle \sum_k \langle\mathbf{P}_{jM}(0) \cdot \mathbf{P}_{kM}(t)\rangle_i\right] \qquad (5.144)$$

$$\left.+ \sum_M \left[\langle_iW(M)\rangle \sum_{M'}\sum_{k'} \langle\mathbf{P}_{jM}(0) \cdot \mathbf{P}_{k'M'}(t)\rangle_i\right]\right\}$$

where i refers to the isotropic or anisotropic phase and, therefore, is I or A, respectively. For the biphasic material, Eqs. (5.8), (5.135), (5.136), and (5.144) lead to

$$\frac{\epsilon^*(\omega) - \epsilon_\infty}{\epsilon_0 - \epsilon_\infty} = \frac{v(_I\Omega)}{v(_I\Omega) + (1 - v)(_A\Omega)}\mathcal{L}\left[\frac{-d\phi_I(t)}{dt}\right]$$

$$+ \frac{1 - v(_A\Omega)}{v(_I\Omega) + (1 - v)(_A\Omega)}\mathcal{L}\left[\frac{-d\phi_A(t)}{dt}\right] \qquad (5.145)$$

where $\mathcal{L}(x)$ represents the Laplace transform of x.

This equation shows that the real and loss components of the complex permittivity in the biphasic region are given by the weighted sum of contributions of the isotropic and anisotropic phases. Any change in the thermodynamic variables of the system changes its composition and, as a consequence, its dielectric response. Summing up, transformations from the isotropic $\rightarrow$ biphasic $\rightarrow$ anisotropic phase lead to reduction in $\Delta\epsilon$ and the average relaxation time. Moreover, the dielectric results show that the dispersion and motions in the anisotropic phase are, respectively, much broader and faster than those in the isotropic phase.

DIELECTRIC BEHAVIOR OF SIDE-CHAIN LIQUID-CRYSTALLINE POLYMERS

An important requirement to obtain thermotropic side-chain liquid-crystalline polymers is the placement of a flexible spacer between the mesogenic group [331–338] and the main chain to partially decouple the mobility of the main chain from that of the mesogenic group. Formation of mesophases without a spacer requires that the polymer backbone be significantly distorted. For this reason, most of the polymers with mesogenic groups directly attached to the backbone are amorphous. However, there are flexible polymers with a polyacrylate backbone that even lacking a spacer

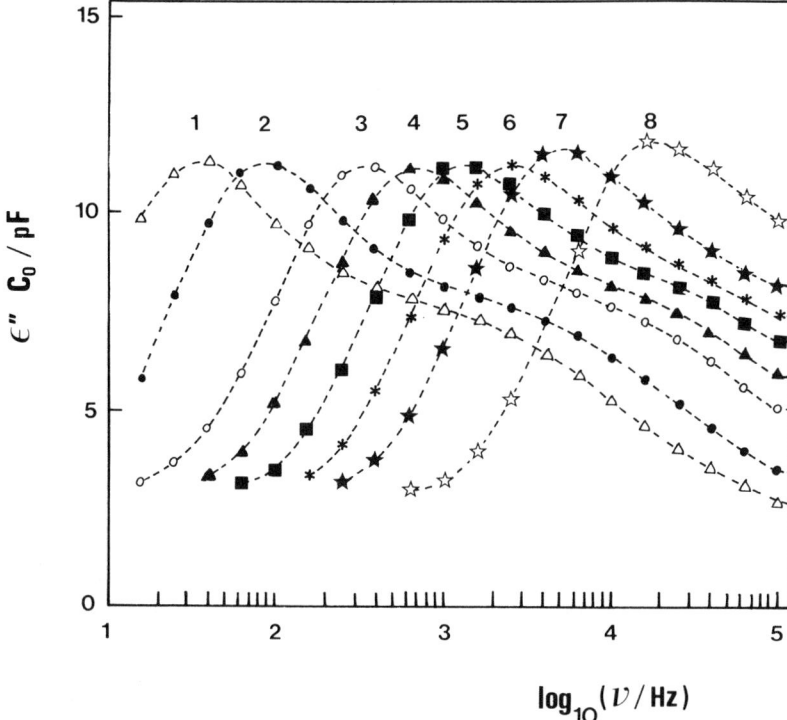

Figure 5.19 $\epsilon''C_0$ versus log frequency for the unaligned smetic polymer, $m = 8$. Curves 1–8 correspond to temperatures of 305.2, 309.2, 316.2, 320.2, 324.2, 328.2, 333.2, 343.2K, respectively. (Redrawn from reference [309] by courtesy of Liquid Crystals.)

have the capability to form liquid-crystalline polymers [338]. Liquid-crystalline side-chain polymers form films 5–100 μm thick that are aligned macroscopically by applied electric or magnetic fields. These polymer films, like low molecular-weight smetics, but in contrast with low molecular-weight nematics, retain their macroscopic alignment after the directing field is removed.

The dielectric relaxation behavior of liquid-crystalline side-chain polymers having siloxane backbones and mesogenic groups $R = (CH_2)_m$—O—C_6H_4 —$COOC_6H_3(CH_3)CN$, where m in the spacer can take the values of 5, 6, or 8, was studied in depth by Attard et al. [339]. Different extents of alignment between fully homeotropic and fully planar (homogeneous) structures can be achieved by slowly cooling the isotropic material back into the liquid-crystalline state in the presence of ac fields of given amplitude and frequency. The loss dielectric curve is bimodal for the isotropic material with $m = 8$ (Figure 5.19), which can be resolved into two broad relaxations, named α and δ; however, the fully homeotropic material has a

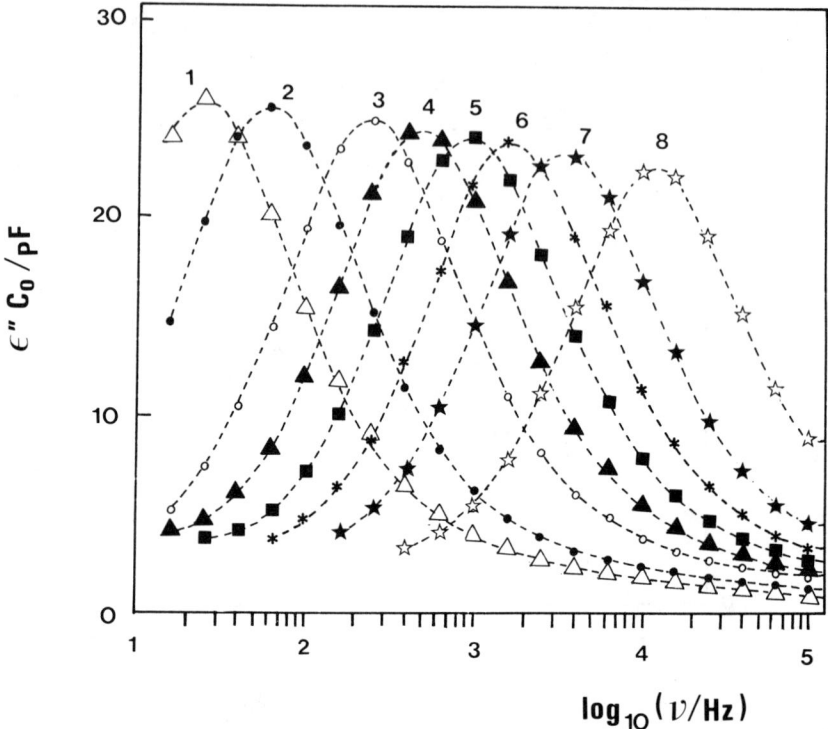

Figure 5.20 $\epsilon''C_0$ versus log frequency for the homeotropically aligned smetic polymer, $m = 8$. Curves 1–8 correspond to the same temperatures indicated in Figure 5.19. (Redrawn from reference [309] by courtesy of Liquid Crystals.)

loss curve that seems to be made up of a δ process, which dominates the spectrum, and a small-intensity α process (Figure 5.20).

Theories [340–342] have been developed for nematic liquid crystals that suggest that the parallel and transverse permittivities, $\epsilon_\parallel$ and $\epsilon_\perp$, are functions of a local-order parameter S and the dipole moment components μ_l and μ_t of the molecule with respect to the local director axis **n**. According to the theories, four relaxation modes are involved in the dielectric relaxation of $\epsilon_\parallel$ and $\epsilon_\perp$, as $\langle D_{ij}^1(0) \cdot D_{ij}^1(t) \rangle$, where ij are 00, 10, 01, and 11; and D indicates a Wigner rotation-matrix element. Thus, $\epsilon_\parallel(\omega)$ is given by linear combinations of 00 and 10 modes, expressed in the frequency domain, and $\epsilon_\perp$ is given by combinations of the 01 and 11 modes. The director-order parameter S_d estimates the average alignment of the local directors **n** with respect to the laboratory axis Z, that is,

$$S_d = \langle 3\cos^2\theta_{Z\mathbf{n}} - 1 \rangle / 2 \qquad (5.146)$$

Accordingly, for a fully homeotropic monodomain ($n\|Z$), $S_d = 1$, whereas for a fully planar monodomain ($n\perp Z$), $S_d = -\frac{1}{2}$. It is obvious that for an unaligned material, $S_d = 0$. For a partially aligned material along the symmetry axis Z, the theory predicts [342] that the relaxation modes ij contribute to $\epsilon_Z(\omega)$ according to the following intensities:

$$A'_{00} = (\mu_t^2/9)(1 + 2S_d)(1 + 2S)$$

$$A'_{10} = (\mu_t^2/9)(1 + 2S_d)(1 - S)$$

$$A'_{01} = (\mu_t^2/9)(1 - S_d)(1 - S) \qquad (5.147)$$

$$A'_{11} = (\mu_t^2/9)(1 - S_d)(1 + S/2)$$

For $S_d = 1$ (homeotropic sample) only the 00 and 10 modes contribute to the dielectric relaxation spectrum and $\epsilon_{\parallel}(\omega)$ is measured; however, for $S_d = -\frac{1}{2}$ only the 01 and 11 modes contribute and $\epsilon_{\perp}(\omega)$ is determined. Calculations of the average relaxation modes τ_{ij} for a small step diffusion in nematic potential suggest that

$$\tau_{00} > \tau_{01} \geq \tau_{11} \geq \tau_{10}$$

In order to reconcile the four relaxation modes that predict the theory with the two processes (α and δ) that appear in the experimental results (Figure 5.19), it seems reasonable to assign the δ process to the 00 relaxation mode and, in addition, to assume that the three remaining modes comprise the broad α process. With these assumptions, reasonable values were obtained for S, S_d, and (μ_t/μ_1) of the smetic polymer in which $m = 8$. The theory was further extended to those cases in which the α and δ processes cannot be resolved and the reader will find more details in reference [343].

Both the δ and α peaks do not obey the Arrhenius equation since the plots of $\log \nu_m$ versus $1/T$ show a strong curvature. Dielectric measurements carried out on the smetic polymer ($m = 8$) in its unaligned and homeotropic aligned states suggest that the δ and α relaxation processes are governed by the free volume. Thus, both loci fit the Vogel equation:

$$\log \nu_m = A - B/(T - T_\infty) \qquad (5.148)$$

finding different values for T_∞ as consequence of the different curvatures of the α and δ processes. It should be pointed out that both processes tend to coalesce as T_g is approached, and, therefore, the crossing of the two processes predicted by the theory does not occur in practice [339].

References for Part One

1. P. W. Atkins, *Physical Chemistry*, Oxford University Press, New York, 1978.
2. A. J. Hopfinger, *Conformational Properties of Macromolecules*, Academic Press, New York, 1973.
3. M. Faraday, *Phil. Trans.*, **128**, 1, 79, 265 (1837/1838).
4. P. F. Massotti, *Bibl. Univ. Modena*, **6**, 193 (1847).
5. R. Clausius, *Die Mechanische Wärmetheory*, Vol. II, Braunschweich, 1879.
6. J. C. Maxwell, *Phil. Trans.*, **155**, 459 (1865); ibid., **158**, 643 (1868).
7. L. Lorentz, *Ann. Phys. Chem.*, **11**, 70 (1880).
8. H. A. Lorentz, *Verh. Kon. Acad. van Wetenschappen, Ann. Phys.*, **9**, 641 (1879).
9. J. J. Berzelius, *Essai sur la Theorie des Proportions Chimique et sur l'Influence Chimique de l'Electricité*, Paris, 1819.
10. P. Debye, *Phys. Z.*, **13**, 97 (1912); *Collected Papers, Polar Molecules*, p. 173, Wiley Interscience, New York, 1954.
11. L. Onsager, *J. Am. Chem. Soc.*, **58**, 1486 (1938).
12. J. G. Kirkwood, *J. Chem. Phys.*, **7**, 911 (1939).
13. P. Debye, *Ber. D. Phys. Ges.*, **15**, 777 (1913); *Collected Papers*, p. 158, Wiley Interscience, New York, 1954.

14. H. Staudinger, *Ber.*, **53**, 1073 (1920); **57**, 1203 (1924); H. Staudinger and J. Fritschi, *Helv. Chim. Acta.*, **5**, 785 (1922).

15. W. Kuhn, *Kolloid-Z.*, **68**, 2 (1934).

16. E. Guth and H. Mark, *Monatsh. Chem.*, **65**, 93 (1934).

17. M. V. Volkenstein, *Configurational Statistics of Polymeric Chains*, trans. from the 1959 Russian edition by S. N. Timasheff and M. J. Timasheff, Wiley Interscience, New York, 1963.

18. T. M. Birshtein and O. B. Ptitsyn, *Conformations of Macromolecules*, trans. from the 1964 Russian edition by S. N. Timasheff and M. J. Timasheff, Wiley Interscience, New York, 1966.

19. P. J. Flory, *Statistical Mechanics of Chain Molecules*, Wiley Interscience, 1969.

20. I. F. Halverstadt and W. D. Kumler, *J. Am. Chem. Soc.*, **64**, 2988 (1942).

21. E. A. Guggenheim, *Trans. Faraday Soc.*, **45**, 714 (1949); **47**, 573 (1951).

22. J. W. Smith, *Trans. Faraday Soc.*, **46**, 394 (1950).

23. J. D. Ferry, *Viscoelastic Properties of Polymers*, 3rd ed., Wiley Interscience, New York, 1980.

24. P. E. Rouse, Jr., *J. Chem. Phys.*, **21**, 1272 (1953); B. Zimm, *J. Chem. Phys.*, **24**, 269 (1956); M. Fixman, *J. Chem. Phys.*, **42**, 3831 (1965); C. W. Pyun and M. Fixman, *J. Chem. Phys.*, **42**, 3838 (1965).

25. P. G. de Gennes, *J. Chem. Phys.*, **55**, 572 (1971).

26. P. G. de Gennes, *Scaling Concepts in Polymer Physics*, Cornell University Press, Ithaca, N.Y., 1979.

27. S. F. Edwards, *Proc. Phys. Soc.*, **92**, 9 (1967).

28. M. Doi and S. F. Edwards, *J. Chem. Soc. Faraday 2*, **74**, 1789 (1978).

29. M. Doi and S. F. Edwards, *The Theory of Polymer Dynamics*, Clarendon Press, Oxford, 1986.

30. W. H. Stockmayer, *Pure Appl. Chem.*, **15**, 539 (1967); W. H. Stockmayer and J. J. Burke, *Macromolecules*, **2**, 647 (1969).

31. D. Boese, F. Kremer, and L. J. Fetters, *Makromol. Chem. Rap. Comm.*, **9**, 367 (1988).

32. E. Helfand, *Science*, **226**, 647 (1984).

33. C. P. Smyth, *Dielectric Behavior and Structure*, McGraw-Hill, New York, 1955.

34. P. Langevin, *J. Phys.*, **4**, 678 (1905); *Ann. Chim. Phys.*, **5**, 70 (1905).

35. P. Debye, *Handbuch der Radiologie*, Akademische Verlagsgesellschaft m.b.H., VI, pp. 597–653, Leipzig, 1925; *Polar Molecules*, Chemical Catalog, New York, 1929.

36. W. Kauzmann, *Revs. Mod. Phys.*, **14**, 12 (1942).

37. C. J. F. Böttcher, *Theory of Electric Polarization*, 2nd ed. rev. by O. C. Van Belle, P. Bordewijk, and A. Rip, Elsevier, Amsterdam, 1973.

38. J. G. Kirkwood, *Ann. N. Y. Acad. Sci.*, **40**, 315 (1940); *Trans. Faraday Soc.*, **42A**, 7 (1946).

39. H. Fröhlich, *Trans. Faraday Soc.*, **44**, 238 (1948); *Theory of Dielectrics*, Oxford University Press, London, 1958.

40. W. H. Heston, Jr., E. J. Hennelly, and C. P. Smyth, *J. Am. Chem. Soc.*, **72**, 2071 (1950).

41. P. Madden and D. Kivelson, *Ann. Rev. Phys. Chem.*, **35**, 75 (1984).

42. H. Block, *Adv. Polym. Sci.*, **39**, 94 (1979).

43. J. Marchal and H. Benoit, *J. Chim. Phys. Phys. Chim. Biol.*, **52**, 818 (1955); *J. Polym. Sci.*, **23**, 223 (1957).

44. K. Nagai and T. Ishikawa, *Polym. J.*, **2**, 416 (1971).

45. M. Fixman, *J. Chem. Phys.*, **23**, 1656 (1955).

46. L. de Brouckere, D. Buess, J. de Bock, and J. Versluys, *Bull. Soc. Chim. Belges*, **64**, 669 (1955).

47. L. de Brouckere, D. Buess, and L. K. H. van Beek, *J. Polym. Sci.*, **23**, 233 (1957).

48. L. de Brouckere and L. Lecocq-Robert, *Bull. Soc. Chim. Belges*, **70**, 549 (1961).

49. M. Doi, *Polym. J.*, **3**, 252 (1972).

50. W. L. Mattice and D. K. Carpenter, *Macromolecules*, **17**, 625 (1984).

51. W. L. Mattice, D. K. Carpenter, M. D. Barkley, and N. R. Kestner, *Macromolecules*, **18**, 2236 (1985).

52. E. Riande, *Makromol. Chem.* **178**, 2001 (1977).

53. S. C. Liao and J. E. Mark, *J. Chem. Phys.*, **59**, 3825 (1973).

54. L. L. Burshtein and T. P. Stepanova, *Polym. Sci. USSR*, **11**, 2885 (1969).

55. F. Blasco and E. Riande, *J. Polym. Sci., Polym. Phys. Ed.*, **21**, 835 (1983).

56. E. Riande, S. Boileau, P. Hemery, and J. E. Mark, *Macromolecules*, **12**, 702 (1979).

57. E. Riande, S. Boileau, P. Hemery, and J. E. Mark, *J. Chem. Phys.*, **71**, 4206 (1979).

58. N. G. McCrum, B. E. Read, and G. Williams, *Anelastic and Dielectric Effects in Polymeric Solids*, Wiley, New York, 1967.

59. L. Hartshorn, J. V. L. Parray, and L. Essen, *Proc. Phys. Soc. (London)*, **68B**, 422 (1955).

60. R. D. McCammon and R. N. Work, *Rev. Sci. Instrum.*, **36**, 1169 (1955).

61. R. H. Cole, *Ann. Rev. Phys. Chem.*, **28**, 283 (1977).

62. R. H. Cole, *J. Phys. Chem.*, **79**, 1459, 1469 (1975).

63. R. H. Cole, S. Mashimo, and P. Winsor, *J. Phys. Chem.*, **84**, 786 (1980).

64. H. Nakamura, S. Mashimo, and A. Wada, *Japan. J. Appl. Phys.*, **21**, 467 (1982).

65. P. J. Flory, *Macromolecules*, **7**, 381 (1974).

66. S. Mizushima, *Structure of Molecules and Internal Rotation*, Academic Press, New York, 1954.

67. W. J. Orville-Thomas, Ed., *Internal Rotation in Molecules*, Wiley, New York, 1974.

68. J. Ladik, J. M. André, and M. Seel, Eds., *Quantum Chemistry of Polymers*, NATO ASI Series, Series C, Vol. 123, 1984.

69. A. Abe and J. E. Mark, *J. Am. Chem. Soc.*, **98**, 6468 (1976).

70. W. L. Bragg, *Phil. Mag.*, **40**, 169 (1920).

71. S. B. Hendricks, *Chem. Rev.*, **7**, 431 (1930).

72. J. C. Slater and J. G. Kirkwood, *Phys. Rev.*, **37**, 682 (1931).

73. R. A. Scott and H. A. Scheraga, *J. Chem. Phys.*, **45**, 2091 (1966).
74. K. S. Pitzer, *Adv. Chem. Phys.*, **2**, 59 (1959).
75. A. Abe, R. L. Jernigan and P. J. Flory, *J. Am. Chem. Soc.*, **88**, 631 (1966).
76. K. Matsuzaki and H. Ito, *J. Polym. Sci., Polym. Phys. Ed.*, **12**, 2509 (1974).
77. T. M. Connor and K. A. McLauchlan, *J. Phys. Chem.*, **69**, 1888 (1965); T. M. Connor, *Mol. Phys.*, **19**, 253 (1970).
78. D. A. Brant, W. G. Miller, and P. J. Flory, *J. Mol. Biol.*, **23**, 47 (1967).
79. I. Tvarovska and T. Bleha, *J. Mol. Struct.*, **24**, 249 (1975).
80. G. A. Jeffrey, J. A. Pople, and L. A. Radom, *Carbohydr. Res.*, **25**, 117 (1972); **38**, 81 (1974).
81. E. R. Chan, E. Clementi, J. E. Mark, and P. J. Flory, unpublished results.
82. W. J. Leonard, R. L. Jernigan, and P. J. Flory, *J. Chem. Phys.*, **43**, 2256 (1965).
83. H. J. G. Hayman and I. Eliezer, *J. Chem. Phys.*, **35**, 644 (1961).
84. C. H. Porter, J. H. L. Lawler, and R. H. Boyd, *Macromolecules*, **3**, 308 (1970).
85. J. Marchal and H. Benoit, *J. Polym. Sci.*, **23**, 223 (1957).
86. K. Bak, G. Elefante, and J. E. Mark, *J. Phys. Chem.*, **71**, 4007 (1967).
87. E. Riande, *J. Polym. Sci., Polym. Phys. Ed.*, **14**, 2231 (1976); E. Riande, J. Guzmán, and L. Garrido, *Macromolecules*, **17**, 1234 (1984).
88. J. E. Mark and D. S. Chiu, *J. Chem. Phys.*, **66**, 1901 (1977).
89. T. Uchida, Y. Kurita, and M. Kubo, *J. Polym. Sci.*, **19**, 365 (1956).
90. K. Kimura and R. Fugishiro, *Bull. Chem. Soc. Jpn.*, **39**, 668 (1966).
91. R. J. W. Le Fevre and K. M. Sundaram, *J. Chem. Soc., Perkin Trans.*, **2**, 2323 (1972).
92. A. Abe, T. Hirano, and T. Tsuruta, *Macromolecules*, **12**, 1092 (1979).
93. E. Riande, J. G. de la Campa, J. Guzmán, and J. de Abajo, *Macromolecules*, **17**, 1891 (1984).
94. E. Riande, J. G. de la Campa, and J. de Abajo, *Macromolecules*, **17**, 1431 (1984).
95. E. Saiz, E. Riande, J. Guzmán, and J. de Abajo, *J. Chem. Phys.*, **73**, 958 (1980).
96. L. Garrido, E. Riande, and J. Guzmán, *Macromolecules*, **20**, 1805 (1982).
97. E. Riande and J. E. Mark, *Macromolecules*, **11**, 956 (1978).
98. E. Riande and J. E. Mark, *J. Polym. Sci., Polym. Phys. Ed.*, **17**, 2013 (1979).
99. E. Riande and J. E. Mark, *Polymer*, **20**, 1188 (1979).
100. E. Riande, J. Guzmán, and E. Saiz, *Polymer*, **22**, 465 (1981).
101. A. de Chirico and L. Zotteri, *Eur. Polym. J.*, **11**, 487 (1975).
102. D. Bhaumik and J. E. Mark, *Macromolecules*, **14**, 162 (1981).
103. W. J. Welsh and E. Riande, *Polym. J.*, **12**, 467 (1980).
104. E. Riande, J. Guzmán, W. J. Welsh, and J. E. Mark, *Makromol. Chem.*, **183**, 2555 (1982).
105. Y. Takahasi, H. Tadokoro, and Y. Chatani, *J. Macromol. Sci., Phys.*, **2**(2), 361 (1968).
106. W. J. Welsh, J. E. Mark, J. Guzmán, and E. Riande, *Makromol. Chem.*, **183**, 2562 (1982).
107. J. L. de la Peña, E. Riande, and J. Guzmán, *Macromolecules*, **18**, 2739 (1985).

108. A. Abe, *Macromolecules*, **13**, 546 (1980).

109. E. Riande and J. Guzmán, *Macromolecules*, **14**, 1234 (1981).

110. P. Guerin, S. Boileau, F. Subira, and P. Sigwalt, *Eur. Polym. J.*, **11**, 337 (1975).

111. A. Abe, *Macromolecules*, **13**, 541 (1980).

112. J. Guzmán, E. Riande, W. J. Welsh, and J. E. Mark, *Makromol. Chem.*, **183**, 2573 (1982).

113. E. Riande, J. Guzmán, E. Saiz, and J. de Abajo, *Macromolecules*, **14**, 608 (1981).

114. E. Riande and J. Guzmán, *Macromolecules*, **19**, 2956 (1986).

115. E. Riande and J. Guzmán, *Macromolecules*, **12**, 952 (1979); **12**, 1117 (1979).

116. E. Riande and J. Guzmán, *Macromolecules*, **14**, 1511 (1981).

117. E. Riande, *J. Polym. Sci., Polym. Phys. Ed.*, **15**, 1397 (1977).

118. E. Riande and J. Guzmán, *J. Polym. Sci., Polym. Phys. Ed.*, **23**, 1235 (1985).

119. C. González, E. Riande, A. Bello, and J. M. Pereña, *Macromolecules*, **21**, 3230 (1988).

120. E. Riande, J. Guzmán, J. G. de la Campa, and J. de Abajo, *J. Polym. Sci., Polym. Phys. Ed.*, **25**, 2403 (1987).

121. R. Díaz-Calleja, E. Riande, and J. Guzmán, *Macromolecules*, **22**, 3654 (1989).

122. E. Saiz, J. P. Hummel, P. J. Flory, and M. Plavsic, *J. Phys. Chem.*, **85**, 3211 (1981).

123. T. Miyazwa, *Bull Chem. Soc. Jpn.*, **34**, 691 (1961).

124. A. D. Williams and P. J. Flory, *J. Polym. Sci. A-2*, **5**, 417 (1967).

125. J. San Román, J. Guzmán, E. Riande, J. Santoro, and M. Rico, *Macromolecules*, **15**, 609 (1982).

126. F. Mendicuti, M. M. Rodrigo, M. P. Tarazona, and E. Saiz, *Macromolecules*, **23**, 1139 (1990).

127. E. Riande, J. Guzmán, M. P. Tarazona, and E. Saiz, *J. Polym. Sci., Polym. Phys. Ed.*, **22**, 917 (1984).

128. E. Riande, J. Guzmán, and J. San Román, *J. Chem. Phys.*, **73**, 958 (1980).

129. E. Riande, J. Guzmán, and J. de Abajo, *Makromol. Chem.*, **185**, 1943 (1984).

130. E. Riande, J. G. de la Campa, D. D. Schlereth, J. de Abajo, and J. Guzmán, *Macromolecules*, **20**, 1641 (1987).

131. E. Riande, J. Guzmán, J. G. de la Campa, and J. de Abajo, *Macromolecules*, **18**, 2739 (1985).

132. E. Riande, J. Guzmán, and J. de Abajo, *Macromolecules*, **22**, 4026 (1989).

133. E. Riande, J. Guzmán, and J. G. de la Campa, *Macromolecules*, **21**, 2128 (1988).

134. S. Dirlikov, J. Stokr, and B. Schneider, *Collect. Czech. Chem. Commun.*, **36**, 3028 (1971).

135. R. M. Moravie and J. Corset, *Chem. Phys. Lett.*, **26**, 210 (1974); *J. Mol. Struct.*, **24**, 91 (1975).

136. A. J. Bowles, W. O. George, and D. B. Cunnliffe-Jones, *Chem. Comm.*, 103 (1970); W. O. George, D. V. Hassid, and W. F. Madams, *J. Chem. Soc., Perkin Trans.*, **2**, 1029 (1972).

137. A. Abe, *J. Am. Chem. Soc.*, **106**, 14 (1984).

138. E. Riande, J. Guzmán, and H. Addabo, *Macromolecules*, **19**, 2567 (1986).

139. E. Riande and J. Guzmán, *J. Chem. Soc., Perkin Trans.*, **2**, 299 (1988).

140. A. Abe, *J. Phys. Chem.*, **91**, 6496 (1987).

141. W. P. Purcell and J. A. Singer, *J. Phys. Chem.*, **69**, 691 (1965).

142. M. M. Kopecnic, R. J. Laub, and D. M. Petkovic, *J. Phys. Chem.*, **85**, 1595 (1981).

143. N. Kulevsky, *Molecular Associations*, Chapter 2, R. Foster (Ed.), Academic Press, London, 1975.

144. J. P. Hummel and P. J. Flory, *Macromolecules*, **13**, 479 (1980).

145. L. A. La Planche and T. Rogers, *J. Am. Chem. Soc.*, **86**, 337 (1964).

146. L. L. Shipman and R. E. Christoffersen, *J. Am. Chem. Soc.*, **95**, 1408 (1973).

147. G. Khanarian, P. Msek, and W. J. Moore, *Biopolymers*, **20**, 1191 (1981).

148. A. López Piñeiro and E. Saiz, *Int. J. Biol. Macromol.*, **5**, 37 (1983).

149. J. F. Yan, F. A. Momany, R. Hoffmann, and H. A. Scheraga, *J. Phys. Chem.*, **74**, 420 (1970).

150. M. M. Rodrigo, M. P. Tarazona, and E. Saiz, *J. Phys. Chem.*, **90**, 2236 (1986).

151. R. E. Cater and J. Sandström, *J. Phys. Chem.*, **76**, 642 (1972).

152. P. C. Chieh, E. Subramanian, and J. Trotter, *J. Chem. Soc.*, **A**, 179 (1970).

153. E. Eliel, *Stereochemistry of Carbon Compounds*, McGraw-Hill, New York, 1962.

154. D. A. Brant and P. J. Flory, *J. Am. Chem. Soc.*, **87**, 2791 (1965).

155. J. E. Mark, *Account Chem. Res.*, **7**, 218 (1974).

156. M. Salmerón, J. M. Barrales-Rienda, E. Riande, and E. Saiz, *Macromolecules*, **17**, 2728 (1984).

157. A. Roig and I. Hernández-Fuentes, *An. Quim.*, **70**, 668 (1974).

158. B. Baysal, H. Yu, and W. H. Stockmayer, *Dielectric Properties of Polymers*, F. E. Karasz (Ed.), Plenum, New York, 1972; F. H. Smith, L. H. Corrado, and N. Work, *Polym. Prep. Am. Chem. Soc. Div. Polym. Chem.*, **12**, 64 (1971).

159. N. Yamaguchi, M. Sato, E. Ogawa, and M. Shima, *Polymer*, **22**, 1464 (1981).

160. A. E. Tonelli and L. A. Belfiore, *Macromolecules*, **16**, 1740 (1983).

161. D. Y. Yoon, P. R. Sundararajan, and P. J. Flory, *Macromolecules*, **8**, 776 (1975).

162. E. Saiz, J. E. Mark, and P. J. Flory, *Macromolecules*, **10**, 967 (1977).

163. P. R. Sundararajan, *Macromolecules*, **13**, 512 (1982).

164. R. M. Masegosa, I. Hernández-Fuentes, E. A. Ojalvo, and E. Saiz, *Macromolecules*, **12**, 862 (1979).

165. E. A. Ojalvo, E. Saiz, R. M. Masegosa, and I. Hernández-Fuentes, *Macromolecules*, **12**, 865 (1979).

166. M. P. Tarazona and E. Saiz, *Macromolecules*, **16**, 1128 (1983).

167. Y. Yarim-Agaev, M. Plavsic, and P. J. Flory, *Prepr. Am. Chem. Soc., Div. Polym. Chem.*, **24**(1), 233 (1983).

168. J. San Román, E. Riande, E. L. Madruga, and E. Saiz, *Macromolecules*, **23**, 1923 (1990).

169. E. Saiz, E. Riande, J. San Román, and E. L. Madruga, *Macromolecules*, **23**, 785 (1990).

170. E. Saiz, J. San Román, E. L. Madruga, and E. Riande, *Macromolecules*, **22**, 1330 (1989).

171. E. Riande, E. L. Madruga, J. San Román, and E. Saiz, *Macromolecules*, **21**, 2807 (1988).

172. P. R. Sundarajan and P. J. Flory, *J. Am. Chem. Soc.*, **96**, 5025 (1974).

173. A. Horta, I. Hernández-Fuentes, L. Gargallo, and D. Radic, *Makromol. Chem. Rapid Comm*, **8**, 523 (1988).

174. L. Gargallo, M. Yazdani-Pedram, D. Radic, I. Hernández-Fuentes, and A. Horta, *Makromol. Chem.*, **189**, 145 (1988).

175. E. Saiz, A. Horta, L. Gargallo, I. Hernández-Fuentes, and D. Radic, *Macromolecules*, **31**, 1736 (1988).

176. E. Saiz, A. Horta, L. Gargallo, I. Hernández-Fuentes, C. Abradelo, and D. Radic, *J. Chem. Res.*, **280** (1988).

177. R. J. W. Le Fèvre and K. M. S. Sundaran, *J. Chem. Soc.*, 1494 (1967).

178. F. Blasco-Cantera, E. Riande, J. P. Almendro, and E. Saiz, *Macromolecules*, **14**, 138 (1981).

179. R. J. W. Le Fèvre and K. M. S. Sundaran, *J. Chem. Soc.*, 4003 (1962).

180. E. Saiz, E. Riande, M. P. Delgado, and J. M. Barrales-Rienda, *Macromolecules*, **15**, 1152 (1982).

181. J. E. Mark, *J. Chem. Phys.*, **56**, 451 (1971); **57**, 2541 (1972).

182. P. J. Flory, P. R. Sundararajan, and L. C. DeBolt, *J. Am. Chem. Soc.*, **96**, 5015 (1974).

183. P. J. Flory and A. D. Williams, *J. Am. Chem. Soc.*, **91**, 3118 (1969).

184. E. Riande, M. Garcia, and J. E. Mark, *J. Polym. Sci., Polym. Phys. Ed.*, **19**, 1739 (1981).

185. M. S. Beevers, S. J. Mumby, S. J. Carlson, and A. J. Semlyen, *Polymer*, **24**, 1565 (1983).

186. C. Sutton and J. E. Mark, *J. Chem. Phys.*, **54**, 5011 (1971); S. Dasgupta and C. P. Smyth, *J. Chem. Phys.*, **47**, 2911 (1967).

187. E. Riande and J. E. Mark, *Eur. Polym. J.*, **20**, 517 (1984).

188. C. J. C. Edwards, S. Bantle, W. Burchard, R. J. Stepto, and J. A. Semlyen, *Polymer*, **22**, 873 (1982).

189. P. J. Flory, V. Crescenzi, and J. E. Mark, *J. Am. Chem. Soc.*, **86**, 146 (1964).

190. C. Salom and I. Hernández-Fuentes, *Eur. Polym. J.*, **25**, 203 (1989).

191. C. Salom, J. J. Freire, and I. Hernández-Fuentes, *Polym. J.*, **20**, 1109 (1988).

192. W. Bates and W. H. Stockmayer, *Macromolecules*, **1**, 1217 (1978).

193. A. Tonelli, *Macromolecules*, **10**, 153 (1977); J. E. Mark, *Characterization of Materials in Research, Ceramics and Polymers*, Chapter 11, Syracuse University Press, Syracuse N.Y., 1975.

194. J. E. Mark, *J. Am. Chem. Soc.*, **94**, 6645 (1972).

195. J. E. Mark, *J. Polym. Sci., Polym. Phys. Ed.*, **11**, 1375 (1973).

196. Z. Kücükyavuz and B. Baysal, *Macromolecules*, **21**, 2287 (1988).

197. L. Boltzmann, *Pogg. Ann. Physik.*, **7**, 108 (1876).

198. J. Hopkinson, *Phil. Trans. Roy. Soc. (London)*, **167**, 599 (1877); *The Feynmam Lectures on Physics*, Addison-Wesley, Reading, Mass., 1963.

199. J. J. O'Dwyer and E. Harting, *Progr. Dielect.*, **5**, 5143 (1963).

200. M. F. Manning and M. E. Bell, *Rev. Mod. Phys.*, **12**, 215 (1940).

201. G. Williams, *Chem. Rev.*, **72**, 55 (1972).

202. B. K. P. Scaife, *Progr. Dielect.*, **5**, 143 (1963).

203. M. Cook, D. C. Watts, and G. Williams, *Trans. Faraday Soc.*, **66**, 2503 (1970).

204. R. H. Cole and K. S. Cole, *J. Chem. Phys.*, **10**, 98 (1942).

205. D. W. Davidson, *Can. J. Chem.*, **39**, 571 (1961).

206. G. Williams and D. C. Watts, *Trans. Faraday Soc.*, **66**, 80 (1970).

207. G. Williams and D. C. Watts, *Trans. Faraday Soc.*, **67**, 2791 (1971).

208. G. Williams and D. C. Watts, *Trans. Faraday Soc.*, **67**, 1971 (1971).

209. G. Williams, D. C. Watts, S. Dev, and A. M. North, *Trans. Faraday Soc.*, **67**, 1323 (1971).

210. R. H. Cole and H. S. Cole, *J. Chem. Phys.*, **9**, 341 (1941).

211. S. Havriliak and S. Negami, *J. Polym. Sci.*, **14**, 99 (1966).

212. S. Havriliak and S. Negami, *Brit. J. Appl. Phys*, **2**, 1301 (1969).

213. S. Havriliak and S. Negami, *Polymer*, **8**, 161 (1967); **10**, 859 (1969).

214. D. W. Davidson and R. H. Cole, *J. Chem. Phys.*, **19**, 1484 (1951).

215. G. Williams, *Chem. Soc. Rev.*, **7**, 89 (1978).

216. R. Zwanzig, *Ann. Rev. Phys.*, **16**, 67 (1965).

217. R. G. Gordon, *Adv. Magn. Resonance*, **3**, 1 (1968).

218. B. J. Berne and G. D. Harpe, *Adv. Chem. Phys.*, **17**, 63 (1970).

219. B. J. Berne and D. Forster, *Ann. Rev. Phys. Chem.*, **22**, 563 (1971).

220. G. Williams, *Adv. Polym. Sci.*, **39**, 59 (1979).

221. A. Jones, W. H. Stockmayer, and R. J. Molinary, *J. Polym. Sci., Polym. Symp.*, **54**, 227 (1976).

222. W. H. Stockmayer and M. E. Baur, *J. Am. Chem. Soc.*, **43**, 4319 (1964).

223. K. Adachi and T. Kotaka, *Macromolecules*, **17**, 120 (1984).

224. K. Adachi and T. Kotaka, *Macromolecules*, **18**, 466 (1985).

225. K. Adachi, H. Okazaki, and T. Kotaka, *Macromolecules*, **18**, 1687 (1985).

226. K. Adachi, H. Okazaki, and T. Kotaka, *Macromolecules*, **18**, 1486 (1985).

227. K. Adachi and T. Kotaka, *Macromolecules*, **20**, 2018 (1987).

228. K. Adachi and T. Kotaka, *J. Mol. Liquid*, **36**, 75 (1987).

229. K. Adachi and T. Kotaka, *Macromolecules*, **21**, 157 (1988).

230. S. Mashimo, P. Winsor IV, R. H. Cole, and W. H. Stockmayer, *Macromolecules*, **16**, 965 (1983).

231. H. A. Kramers, *Physica*, **7**, 284 (1940).

232. S. Mashimo, S. Yagihara, and A. Chiba, *Macromolecules*, **17**, 630 (1984).

233. P. H. Verdier and W. H. Stockmayer, *J. Chem. Phys.*, **36**, 227 (1962).

234. P. H. Verdier, *J. Chem. Phys.*, **45**, 2118 (1966).

235. L. Monnerie and F. Gény, *J. Chim. Phys.*, **66**, 1961 (1969).

236. F. Gény and L. Monnerie, *J. Chim. Phys.*, **66**, 1708 (1969).

237. F. Gény and L. Monnerie, *J. Polym. Sci., Polym. Phys. Ed.*, **17**, 131, 147 (1979).

238. E. Helfand, *J. Chem. Phys.*, **69**, 1010 (1978).

239. E. Helfand, Z. Wassermann, and T. A. Weber, *Macromolecules*, **13**, 526 (1980).

240. C. K. Hall and E. Helfand, *J. Chem. Phys.*, **77**, 3275 (1982).

241. E. Helfand, *J. Chem. Phys.*, **54**, 4651 (1971).

242. J. Skolnick and E. Helfand, *J. Chem. Phys.*, **72**, 5489 (1980).

243. E. Helfand and J. Skolnick, *J. Chem. Phys.*, **77**, 5714 (1982).

244. H. Helfand, *Science*, **226**, 647 (1984).

245. A. Perico and M. Guenza, *J. Chem. Phys.*, **83**, 3103 (1985).

246. A. Perico and M. Guenza, *J. Chem. Phys.*, **84**, 510 (1986).

247. A. Perico, F. Ganazzoli, and G. Allegra, *J. Chem. Phys.*, **87**, 3677 (1987).

248. A. Perico, *J. Chem. Phys.*, **88**, 3996 (1988).

249. A. Perico, *Acc. Chem. Res.*, **22**, 336 (1989).

250. I. Bahar and B. Erman, *Macromolecules*, **20**, 1368 (1987).

251. I. Bahar, *J. Chem. Phys.*, **91**, 6525 (1989).

252. R. L. Jernigan, *Dielectric Properties of Polymers*, F. E. Karasz (Ed.), p. 99, Plenum, New York, 1972.

253. H. Eyring, *J. Chem. Phys.*, **3**, 107 (1935).

254. W. F. K. Wynne-Jones and H. Eyring, *J. Chem. Phys.*, **3**, 492 (1935).

255. M. Mandel and T. Odijk, *Ann. Rev. Phys. Chem.*, **35**, 75 (1984).

256. R. H. Boyd, *Polymer*, **26**, 323 (1985).

257. R. H. Boyd, *Polymer*, **26**, 1123 (1985); R. Diaz Calleja, J. L. Gómez, A. Ribes, E. Riande, and J. Guzmán, *Macromolecules*, **21**, 2121 (1988).

258. M. Doi and S. F. Edwards, *J. Chem. Soc., Faraday Trans. 2*, **74**, 1789, 1802, 1818 (1978).

259. P. G. DeGennes, *J. Chem. Phys.*, **55**, 572 (1971).

260. G. Williams and D. A. Edwards, *Trans. Faraday Soc.*, **62**, 1329 (1966).

261. G. Williams, *Trans. Faraday Soc.*, **62**, 2091 (1966).

262. H. Sasabe and S. Saito, *J. Polym. Sci., A-2*, **6**, 1401 (1968).

263. G. Williams and D. C. Watts, *NMR, Basic Principles and Progress, Vol. IV, NMR of Polymers*, p. 271, Springer, Berlin, 1971.

264. G. Williams, M. Cook, and P. J. Hains, *J. Chem. Soc., Faraday Trans. 2*, **68**, 1045 (1972).

265. G. Williams, *J. Chem. Soc., Faraday Symp.*, **6**, 44, (1972).

266. G. Williams, *Dielectric Properties of Polymers*, F. E. Karasz (Ed.), p. 17, Plenum, New York, 1972.

267. G. Williams, D. C. Watts, and J. P. Nottin, *J. Chem. Soc., Faraday Trans. 2*, **68**, 16 (1972).

268. K. L. Ngai, R. W. Rendell, A. K. Rajagopal, and S. Teitler, *Dynamic Aspects of Structural Change in Liquid and Glasses*, C. A. Angell and M. Goldstein (Eds.), Vol. 484, p. 150, New York Academy of Sciences, New York, 1987.

269. K. L. Ngai, A. K. Rajagopal, R. W. Rendell, and S. Teitler, *IEEE Trans. Elect. Insul.*, **EI21**, 313 (1986).

270. K. L. Ngai, A. K. Rajagopal, and S. Teitler, *J. Chem. Phys.*, **88**, 5086 (1988).

271. K. L. Ngai, S. Mashimo, and G. Fytas, *Macromolecules*, **21**, 3030 (1988).

272. H. Goldstein, *Classical Mechanics*, 2nd ed., Addison-Wesley, Reading, Mass., 1980.

273. E. C. G. Sudarshan and N. Mukunda, *Classical Dynamics: A Modern Perspective*, edited by Wiley Publishers, Academic Press, New York, 1986.

274. P. A. M. Dirac, *Lectures on Quantum Mechanics*, Belfer Graduate School of Science Monograph Series No. 2, Yeshiva University, New York, 1964.

275. D. Sahoo and G. Venkataraman, *Pramana*, **185** (1975).

276. K. L. Ngai and D. J. Plazek, *J. Polym. Sci., Polym. Phys. Ed.*, **28**, 2159 (1985).

277. G. B. McKenna, K. L. Ngai, and D. J. Plazek, *Polymer*, **26**, 1651 (1985).

278. D. Boese, F. Kremer, and L. Fetters, *Macromolecules*, **23**, 1826 (1990).

279. A. K. Doolittle and B. D. Doolittle, *J. Appl. Phys.*, **28**, 901 (1957).

280. M. H. Cohen and D. Turnbull, *J. Chem. Phys.*, **31**, 1164 (1954).

281. H. Vogel, *Physik.*, **22**, 645 (1921); R. Diaz Calleja, E. Riande, and J. San Román, *Macromolecules*, **24**, 264 (1991).

282. J. D. Plazek and V. M. O'Rourke, *J. Polym. Sci., A-2*, **9**, 208 (1971).

283. L. Mandelkern, *Crystallization of Polymers*, McGraw-Hill, New York, 1964.

284. J. C. Coburn, *Diss. Abstr. Int. B*, **45**(6), 1978 (1984).

285. Y. Ishida, M. Matsuo, and K. Yamafuji, *Kolloid Z.*, **180**, 108 (1962).

286. Y. Ishida, *Kolloid Z.*, **168**, 29 (1960).

287. H. Yamakawa, *Modern Theory of Polymer Solutions*, Harper & Row, New York, 1971.

288. G. Porod, *J. Polym. Sci.*, **10**, 157 (1953).

289. S. Heine, O. Kratky, and P. J. Schmitz, *Makromol. Chem.*, **44–46**, 682 (1961).

290. J. J. Hermans and R. Ullman, *Physica*, **18**, 951 (1952).

291. K. F. Freed, *Renormalization Group Theory of Macromolecules*, Wiley Interscience, New York, 1987.

292. C. Robinson, *Trans. Faraday Soc.*, **52**, 571 (1956).

293. E. Iizuka, *Adv. Polym. Sci.*, **20**, 79 (1976).

294. E. T. Samulsi, *Liquid Crystalline Order In Polymers*, A. Blumstein (Ed.)., Chapter 5, Academic Press, New York, 1978.

295. R. S. Werbowyj and D. G. Gray, *Mol. Cryst. Liq. Cryst.*, **34**, 1 (1976).

296. R. S. Werbowyj and D. C. Gray, *Macromolecules*, **13**, 69 (1980).

297. S. M. Aharoni, *Mol. Cryst. Liq. Cryst.*, **56**, 237 (1980).

298. P. Navard, J. M. Haudin, S. Dayan, and P. Sixou, *Polym. Lett.*, **19**, 379 (1981).

299. S. N. Bhadani and D. G. Gray, *Makromol. Chem. Rapid Commun.*, **3**, 449 (1982).

300. P. Zugenmeier and U. Vogt, *Makromol. Chem.*, **184**, 1749 (1983).

301. S. Bhadani, So-Lang Tseng, and D. G. Gray, *Makromol. Chem.*, **184**, 1727 (1983).

302. J. Preston, *Liquid Crystalline Order in Polymers*, A. Blumstein (Ed.), Chapter 4, Academic Press, New York, 1978.

303. P. W. Morgan, *Macromolecules*, **10**, 1381 (1977).

304. *Polymer Preprints, Am. Chem. Soc. Div. Polym. Chem.*, **17**(1), 41–78 (1976).

305. *J. Polym. Sci., Polym. Symp.*, **65** (1978), special volume devoted to rigid polymer chains.

306. *Macromolecules*, **14**, 900–953 (1981), articles devoted to rigid polymer chains.

307. V. N. Tsvetkov, *Rigid-Chain Polymers (Hydrodynamic and Optical Properties in Solution)*, Consultants Bureau, New York, 1989.

308. S. M. Aharoni and E. K. Walsh, *J. Polym. Sci., Polym. Lett. Ed.*, **17**, 321 (1979).

309. S. M. Aharoni and E. K. Walsh, *Macromolecules*, **12**, 271 (1979).

310. J. K. Moscicki, G. Williams, and S. M. Aharoni, *Macromolecules*, **15**, 642 (1982).

311. P. J. Flory, *Proc. R. Soc. London, Ser. A*, **73**, 234 (1956).

312. P. J. Flory, *Ber. Bunsenges. Phys. Chem.*, **81**, 885 (1977).

313. P. J. Flory and A. Abe, *Macromolecules*, **11**, 1119 (1978).

314. A. Abe and P. J. Flory, *Macromolecules*, **11**, 1122 (1978).

315. P. J. Flory and R. S. Frost, *Macromolecules*, **11**, 1126 (1978).

316. R. S. Frost and P. J. Flory, *Macromolecules*, **11**, 1134 (1978).

317. H. Yu, A. J. Bur, and L. Fetters, *J. Chem. Phys.*, **44**, 2568 (1966).

318. A. J. Bur and D. E. Roberts, *J. Chem. Phys.*, **51**, 406 (1969).

319. S. B. Dev, R. Y. Lockhead, and A. M. North, *Discuss. Faraday Soc.*, **49**, 244 (1970).

320. J. Pierre and E. Marchal, *J. Polym. Sci., Polym. Lett. Ed.*, **13**, 11 (1975).

321. A. J. Bur and L. Fetters, *Chem. Rev.*, **76**, 727 (1976).

322. M. S. Beevers, D. C. Garrington, and G. Williams, *Polymer*, **18**, 540 (1977).

323. H. J. Coles, A. K. Gupta, and E. Marchal, *Macromolecules*, **10**, 182 (1977).

324. M. Doi, *J. Polym. Sci., Polym. Phys. Ed.*, **20**, 1963 (1982).

325. M. Doi and S. F. Edwards, *J. Chem. Soc. Faraday Trans. 2*, **74**, 560 (1978).

326. M. Doi and S. F. Edwards, *J. Chem. Soc. Faraday Trans. 2*, **74**, 718 (1978).

327. G. Williams, *J. Polym. Sci., Polym. Phys. Ed.*, **21**, 2037 (1983).

328. M. P. Warchol and W. E. Vaughan, *Adv. Mol. Relaxation Interaction Processes*, **13**, 317 (1978).

329. C. C. Wang and R. Pecora, *J. Chem. Phys.*, **72**, 5333 (1980).

330. J. K. Moscicki and G. Williams, *J. Polym. Sci., Polym. Phys. Ed.*, **21**, 213 (1983).

331. H. Finkelmann, H. Ringsdorf, and J. H. Wendorff, *Makromol. Chem.*, **179**, 273 (1978).

332. H. Finkelmann, M. Happ, M. Portugal, and H. Ringsdorf, *Makromol. Chem.*, **179**, 2541 (1978).

333. H. Finkelmann and G. Rehage, *Adv. Polym. Sci.*, **60/61**, 99 (1984).

334. M. Engel, B. Hisgen, R. Keller, W. Kreuder, B. Reck, H. Ringsdorf, W. Schmidt, and P. Tschirner, *Pure Appl. Chem.*, **57**, 1009 (1985).

335. V. P. Shibaev and N. A. Plate, *Adv. Polym. Sci.*, **60/61**, 173 (1984).

336. A. C. Griffin, A. M. Bhatti, and R. L. S. Hung, *Nonlinear Optical and Electroactive Polymers*, P. Prasad and D. R. Ulrich (Eds.), p. 375, Plenum Press, London, 1988.

337. R. N. De Martino, E. W. Choe, G. Khanarian, D. Haas, T. Leslie, G. Elson, J. Stamatoff, D. Suetz, C. C. Teng, and H. Yoon, *Nonlinear Optical and Electroactive Polymers*, P. N. Prasa and D. R. Ulrich (Eds.), p. 169, Plenum Press, London, 1988.

338. C. Pugh and V. Perce, *Chemical Reactions on Polymers*, ACS Symposium Series 364, J. L. Benham and J. F. Kinstle (Eds.), p. 97, American Chemical Society, Washington, D.C., 1988.

339. G. S. Attard, K. Araki, J. J. Moura-Ramos, and G. Williams, *Liquid Crystals*, **3**, 861 (1988).

340. A. J. Martin, G. Meier, and A. Saupe, *J. Chem. Soc., Faraday Symp.*, **5**, 119 (1971).

341. P. L. Nordio, G. Rigatti, and U. Segre, *Mol. Phys.*, **25**, 129 (1973).

342. G. S. Attard, *Mol. Phys.*, **58**, 1087 (1986).

343. G. S. Attard, K. Araki, and G. Williams, *Brit. Polym. J.*, **19**, 119 (1987).

6

Introduction

A crystal is formed by molecules, atoms, or ions ordered according to a well-defined pattern. Because our main interest is centered on polymers whose chemical species are macromolecules, we will always refer to molecules even in cases like sodium chloride or diamond crystals. A macroscopic consequence of this microscopic order is that many properties have different values, depending on the direction in which they are measured. Thus, it is a well-known fact that the toughness or the electric conductivity of graphite is quite different when measured along the planes containing pseudomolecular arrangements of carbon atoms than in the direction normal to those planes. On the contrary, the molecules of an amorphous sample are disordered (or randomly oriented) and the macroscopic properties of the material do not depend on the direction in which they are measured; for instance, the electric conductivity of water changes with temperature and is very sensitive to the purity of the sample, but no change with the direction of measurement has yet been found. Then, amorphous substances are said to be *isotropic* (from *isos*, the same, and *tropos*, direction), meaning that their properties are the same in any direction, whereas crystals are *anisotropic*.

A beam of polarized light is a good probe for testing anisotropy, and, therefore, the degree of molecular order of a sample. The reason is that the interaction between the sample and light depends on the orientation of the molecules with respect to the plane of polarization of the light. Thus, in an amorphous sample, these interactions, on average, will be the same for any direction and any plane of polarization of the light. Therefore, characteristics like intensity, state of polarization, phase, etc., of the emerging light as compared with those of the incident beam do not depend on the direction in which the light travels through the sample. On the contrary, if the sample is anisotropic, the interaction of matter to light will depend on the relative situation of the plane of polarization of the light with the direction of preferential orientation of the molecules. Thus, a beam of light traveling through the sample with a plane of polarization that contains the direction in which the molecules are oriented will "see" a very different microscopic situation than an identical beam traveling in the same direction but with a plane of polarization normal to the direction of molecular orientation. These two beams will interact with the sample in quite a different way and may travel with different speed, be absorbed to a different degree, change their plane of polarization in a different amount, etc. Therefore, two radiations that enter the sample, differing only in the orientation of their plane of polarization, may differ in many other properties upon emerging from the sample. Many optical applications of crystals arise from this behavior, which is also the basis for several techniques used to study the microscopic structure of matter.

Among all the features named in the preceding paragraph, we will focus our attention on the speed of the light inside the sample that will be characterized by its refractive index n. As explained before, an isotropic sample has the same refractive index regardless of the direction or plane of polarization of the light passing through it. On the contrary, a sample whose molecules are preferentially aligned in a given direction (this direction is the optical axis of the sample) may exhibit different refractive indices for a beam of light traveling in a direction perpendicular to the optical axis, depending on the plane of polarization of the radiation. Such a substance is called *birefringent* and transforms plane polarized light into elliptically polarized radiation.

Amorphous polymers above the glass-transition temperature or in liquid solutions at any temperature are good examples of substances having isotropic properties. The reason is that under these circumstances, the macromolecular chains have a certain freedom of movement, and, by effect of random thermal motion, they orientate in any possible direction with the same probability. Thus, flexible polymers are constantly changing their conformation; consequently, provided that the chain is long enough, the average distribution of segments over the center of gravity of each chain is given by a Gaussian function having spherical symmetry. Moreover, the location of those centers changes randomly by thermal motions. Rigid polymers have no conformational freedom, but they are still subject to Brownian motions that randomize both position and orientation of the molecules. A beam of light traveling through such a sample has, on average, the same interactions regardless of its direction or plane of polarization.

However, this random orientation of the macromolecules and the consequent isotropy of polymeric samples can be modified by application of an external perturbation that could orientate the molecules along a given direction. Under such a perturbation, the sample behaves like a uniaxial crystal with its optical axis in the direction of preferential orientation of the molecules and exhibits anisotropy in several of its properties, for example, in the refractive indices, among others. This kind of perturbation, therefore, is able of inducing birefringence in a sample that, when unperturbed, is isotropic.

There are several kinds of perturbations that can be applied to a sample in order to produce molecular reorientations, thus inducing birefringences that are named after the type of perturbation that produces them. The most common kinds are as follows:

(a) *Electric Birefringence.* The driving force orientating the molecules is the interaction with an externally applied electric field.

(b) *Optical Birefringence.* Strictly speaking, this is a kind of electric birefringence. It is called optical when the electric field orientating the sample is produced by a powerful laser beam with frequencies in the optical range.

(c) *Magnetic Birefringence.* In this case, the molecules are oriented by the effect of a magnetic field.

(d) *Stress Birefringence.* The molecules are oriented by the effect of a mechanical force producing stress in the sample.

(e) *Flow Birefringence.* When the sample is under laminar flow, the molecules are forced to orientate and, therefore, birefringence is induced.

There are several differences among these processes; for instance, stress birefringence is measured in solid amorphous samples, whereas the other four are usually studied in solution, and, of course, the experimental setup required to produce and measure these birefringences are quite different. However, because all of them have the same molecular origin, they also have some basic similarities that can be explained, at a qualitative level, in the following way.

Let us define a laboratory coordinate system with the x axis in the direction of the external perturbation (direction of the electric field, flow, etc.); the z axis will be taken in the direction of propagation of the beam of light used as a probe in the measurement; finally, the y axis will complete a right-handed Cartesian frame (see Figure 6.1). The incident radiation is polarized, by means of an adequately oriented polarizer, in the plane bisecting the first quadrant, that is, the plane of polarization makes angles of 45° with both the xz and yz planes. The reason for this orientation is that the radiation can then be taken as the sum of two components with the same amplitude, one polarized in the direction of the perturbation, A_x, and a second one polarized in a direction normal to the perturbation, A_y. When entering the sample, both components have the same phase and, therefore, can be written as

$$A_x = A_y = A_0 \cos \omega t \qquad (6.1)$$

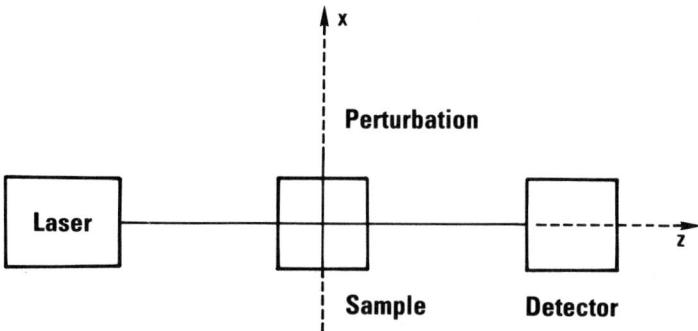

Figure 6.1   Coordinate system used in the analysis of induced birefringence. The x axis is taken in the direction of the external perturbation; z coincides with the direction of propagation of the probe beam of light; and y completes a right-handed Cartesian frame.

where ω is the angular frequency of the radiation. The only difference between these two components is that A_x oscillates in the xz plane and A_y oscillates in the yz. The amplitude A_0 of each component is given by $2A_0^2 = A_t^2$, where A_t is the amplitude of the total incident radiation.

If the sample is isotropic (the external perturbation is not applied), both A_x and A_y components travel through the sample with the same speed, and, therefore, they emerge with the same phase; the intensity of the emerging radiation may be smaller than the incident due to absorption, scattering, etc., but both components are affected in the same way. The emerging radiation is then formed by two components of equal intensity and with the same phase; therefore, it is polarized in the same plane as the incident light and it is extinguished if an analyzer, crossed with the polarizer, is mounted after the sample.

Upon application of the external perturbation in the x direction, the molecules tend to orientate. The orientation can be either along the x or y axis, depending on the kind of interactions produced by the perturbation. On the other hand, the fraction of molecules having the preferred orientation is determined by the balance between the energy of the interaction and the thermal random motion. But, at any rate, the important feature is that the number of molecules oriented over the xz and yz planes becomes different and, therefore, the A_x and A_y components of the radiation interact in a different way with the sample and, as a consequence, they travel with different speeds. The sample becomes birefringent and exhibits a difference in refractive indices, $\Delta n = n_{xz} - n_{yz} \neq 0$, between the components polarized in the xz and yz planes.

Let us assume, for instance, that $\Delta n > 0$ (i.e., the A_y component travels faster than the A_x). If the sample has a thickness l, when the slow component emerges from the sample, the fast one has already traveled a distance $l\,\Delta n$ out of the sample.

This difference in optical path between the two components can be written in units of length, in number of waves of the used radiation, or as the angle of phase difference using the relationships:

$$\text{Retardation: } R = l\,\Delta n \text{ (meters)} \tag{6.2}$$

$$\text{Number of waves: } v = R/\lambda = l\,\Delta n/\lambda \tag{6.3}$$

$$\text{Phase difference: } \delta = 2\pi v = 2\pi l\,\Delta n/\lambda \text{ (radians)} \tag{6.4}$$

The emerging radiation, therefore, has two components described by

$$A_x = A_0 \cos \omega t \quad \text{and} \quad A_y = A_0 \cos(\omega t + \delta) \tag{6.5}$$

The combination of these two components gives an elliptically polarized radiation that will not be completely extinguished by the analyzer. Of course, the radiation described by Eq (6.5) is polarized in the original plane of the incident beam when $\delta = 2m\pi$, with $m = 0, 1, 2, \ldots$; $m = 0$ represents the trivial case of an isotropic sample and the values $m \geq 1$ are not reached in the usual experimental determinations.

From the experimental point of view, the study of any of these induced birefringences consists in applying the external perturbation (for example, the electric field), and measuring the birefringence produced (expressed as Δn, v, or δ) as a function of the strength of the perturbation; thermodynamic variables like p, T, and concentration; and the chemical characteristics of the sample. A theoretical analysis requires the use of a model that will allow the calculation of the birefringence in terms of microscopic parameters of the sample like electric or magnetic characteristics, size, shape, etc. The comparison of theoretical results with the experimental values gives valuable information about those molecular parameters. In the next sections, we will be mainly concerned with these kinds of applications of induced birefringences, and, for this reason, we treat with more detail the electric and stress birefringences, which are the most commonly used in the investigation of the microstructure of polymers.

However, there are at least two other applications of induced birefringences. Probably the most obvious, and certainly the less used, one is a light switch. In the unperturbed state, the sample is isotropic and the combination polarizer-analyzer gives complete extinction. Upon application of the perturbation, part of the light passes through the analyzer. Probably the best choice for this application is electric birefringence since electric fields can be easily applied and some low molecular-weight substances give strong and very fast responses. Some attempts have been made to use an electric birefringence ensemble as a shutter for high-speed photography or for modulation of laser beams, although the application is not widespread. Some slightly different electrooptical devices, based upon the response of liquid crystals to electric fields, seem to be more adequate for that particular purpose. For instance, polymer dispersed liquid crystals (PDLCs) are formed by liquid crystals dispersed in a polymer binder by means of a process of phase separation from a homogeneous solution [1]. Their microscopic behavior is very similar to that explained before for the case of induced birefringences, namely, the difference in the refractive index of the liquid crystal when its molecules are disordered as compared with the state in which they are oriented by means of an externally applied electric field. The only difference is that the two components (liquid crystal and polymer) are chosen in such a way that their refractive indices match when the liquid crystal has its molecules oriented (field ON), whereas they are noticeably different when the field is OFF and the liquid crystal is disordered. In the former state, the system is transparent to the radiation, whereas in the latter, the difference in refractive indices of the two components produces a very strong scattering and the sample becomes opaque. The response of the system to the applied field is very fast and does not require the use of polarizers [2–4].

The last application of the measurement of induced birefringence is to obtain information on the applied perturbation. The most interesting case is that of stress birefringence. The measurement can be as simple as examining the sample between crossed polarizers; the areas of the sample that are not stressed appear dark, whereas the brighter that a given area appears, the more stressed it is. Thus, a stress map of the whole sample can be easily obtained. The stress-producing birefringence can be originated by a mechanical force externally applied to the sample or it can be internal stress due to inhomogeneities or lack of thermodynamical equilibrium in the sample. The typical example is a piece of polymeric material obtained by extrusion; the combination of high shear stresses and temperatures during the processing makes the macromolecules adopt highly extended conformations and almost crystalline packing. Once the pressure is removed and the temperature lowered, the polymeric chains tend to reach their equilibrium random-coil distribution. However, the changes required to achieve the equilibrium could be quite slow if the temperature is lowered close to or below the glass transition. Thus, an internal stress can build up in the sample by the differences between the areas in which the equilibrium distribution has been reached and those in which it has not. The measurement of birefringence is a very sensitive technique for detecting and measuring this kind of stress. A lot of work has been done on this subject, mainly in the area of applied polymer science; the interested reader can find many examples and the description of more sophisticated techniques on some on the revisions in the literature [5, 6].

7
Electric Birefringence

HISTORY

"The thought which led me to the following inquiry was briefly this: that if a transparent and optically isotropic insulator were subjected properly to intense electrostatic force, it should act no longer as an isotropic body upon light sent through it." This is the first sentence of a paper published in 1875 by John Kerr [7].

Following this thought, Kerr prepared an apparatus consisting of a paraffin flame as a light source, two Nicol's prisms as a polarizer and analyzer, and Ruhmkorff's induction machine to provide the required electric field. He placed a plate of glass between the Nicol's prisms and observed that the analyzer gave complete extinction. However, when a electric field was applied to the glass in the direction perpendicular to the beam of light, some light passed through the analyzer, and moreover, using his own words: *"The light thus restored by electric action cannot be extinguished again, at any stage of the experiment, by any rotation of the analyzer either way."*

The isotropic glass becomes birefringent by the action of the electric field. Thus, a difference in refractive index, $\Delta n = n_p - n_v \neq 0$, between the directions parallel and perpendicular (vertical) to the electric field is induced in the glass. When l is the length of sample traversed by light of wavelength λ, this difference in refractive indices produces a retardation δ between the two components of the incident-plane polarized light, which is given by Eq. (6.4) in Chapter 6. The radiation emerging from such a birefringent media contains two components that are out of phase by an angle δ. It is no longer plane but elliptically polarized, and, consequently, it cannot be extinguished at any position of the analyzer.

During 20 years, Kerr studied many substances, both solid and liquid. He was able to perform quantitative measurements and obtained the following empirical law relating the electric field applied to the sample E with the induced birefringence δ:

$$\delta = 2\pi l \, \Delta n/\lambda = 2\pi l B E^2 \tag{7.1}$$

where B represents a constant that is characteristic of the measured sample. Thus, some substances such as benzene, toluene, chlorobenzene, etc., have a positive birefringence (i.e., $B > 0$ or $n_p > n_v$), whereas some others like chloroform or olive oil exhibit negative values of B. The dimensions of B are [length $\times$ (electric field)^2]$^{-1}$, so that its SI units are m V^{-2}.

Of course, the induction of birefringence in an isotropic sample by application of an electric field is known as the Kerr effect, Eq. (7.1) is Kerr's law, and B is usually called the Kerr constant.

The work of Kerr was soon followed by many other researchers [8–23] who measured several hundreds of compounds, mainly in solid or liquid state. In most of these early works, relative values of the Kerr constant, using carbon disulfide (CS_2) as reference, were obtained. In 1915, Chaumont [23] measured the first reliable absolute value of B for CS_2, thus opening the way for absolute determinations of B; he measured at $\lambda = 589$ nm in the temperature range $18°C \leq t \leq 28°C$, and obtained $B = (3.58 \pm 0.005) \, (t - 20)10^{-14}$ m V^{-2}.

The Kerr effect of gases is much smaller than in liquids and, therefore, it is harder to measure with precision. However, the theoretical interpretation is easier in gases and was developed earlier than for liquids. Thus, for some time, the only available experimental data were obtained in liquids or solids, whereas the theories were strictly applicable only to gases. The first measurements performed in the gas state [24] were reported in 1911, but a systematic study of gases, including dependence of B with pressure, wavelength, and temperature required some improvements of the experimental apparatus [25–35], mainly photocells as detectors and more accurate measuring devices, which started to be introduced in the early 1920s.

In the case of solutions, the magnitude B contains contributions from both solvent and solute. Although molar fractions were first used as a unit of concentration [32, 36–39] in equations relating Kerr constants of solute and solvent, nowadays volume fractions ϕ are commonly used [40]. Thus, representing magnitudes of solvent and solute by subscripts 1 and 2, respectively,

$$B = B_1\phi_1 + B_2\phi_2 = B_1 - B_1\phi_2 + B_2\phi_2$$
$$B_2 = \Delta B/\phi_2 + B_1 \tag{7.2}$$

where ΔB represents the difference between the Kerr constant of solution and that of the pure solvent. Usually, an extrapolation to zero concentration is taken in order to avoid intermolecular interactions between the molecules of solute (see what follows).

The use of sinusoidal [41, 42] and pulsed electric fields [43–57] opened two interesting perspectives. In the first place, conducting substances such as water solutions could be measured; in this regard, the pioneer work on Kerr constants of polymers [45, 46], carried out around 1950, was performed mainly in solutions of biopolymers, such as the tobacco mosaic virus, using this kind of technique. But, on the other hand, if a fast-response recording device such as an oscilloscope or a computer is used to store the signal from the detector, the dynamic birefringence (i.e., the variation of birefringence with time) of the sample can also be studied; thus, information of magnitudes such as the dipole moment and the rotational diffusion coefficient can also be obtained. Therefore, electric birefringence became a valuable tool for the determination of the size and shape of macromolecules and it is commonly used in the analysis of their hydrodynamic properties [58, 59].

Any historical summary, even a very short one like the present, should give a special mention to the work of the group leaded by Le Fevre, who during more than 30 years has reported Kerr constants of well over a thousand compounds in about 300 publications. Some of their reviews on this subject are given in references [60–62]. Lists of Kerr constants have also been published by some other authors [63–67].

Probably the latest significant improvement in the instrumentation of electric birefringence is the use of intense beams of light to provide the electric field required to perform a Kerr-constant measurement. The possibility of carrying out such an experiment was suggested by Buckingham [68] more than 30 years ago. However, experimental measurements on pure liquids [69–71] and mixtures [72, 73] were not performed for more than 10 years afterwards, and only in the mid-1970s was it possible to measure polymer solutions [74, 75]. The phenomenon is known as *optical birefringence,* or the optical Kerr effect, and presents some advantages over the conventional static or low-frequency field techniques because it produces neither heating of conducting solutions nor electrophoretic polarization and, therefore, it is specially suitable for biopolymers in solutions having high salt concentration. In a typical experiment [76], pulses of a high-power laser beam (provided, for instance, by a neodymium or ruby laser) are used to produce the electric field that orientates the molecules of the sample; the width of the pulses are smaller than 1 μs, and the power may be as high as 20 MW, producing electric fields up to 4×10^7 V/m.

The first theoretical interpretation of the Kerr effect was published by Langevin in 1910 [77] and was later improved by others [60, 78–80]. Quantum mechanics theory has also been used to relate the Kerr and Stark effects [79, 81–83], although the expression obtained for B contains some parameters that are difficult to evaluate. A more recent quantum mechanics treatment was published by Tobias and Balazs [84], who studied the interactions between a molecule in an external constant field with a linearly polarized photon. They found that when the field chosen is magnetic and directed along the wave vector of the photon (longitudinal), the time dependence of the state function implies a rotation of the plane of polarization of the photon, characteristic of the Faraday effect. Similarly, when the field is transverse, the time evolution of the state corresponds to the Kerr effect in the case of an electric field and the Cotton–Mouton effect in the case of a magnetic field.

Modern theories of electric birefringence use different treatments depending on the strength of the electric field. Under very strong electric fields, Eq. (7.1) does not hold; instead, a more complicated dependence of Δn with E is found. The results can then be expressed by a power series of the field strength [85] or obtained by a Monte Carlo simulation procedure [86]. A good revision of different possible treatments for the region of strong fields was published by Watanabe and Morita [87].

When applied to polymers, the flexibility of the macromolecule also plays an important role on the difficulty of making the theoretical analysis. Thus, rigid polymers move inside the solution in order to orientate by the effect of the electric field, whereas flexible molecules not only move, but they also can change its shape. Consequently, the treatment becomes more complicated as the flexibility of the polymer increases. A complete treatment of rigid polymers has been published by Wegener and co-workers [88, 89]. Moderately flexible molecules such as bent rods required the use of either approximate [90, 91] or simulation procedures [92]. The realistic treatment of flexible polymers was started by Nagai and Ishikawa [93] to the point that all the calculations carried out before were of simplified models such as freely jointed or freely rotating chains. The application of the isomeric rotational scheme [94] and Flory's matrix-multiplication procedure [95, 96] allows performing calculations for any polymer with the desired degree of microscopic detail, although, in general, these calculations are only undertaken for the simplest possible electric field (i.e., dc fields with low strength).

There is a very large number of publications on electric birefringence, mainly for biological systems. Some interesting reviews are given in references [60–62] and [97–101].

MOLECULAR ORIGIN OF THE KERR CONSTANT

At the time that Kerr performed his experiments, it was known that glass becomes birefringent under strain. Thus, the first interpretation of the Kerr effect was that the electric field heated the glass, producing strain that was then responsible for the observed birefringence. In support of this interpretation were the time lags between the application of the field and the observation of light emerging from the analyzer and between the removal of the field and the extinction of light. These two time lags were associated with the time required for the sample to heat up or to cool down.

Kerr argued against this theory that the effect was not exclusive of solid samples; on the contrary, liquids also become birefringent by the action of the electric field. Therefore, the effect should have a molecular origin; it should be produced by interaction of the molecules of the sample, regardless of their physical state, with the electric field.

How can the molecules of an isotropic sample interact with an electric field in order to produce birefringence? Let us think, first, at a qualitative level. It is well known that when a molecule is placed inside an electric field, its electronic structure is distorted so that a dipole moment is induced in the molecule. In most cases, the effect is linear and the induced dipole moment $\mathbf{m}$ is proportional to the applied electric field $\mathbf{E}$:

$$\mathbf{m} = \boldsymbol{\alpha} \cdot \mathbf{E} \qquad (7.3)$$

where $\boldsymbol{\alpha}$ is the polarizability tensor of the molecule. If the induced moment has the same direction as the field, $\boldsymbol{\alpha}$ is a scalar that can be qualitatively related with the "looseness" of the electric charges (mostly electrons) in the molecule. But, in general, the directions of $\mathbf{m}$ and $\mathbf{E}$ do not coincide; thus, $\boldsymbol{\alpha}$ is a tensorial magnitude that can be written as a 3×3 matrix representing the polarizability in every direction of the space. Thus, by choosing the xyz coordinate system, Eq. (7.3) can be written as

$$\begin{bmatrix} m_x \\ m_y \\ m_z \end{bmatrix} = \begin{bmatrix} \alpha_{xx} & \alpha_{xy} & \alpha_{xz} \\ \alpha_{yx} & \alpha_{yy} & \alpha_{yz} \\ \alpha_{zx} & \alpha_{zy} & \alpha_{zz} \end{bmatrix} \begin{bmatrix} E_x \\ E_y \\ E_z \end{bmatrix} = \begin{bmatrix} \alpha_{xx}E_x + \alpha_{xy}E_y + \alpha_{xz}E_z \\ \alpha_{yx}E_x + \alpha_{yy}E_y + \alpha_{yz}E_z \\ \alpha_{zx}E_x + \alpha_{zy}E_y + \alpha_{zz}E_z \end{bmatrix} \qquad (7.4)$$

The $\boldsymbol{\alpha}$ tensor can be diagonalized by rotating the xyz system until a reference frame XYZ is found in which $\alpha_{ij} = 0$ for $i \neq j$. Using the XYZ axes, Eq. (7.4) simplifies to

$$\begin{bmatrix} m_X \\ m_Y \\ m_Z \end{bmatrix} = \begin{bmatrix} \alpha_X & 0 & 0 \\ 0 & \alpha_Y & 0 \\ 0 & 0 & \alpha_Z \end{bmatrix} \begin{bmatrix} E_X \\ E_Y \\ E_Z \end{bmatrix} = \begin{bmatrix} \alpha_X E_X \\ \alpha_Y E_Y \\ \alpha_Z E_Z \end{bmatrix} \qquad (7.5)$$

XYZ are the main axes of the molecule and α_X, α_Y, and α_Z are the main polarizabilities; they represent the dipole moment induced, respectively, in the X, Y, and Z directions by application of a unit of electric field in those directions. Sometimes the main polarizabilities are visualized by means of a polarizability ellipsoid, which is defined as an ellipsoid having α_X, α_Y, and α_Z as semiaxes.

The main axes of a molecule can be obtained by applying any mathematical procedure to diagonalize $\boldsymbol{\alpha}$. However, chemical knowledge of the molecule may help to find them since a symmetry axis should coincide with one of the main axes and a symmetry plane should contain two of them. Thus, as Figure 7.1(a) indicates, the molecule of benzene has two of the axes contained in the plane of the ring and the third one is perpendicular to that plane.

Molecules having $\alpha_X = \alpha_Y = \alpha_Z$ are called *isotropically polarizable* and their polarizability ellipsoid is just a sphere; this is the case of molecules with spherical symmetry such as CCl_4. Application of an electric field to this kind of molecule

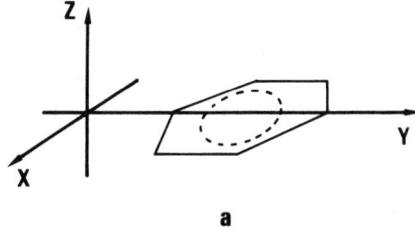

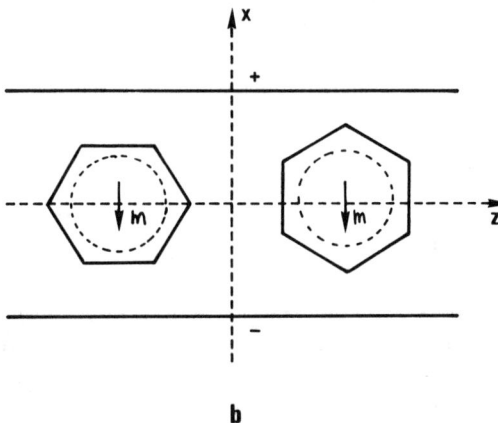

Figure 7.1 (a) Main polarizability axes for the molecule of benzene. (b) Preferred orientations of the molecules of benzene in an electric field.

produces a dipole moment that is parallel to the field regardless of the relative orientation of the molecule and field.

If a molecule had $\alpha_X \neq 0$ and $\alpha_Y = \alpha_Z = 0$, its polarizability ellipsoid would be a straight line. Such a molecule would not interact with electric fields applied in either Y or Z directions, whereas it will develop a induced dipole moment under the effect of a field having the X direction. The experimental situation when a electric field is applied to a macroscopic sample is that the molecules are randomly orientated with respect to the field. Molecules having orientations in which Y or Z is aligned with the field will not interact with it; molecules with any other orientation will induce a dipole moment that will depend on its orientation, being maximum for those molecules having X in the direction of the field. Taking into account that the interaction between an electric field and a dipole moment parallel to it and with opposite sense is attractive, it comes out that molecules having X in the direction of the field have less energy than those in any other orientation. In other words, the electric field tends to orientate the molecules of the sample in the direction in which the nonzero polarizability is parallel to the field. If a beam of polarized light travels through the sample, when its electrical vector is parallel to the external field, it will interact with molecules aligned with their largest polarizability parallel to it. Therefore, it will travel faster than if it were perpendicular (vertical) to the external field, in which case it would find the molecules aligned in the nonpolarizable directions. Thus, the sample becomes birefringent with $n_p > n_v$, or $\delta > 0$ and $B > 0$.

There are no molecules such as those described in the preceding paragraph. The most common situation, however, is that in which either the three main polarizabilities are different or at least one of them is different from the other two. Benzene is a example of the second kind of molecules; the electron clouds can be more easily deformed in the plane of the ring than in the direction perpendicular to it. Thus, $\alpha_X = \alpha_Y > \alpha_Z$, and its polarizability ellipsoid looks like a flying saucer. Let us assume that an electric field is applied to a sample of benzene in a direction x while a beam of light passes through the sample in direction z; see Figure 7.1(b). There will be four preferred and equivalent orientations of the molecules in the field: those having the ring in the xz plane with either X or Y parallel to the direction of the field x, as shown in Figure 7.1(b), and those having the ring in the xy plane, again with either X or Y in the x direction (these two orientations can be obtained by rotating the orientations shown in Figure 7.1(b) 90° over x until the ring is perpendicular to the plane of the drawing). All four orientations have one of the largest polarizabilities (either α_X or α_Y) in the direction of the electric field x, whereas in the case of y, two of the orientations (those shown in Figure 7.1(b)) align the smallest polarizability along this axis. Thus, if the beam of light is polarized in the xz plane, it will travel faster than if it were polarized in the yz plane. Therefore, the sample develops a positive birefringence although its value should be smaller than in the hypothetical case explained before in which all the preferred orientations favor one of the polarizations.

Up to now, we have considered only molecules that have no permanent dipole moment, but what will happen if the sample molecules have a permanent dipole μ? The situation, as far as inducing a dipole m, is the same; however, μ is usually several orders of magnitude larger than m. Therefore, the orientation in the electric field will be mostly dictated by μ with very little dependence on the polarizability of the molecule. We can imagine two extreme cases (Figure 7.2): first, molecules such as chloromethane in which the largest polarizability is parallel to μ, and, second, molecules like chloroform in which the largest polarizability is perpendicular to μ. In the first example, the polarizability ellipsoid looks like a rugby ball with its longest axis parallel to the Cl—C bond, whereas in the second, the ellipsoid is again like a flying saucer with the smallest axis along the H—C bond. Both kinds of molecules will align μ in the direction of the external field and, therefore, largest/smallest polarizabilities would then be aligned with the field for molecules of

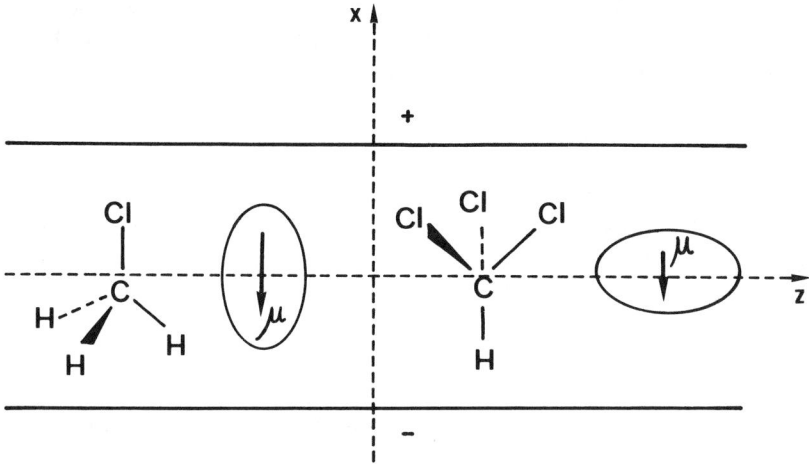

Figure 7.2 Preferred orientation in an electric field of molecules like chloromethane and chloroform having μ parallel and perpendicular, respectively, to the direction of largest polarizability.

the first/second kind. Thus, in the first case, $n_p > n_v$ and, consequently, $B > 0$, whereas molecules like chloroform produce $n_p < n_v$ and exhibit a negative Kerr effect.

Therefore, the preferential orientation of anisotropically polarizable molecules in an electric field, produced by interaction of the field with permanent μ, induced $\mathbf{m}$, or both kinds of dipole moments, is responsible for the Kerr effect. The driving force for the orientation of the molecules is the interaction between the field and the dipole moment; consequently, the birefringence produced should increase with $\mathbf{E}$, μ, and the difference between the main polarizabilities of the sample (the so-called anisotropy of the polarizability). On the other hand, the thermal motion of the molecules tends to randomize their orientations, so that the birefringence should decrease with increasing values of kT. Moreover, it will take a certain time for the molecules both to orientate upon application of the field and to randomize when the field is removed; this explains the time lags experimentally observed and is the basis of the dynamic Kerr-effect experiments.

These ideas were first written in a quantitative form by Langevin [77] in 1910 for molecules without permanent dipole moments and expanded for molecules with dipole moments by Born [78] in 1918. The combination of both treatments is known as the Langevin–Born theory. These authors assumed that the distribution of orientations of the molecules in the electric field is governed by a Boltzmann function of the energy of interaction between the field $\mathbf{E}$ and the dipole moments μ and $\mathbf{m}$. With this distribution, they calculated the polarizability and the refractive indices in directions parallel and perpendicular to $\mathbf{E}$ from which B is immediately obtained, as Eq. (7.1) suggests. Full mathematical treatment has been published elsewhere [60, 79]. Their final equation is

$$B = \frac{\pi \mathcal{N}}{27n\lambda} (n^2 + 2)^2 (\epsilon + 2)^2 (\theta_1 + \theta_2) \qquad (7.6)$$

where $\mathcal{N}$ represents the number of molecules per unit of volume, λ is the wavelength of the light, n is the macroscopic refractive index, and ϵ is the dielectric constant of the sample. The terms θ_1 and θ_2 represent, respectively, the contributions of anistropy and the dipole moment to the value of B and can be written as

$$\theta_1 = \frac{1}{45KT} [(\alpha_X - \alpha_Y)(\alpha_X' - \alpha_Y') + (\alpha_Y - \alpha_Z)(\alpha_Y' - \alpha_Z')$$

$$+ (\alpha_Z - \alpha_X)(\alpha_Z' - \alpha_X')] \qquad (7.7)$$

where α and α' represent the electrooptic and electrostatic polarizabilities, respectively. Thus, only electrons contribute to α, whereas α' contains contributions of both electrons and atoms.

$$\theta_2 = \frac{1}{45k^2T^2} [(\mu_X^2 - \mu_Y^2)(\alpha_X - \alpha_Y) + (\mu_Y^2 - \mu_Z^2)(\alpha_Y - \alpha_Z)$$

$$+ (\mu_Z^2 - \mu_X^2)(\alpha_Z - \alpha_X)] \qquad (7.8)$$

where μ_X, μ_Y, and μ_Z represent the components of the permanent dipole moment written in the XYZ main polarizability axes of the molecule.

Both electrostatic α' and electrooptic α polarizabilities are required to calculate the value of B according to Eq. (7.6). Some approximations are usually employed in order to eliminate α'. Thus, Gans [80] suggested in 1921 to use the relationship

$$\alpha_X'/\alpha_X = \alpha_Y'/\alpha_Y = \alpha_Z'/\alpha_Z = (\epsilon - 1)/(n^2 - 1) \qquad (7.9)$$

and Le Fevre [60] uses

$$\alpha_X'/\alpha_X = \alpha_Y'/\alpha_Y = \alpha_Z'/\alpha_Z = P_d/P_e \qquad (7.10)$$

where P_e and P_d represent the electronic and total (i.e., electronic plus atomic) polarizations, respectively.

This classical theory gives, with the exceptions that will be shown in what follows, good account of the experimental values of B, mainly in what refers to dependence of B on the dipole moment, optical anisotropy, temperature, etc.

However, the Langevin–Born theory has two important failures, namely, the optically isotropic molecules and the condensed (liquid or solid) states.

According to Eqs. (7.6) to (7.8), molecules having $\alpha_X = \alpha_Y = \alpha_Z$ should have $B = 0$. However, experimental determinations on molecules such as CCl_4 prove that they have small but definitively nonzero values of B (for instance, at 25°C and with $\lambda = 589$ nm, B (CCl_4) = 0.09×10^{-14} mV^{-2} [62]). Some attempts were made to explain these values through deformations of the molecular geometry produced by the effect of the electric field [60–62, 102, 103]. However, the explanation accepted nowadays was given by Buckingham [104–108], who assumes that the dipole moment induced in the sample by the electric field **E** is not a linear function of **E**; thus, he writes Eq. (7.3) as a series expansion:

$$\mathbf{m} = \alpha\mathbf{E} + (1/2)\beta\mathbf{E}^2 + (1/6)\gamma\mathbf{E}^3 + \cdots \qquad (7.11)$$

where β, γ, . . . are hyperpolarizability coefficients such that $\alpha \gg \gamma \gg \cdots$, so that their effect in the case of polar or anisotropic molecules is negligible unless the field is very strong (on the order of several million V/m); in that case, Eq. (7.11) explains the observed departure from linearity of δ versus E^2 indicated by Kerr's law. However, most experimental determinations are performed under conditions in which Eq. (7.3) can be used and, therefore, contributions from hyperpolarizabilities are neglected.

The second failure of the Langevin–Born theory is a more serious one. Because neither interactions between neighbor molecules nor modification of the macroscopic field **E** due to the presence of the sample are introduced, the theory is only strictly applicable to gases at low pressures. This situation implies a hard limitation to the applicability of the Kerr effect in chemistry, because many molecules (polymers among others) are difficult to vaporize without decomposition, and, moreover, measurements in gases are in general more difficult and less accurate than in liquids.

There were some attempts to correct the theory in order make it applicable to liquids [109, 110]. However, the procedure employed at present [60, 111] uses a different approach consisting in the utilization of the Lorentz–Lorenz equation relating the refraction of a given sample with the optical polarizability of its molecules. This equation can be written per unit volume as

$$\frac{n^2 - 1}{n^2 + 2} = \frac{4\pi}{3} \sum_{i=1}^{\mathcal{N}} \alpha_i = \frac{4\pi}{3} \mathcal{N}\alpha \qquad (7.12)$$

where $\mathcal{N}$ is the number of molecules per unit volume, and the last equality assumes that all those molecules have the same polarizability α.

Applying Eq. (7.12) to the refractive indices in directions parallel (n_p) and perpendicular (n_v) to the electric field and assuming that the difference $\Delta n = n_p - n_v$ is small compared with the main value n, so that the following approximations can be used:

$$(n_p^2 + 2)(n_v^2 + 2) \approx (n^2 + 2)^2 \quad \text{and} \quad n_p + n_v \approx 2n \qquad (7.13)$$

the value of Δn can be written as

$$\Delta n = \frac{2\pi}{9n} (n^2 + 2)^2 \mathcal{N} (\alpha_p - \alpha_v) = \frac{2\pi}{9n} (n^2 + 2)^2 \mathcal{N} \, \Delta\alpha \tag{7.14}$$

Solving Eq. (7.1) for B, one obtains $B = \Delta n/\lambda E^2$, which together with Eq. (7.14) gives the value of B as

$$B = \frac{2\pi (n^2 + 2)^2 \mathcal{N}}{9n\lambda} \frac{\Delta\alpha}{E^2} \tag{7.15}$$

All that is required then to compute B according to Eq. (7.15) is to evaluate $\Delta\alpha$ as a function of the applied electric field **E**. The procedures used for this calculation with different kinds of applied fields will be explained in the section devoted to theoretical calculations.

However, Eq. (7.15) has the inconvenience of depending on $\mathcal{N}$, and, therefore, the value of B determined for a given substance will change with its physical state (i.e., solid, liquid, gas, solution, etc.). This problem can be obviated by defining a "molar Kerr constant" as the difference in molar refractions parallel and perpendicular to the electric field divided by the square of the effective field:

$$_mK = (R_p - R_v)/E_{\text{eff}}^2 \tag{7.16}$$

Equation (7.12) can be written for a mole of the sample as

$$R = \frac{(n^2 - 1)M}{(n^2 + 2)d} = \frac{4\pi}{3} N\alpha \tag{7.17}$$

where N is Avogadro's number, M is the molecular weight, and d is the density of the sample. The molar Kerr constant can be written as

$$_mK = \frac{4\pi N}{3E_{\text{eff}}^2} (\alpha_p - \alpha_v) = \frac{4\pi N}{3E_{\text{eff}}^2} \Delta\alpha \tag{7.18}$$

which again depends only on $\Delta\alpha$.

The relationship between $_mK$ and B can be obtained by writing Eq. (7.17) for R_p and R_v and using the approximations given in Eq. (7.13) to calculate the difference $R_p - R_v$ as

$$R_p - R_v = \frac{6Mn}{(n^2 + 2)^2 d} \Delta n \tag{7.19}$$

Taking the effective field as

$$E_{\text{eff}} = E(\epsilon + 2)/3 \tag{7.20}$$

and using the relationship $B = \Delta n/\lambda E^2$, Eq. (7.19) can be written as

$$R_p - R_v = \frac{54\lambda Mn}{(n^2 + 2)^2 (\epsilon + 2)^2 d} BE_{\text{eff}}^2 \tag{7.21}$$

which substituted into Eq. (7.16) gives [40, 112, 113]

$$_mK = \frac{54\lambda M n}{(n^2 + 2)^2(\epsilon + 2)^2 d} B = \varphi \frac{M}{d} B \tag{7.22}$$

The SI units of $_mK$ are $m^5 \, V^{-2} \, mol^{-1}$, however, old publications frequently use cgs units. The conversion factor between the two systems is $1 \, cm^5 \, statvolt^{-2} \, mol^{-1} = 1.1126 \times 10^{-15} \, m^5 \, V^{-2} \, mol^{-1}$.

Unfortunately, Eq. (7.22) is not the only definition of the molar Kerr constant used in the literature. Thus, some authors [37, 38, 60–62] (Le Fevre's group, for example) eliminate the factor of 3 in Eq. (7.20) and, therefore, use a $_mK$ that is identical to that given in Eq. (7.22) except for a factor of 6 instead of 54; their values are, therefore, $1/9$ of those obtained by means of Eq. (7.22). Still other groups [30, 32, 35, 65] eliminate the whole numerical factor of Eq. (7.22) and then their values should be multiplied by 54 prior to comparison. Finally, sometimes a "specific" Kerr constant, obtained dividing Eq. (7.22) by M, is used.

In the case of solutions, the molar Kerr constant of solute $_mK_2$ can be obtained by substituting Eq. (7.2) into Eq. (7.22) as

$$_mK_2 = \left(\Delta B \frac{M_2}{d_2 \phi_2} + B_1 \frac{M_2}{d_2} \right) \varphi \tag{7.23}$$

The first fraction of Eq. (7.23) represents the inverse of the molar concentration of the solution, m, and the second is the molar volume of the solute, v_2. Making these two substitutions into Eq. (7.23) and extrapolating to $m \to 0$ to eliminate the intermolecular interactions for the solute, one obtains the molar Kerr constant of a solute at infinite dilution as [40]

$$(mK_2)_\infty = \varphi \left[\lim_{m \to 0} \left(\frac{\Delta B}{m} \right) + v_2 B_1 \right] \tag{7.24}$$

Since the symbol used in Eq. (7.24) to represent the molar Kerr constant is quite complicated, it is usually simplified to $_mK$, and, depending on whether the substance studied is pure or a solute in a solution, its value is given by Eq. (7.22) or Eq. (7.24). On the other hand, the φ constant contains the refractive index and the dielectric constant of the medium, that is, the solution; however, because the measurements are usually performed on very dilute solutions, the effect of solute in both magnitudes is neglected and the final expression for the molar Kerr constant of solute at infinite dilution is commonly written as

$$_mK = \frac{54\lambda n_1}{(n_1^2 + 2)^2(\epsilon_1 + 2)^2} \left[\lim_{m \to 0} \left(\frac{\Delta B}{m} \right) + v_2 B_1 \right] \tag{7.25}$$

When Eq. (7.25) is applied to polymers, m is quite frequently taken as the "molar concentration of repeating units," that is, the molecular weight of the repeating unit M_0 is used instead of that of the polymer $M = x M_0$ to compute m. The result thus obtained is the Kerr constant per repeating unit of the chain, $_mK/x$.

Equation (7.25) is customarily used for flexible synthetic polymers, which usually are not very birefringent, so that, even if a nearly isotropically polarizable solvent such as p-dioxane is used for the measurements, the B_1 contribution cannot be neglected at the low concentrations used.

However, when very birefringent polymers are studied, the contribution of the solvent can be neglected even for extremely dilute solutions. This situation is quite frequent in the case of biopolymers, liquid–crystal systems, colloidal suspensions, etc. Then, Eq. (7.3) can be simplified assuming that the value of B for the solution is due only to the contribution of the solute molecules. Thus,

$$B \approx \phi_2 B_2 \tag{7.26}$$

Then, it is easier to use the value of B_2 as a specific constant and calculate it as

$$B_{sp} = B_2 = \lim_{\phi_2 \to 0} (B/\phi_2) \tag{7.27}$$

Finally, it should be pointed out that many authors [30, 32, 65] use a wavelength-independent constant K related to B by the expression

$$K = B\lambda/n \tag{7.28}$$

where n is the main (unperturbed) refractive index of the media, which usually is approximated by the value for the pure solvent. Substituting Eq. (7.28) into (7.27) gives

$$K_{sp} = (\lambda/n)B_{sp} = \lim_{\phi_2 \to 0} (B\lambda/n\phi_2) \tag{7.29}$$

Substituting the relationship $B = \Delta n/\lambda E^2$ into Eq. (7.29) gives the expression most commonly used as the definition of the Kerr constant in publications dealing with highly birefringent systems [98–101]:

$$K_{sp} = \lim_{\phi_2 \to 0} \left(\frac{\Delta n}{n\phi_2 E^2} \right) \tag{7.30}$$

The value of K_{sp} can be related to $\Delta\alpha$ through substituting Eq. (7.14) into (7.30) to give

$$K_{sp} = \frac{2\pi}{9} \lim_{\phi_2 \to 0} \left[\frac{(n^2 + 2)^2 \mathcal{N} \, \Delta\alpha}{n^2 \phi_2 E^2} \right] \tag{7.31}$$

where $\mathcal{N}$ is taken as the number of molecules of solute per unit volume of solution, because the solvent produces a negligible birefringence.

Obviously from the preceding paragraphs, quite different definitions of the Kerr constant are often used, depending on the kind of systems studied (gas, liquid, solutions, etc.) and even on the particular preference of the authors. Moreover, the possible use of either cgs (mainly in old publications) or SI units is an added difficulty for the comparison of results obtained by different authors.

EXPERIMENT

A typical Kerr-constant measurement consists in applying an electrical field to a given sample and determining the birefringence produced by the field. For this purpose, a beam of monochromatic and plane polarized light is passed through the sample. When the electric field is on, the sample becomes birefringent and transforms the incoming plane polarized light into elliptically polarized light. The ellipticity of this outcoming light is directly related to the birefringence of the sample. Thus, all that the experiment requires is to measure the state of polarization of a beam of light emerging from the sample when it is under a strong electric field.

To date, there is no commercially available Kerr-constant instrument. This means that every group working in this area has built its own personalized equipment, most of which are described in the literature [60–62, 99, 100, 114–119]. A block diagram showing the arrangement for the main components of the experimental setup is shown in Figure 7.3.

The main basic difference between these instruments lies in the kind of electric field applied to the sample or, more precisely, in the time dependence of the

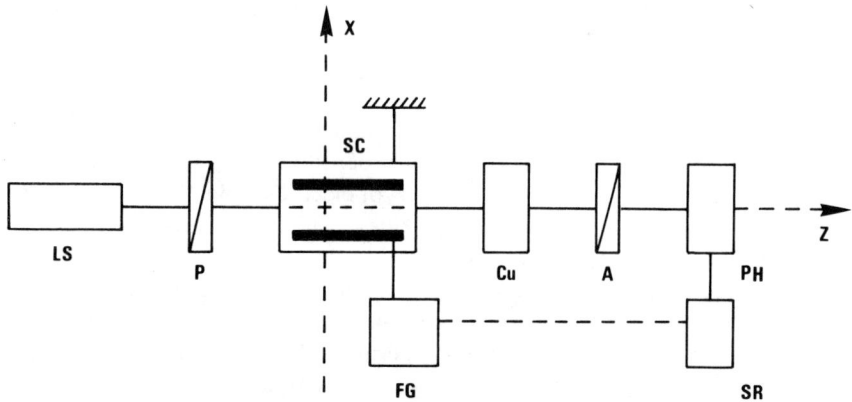

Figure 7.3 Block diagram of the experimental setup used for electric birefringence measurements: (LS) light source, (P) polarizer, (A) analyzer, (SC) sample cell containing the electrodes for application of the electric field, (FG) electric field generator, (CU) compensating unit, (PH) photomultiplier, and (SR) signal recording system.

field that determines the kind of recording system that should be used to store the signal from the photomultiplier. Most instruments used nowadays employ pulsed electric fields of different shape (square wave, bipolar, sawtooth, sinusoidal, etc.), duration, and frequency of repetition. This procedure offers two important advantages over the previous dc field instruments because, in the first place, it minimizes the heating of the sample and the problems arising from electric conductivity, thus allowing the measurement of ionic solutions. Moreover, the evolution of birefringence with time can also be studied and valuable information about the dynamic properties of the polymer can thus be obtained. Both advantages are specially valuable in the case of biopolymers to the point that the measurement of these polymers was practically impossible before the pulsed techniques became available and, of course, almost all the measurements performed nowadays in biological systems are carried out with this kind of field. The use of high-power lasers as electric field generators decreases more the problems with electric conductivity of the sample, but, on the other hand, it complicates both the experimental setup and the processing of the signal and, therefore, this technique is not yet widespreadly used.

The device used to register the signal from the photomultiplier in a pulsed-field experiment should be fast enough to follow the variation of the signal with time. These devices became available in the late 1940s and consisted of a dual-trace oscilloscope that simultaneously registered both the attenuated electric pulses and the photomultiplier signal. Photographic recordings of the oscilloscope traces were then used to measure the intensity of the signals. This procedure is still used, with small variations, to date to reach time scales of the order of nanoseconds. However, for longer times (in the range of microseconds), it is more convenient to use a dual-channel transient recorder that receives and digitizes the signals, feeding them to a computer that can be programmed to control the whole experiment, including the numerical analysis of the results [117].

However, if the apparatus is intended to measure only flexible synthetic polymers and low molecular-weight model compounds, it still makes sense to use a dc field setup. These kinds of substances can be, in general, dissolved in nonconducting solvents and, therefore, the problems of electric conductivity and heating of the samples are minimal. Moreover, they often exhibit very low birefringence so that strong and extremely stable electric fields are required. Finally, equilibrium values of the Kerr constant are much more important for these polymers than their dynamic parameters. Under these conditions, a dc apparatus can perform better than a pulsed-field one with a much simpler experimental setup. A dc instrument [118] that has been working nicely for several years is described in the next paragraphs along with discussions on some alternatives for the different components and explanations of the measuring procedures. Most of what is going to be described here for the dc instrument (components, alignment and testing procedures, state of polarization of the radiation, etc.) is directly applicable to pulsed field setups, keeping in mind the basic difference in the field generator and recording devices.

A block diagram of the instrument is shown in Figure 7.4. It includes the following components: light source, polarizer and analyzer, detection system, sample cell, and compensating unit.

(a) **Light Source.** It must produce a beam of intense and monochromatic light whose intensity should be constant over long periods of time. A laser, for example a He–Ne (λ = 632.8 nm) with a power of about 5 mW, is the best choice to meet these requirements. Fluctuations of the light intensity, measured over time periods of several hours, should be smaller than 0.1%.

(b) **Polarizer and Analyzer.** Their only requirement is to give a good extinction ratio. Glan-type polarizers with extinction ratios of $10^5:1$ can be used. They should be mounted in a rotating device that should allow rotations of about 0.01°.

(c) **Detection System.** It has the role of measuring the intensity of light emerging from the analyzer. Although the easiest detection system is one's own eyes, it is safer and more accurate to use a photomultiplier with a spectral response adequate to the wavelength of the laser (for example, S-20 for the He–Ne). The photomultiplier should be equipped with a narrow-band interference filter centered at the wavelength of the laser and a standard photographic obturator that allows closing the aperture and thus avoids overloading the photomultiplier when the instrument is opened, for instance, to change the cell. The signal from the photomultiplier is fed to a lock-in amplifier equipped with a mechanical chopper working at a frequency of 93 Hz. The output of the amplifier is displayed in a digital voltmeter and fed to a register.

(d) **Sample Cell.** This is the only component of the whole apparatus that is not commercially available. It should be designed and built keeping in mind several characteristics that it has to meet. Basically, the cell has to provide a cavity to hold a liquid without leaking (it can be designed for measurements in gases, but this is not the case in polymers); two conducting plates should be in contact with the liquid, thus allowing the application of an electrical field; a beam of light should be able to pass between the electrodes without changing direction or state of polarization apart from the effect of the field (thus, such things as air bubbles, reflections, strained and therefore birefringent glass, etc., should be avoided); finally, it should be easy to disassemble for cleaning purposes.

The physical dimensions, mainly the length of the electrodes and the gap between them, are also important. The birefringence produced under fixed conditions is proportional to the optical length; therefore, long electrodes produce stronger signals, but, on the other hand, they are more difficult to keep parallel and they increase the volume of sample required for the measurement. If the cell is designed to measure very birefringent samples, lengths of about 1 cm are used; this dimension is very convenient because spectroscopic cuvetts can be used as external containers for the whole cell. However, for most synthetic polymers, lengths of about 10 to 20 cm are required in order to obtain reasonable signals. As for the gap between electrodes, the smaller it is, the stronger the electrical field produced by a given voltage, and the lesser sample is required. Both characteristics are desirable, but if the gap is too small, there may be electrical discharges between the electrodes and the measurement would be ruined. Compromise values between 2 and 4 mm are used for this separation.

Several designs of cells have been published [119–122]. However, a perfect compromise between simplicity and good performance has not yet been reached. Figure 7.5 shows, as an example, front and side views of a cell. The sample cavity has a volume of 8 ml and it is

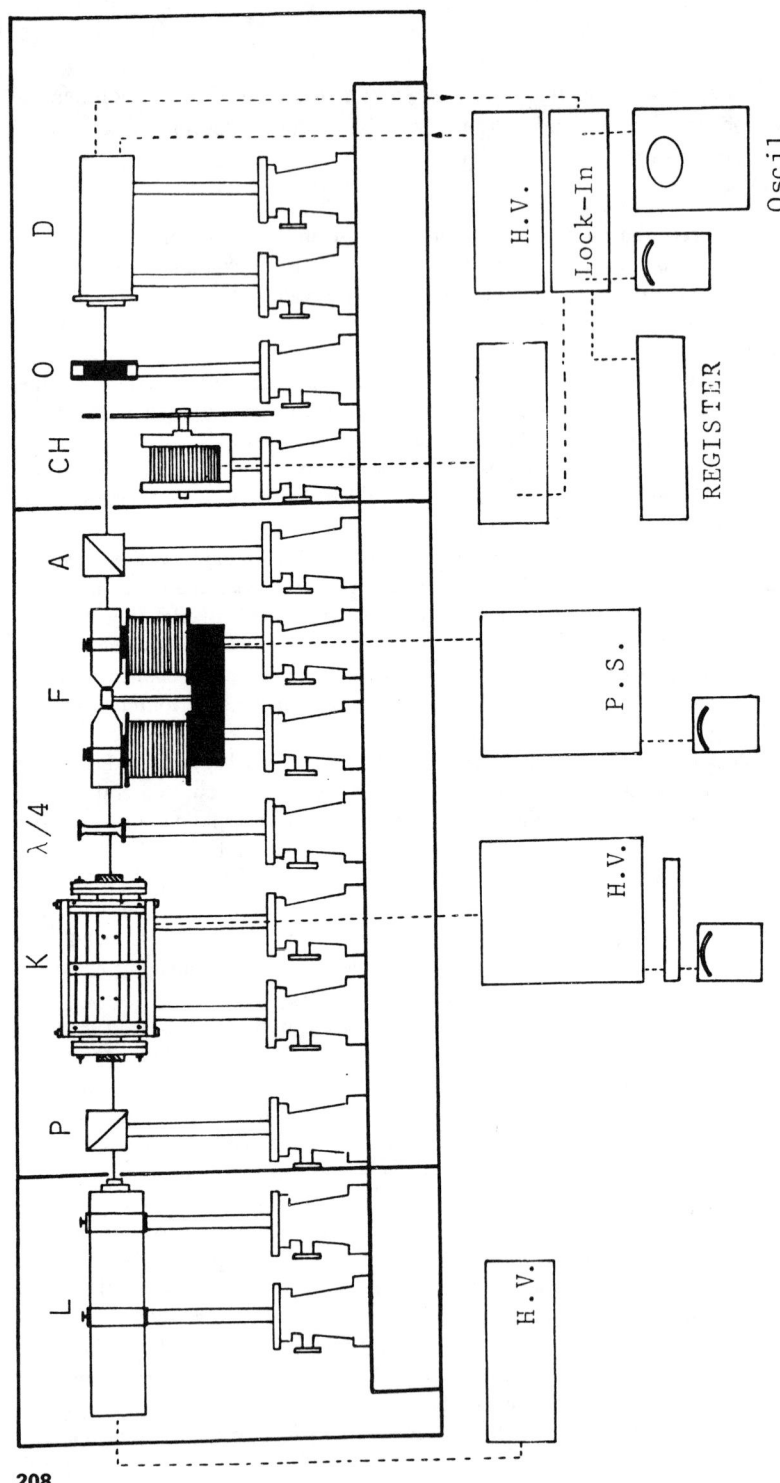

Figure 7.4 Block diagram of the instrument used for dc measurements of electric birefringence: (L) laser, (P) polarizer, (A) analyzer, (K) sample cell, (λ/4) quarter-wave plate, (F) Faraday coils, (CH) light chopper, (O) photographic obturator, and (D) detector. (Redrawn from reference [118] by courtesy of the Real Sociedad Española de Química.)

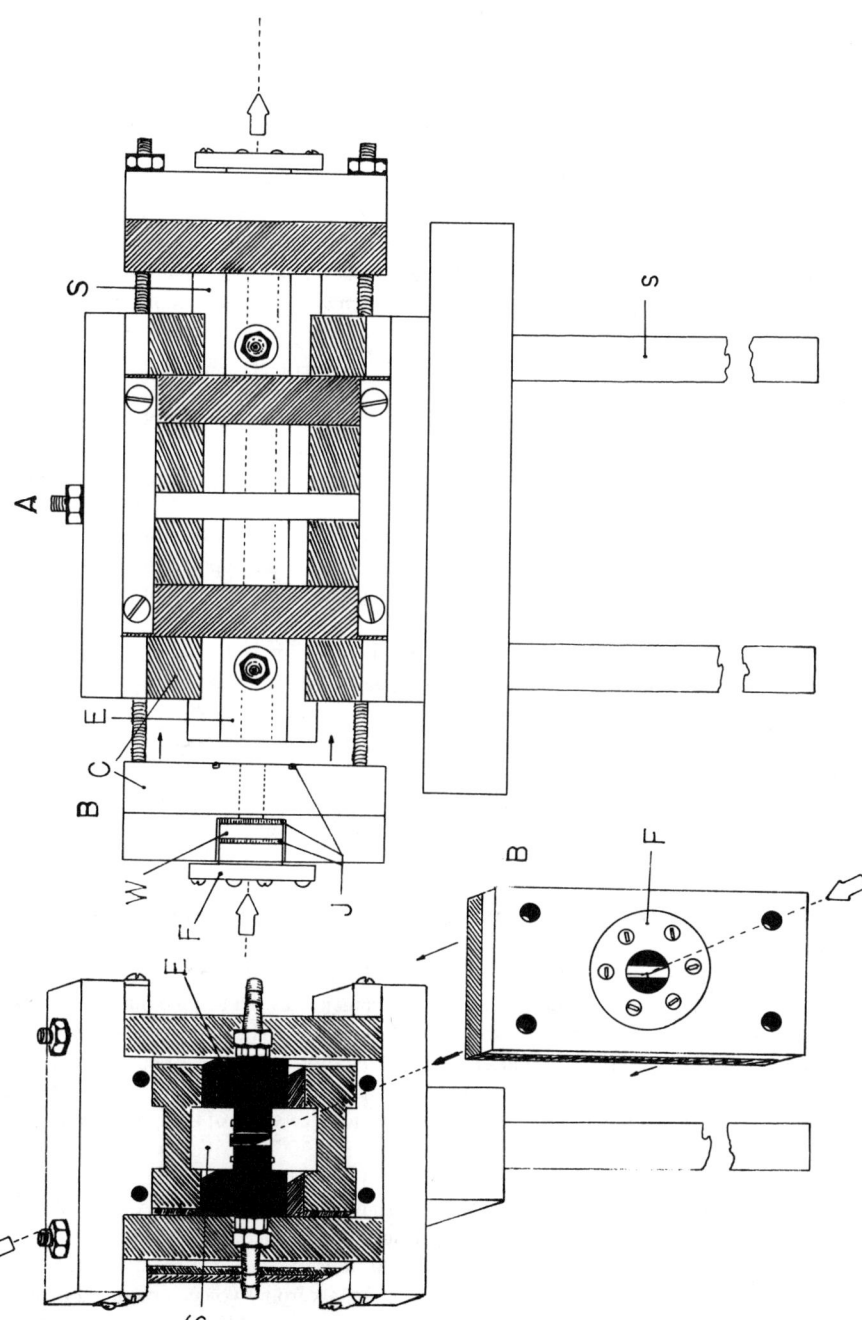

Figure 7.5 Front and side views of the sample cell used for electric birefringence measurements. (Redrawn from reference [119] by courtesy of the American Institute of Physics.)

formed by two hollow electrodes, 15 cm long and 15 mm wide, made out of bronze. Two spacers, fabricated in ceramic with good electrical insulation properties and mechanical rigidity, hold the electrodes parallel with a separation of 3.5 mm. Two aluminum blocks and several pieces of an electrical insulating phenolic material are used to keep together both the electrodes and the spacers. The leakproof condition is provided by four Teflon joints lying in slots placed along the electrodes. The top spacer has a little slot providing a trap for air bubbles, which are thus eliminated from the optical path; it also has two small holes, closed with Teflon stoppers, that allow filling the cell with a syringe. The cell windows are 5 mm thick and are of 25-mm-diameter optical quality glass (Oriel). They are attached to a rigid aluminum body by means of an aluminum flange and six screws. The window assemblies are fixed to the main body of the cell by means of four screws interposing a piece of insulating material and Teflon joints. When the cell is assembled, care should be taken in tightening the six screws of the windows so as to avoid stressing the glass and therefore producing birefringence. The best procedure consists of filling the cell with solvent, placing it in the beam of light between the crossed polarizers, and tightening or loosening the screws until no light passes through the analyzer. It is also important to avoid materials such as rubber for the inner joints of the windows because some solvents may swell those materials, thus producing stress in the windows and giving rise to birefringence; a good test is to fill the cell with solvent and to check that the analyzer gives extinction even after several hours.

The electrical power for the cell is provided by a high-voltage power supply that is able to deliver up to 25 kV. The applied voltage is monitored with a digital voltmeter equipped with a high-voltage probe.

(e) Compensating Unit. The role of this component is to provide a simple and accurate way of determining the state of polarization of the light emerging from the sample cell. There are many ways in which this determination can be performed. According to the kind of modifications produced in the beam of light, the measuring procedures fall into one of the following three groups.

(1) Direct measurement of the ellipticity of the polarization of the light. It consists in rotating the analyzer and measuring the maximum and the minimum of the light intensity. These two values give the long and short axes of the ellipse, and from them, the birefringence produced by the sample can be calculated. This is the simplest procedure from the point of view of instrumentation since no measuring device is required; however, it is also the less accurate.

(2) Introduction of a second birefringent element. In this method, an element is introduced that can produce exactly the same birefringence as the sample but with opposite sign, so that it compensates the effect of the sample cell. The light emerging from the compensator will then have its original plane of polarization and the analyzer gives complete extinction. Obviously, the compensator should be able to be adjusted to produce different birefringences and it has to be calibrated in order to know the value of the birefringence when the condition of compensation is reached.

The most obvious compensator of this kind is a second sample cell (with its own power supply) filled up with a substance whose Kerr constant is known so that the values of applied voltage can be directly transformed into birefringences produced by the compensator. An apparatus of this kind is described in reference [116]; it uses a two-channel signal averager to obtain simultaneously the ac electric birefringence signals from two cells mounted in series. It has the advantage that the ratio of the two signals is not affected by fluctuations of the total light intensity such as those produced by turbulences caused by Joule heating of the samples.

Thus, relative Kerr constants can be measured when one of the cells contains a standard sample.

However, since the problems of alignment, window birefringence, air bubbles, etc., produced by just one cell are important nuisances, this setup is not often used. A modified, and much more used, version of this kind of compensator is the Pockels cell in which the birefringence is produced in a crystal by means of an electrical field applied in the direction of propagation of the radiation. The optical axis of the crystal may be either parallel or perpendicular to the direction of the beam of light; any of these possible orientations has some advantages and some disadvantages, as compared with the other, so that both of them can be used and none is clearly preferable over the other one. The birefringence can also be induced either in solids or in liquids by application of magnetic fields (Cotton–Mouton effect); however, this effect is not used for compensating devices because it requires the use of very large and stable magnetic fields.

A second kind of compensator of this group is mechanical. An excellent description of these compensators is given in reference [62]. The general idea is to use a birefringent material, such as mica or quartz, cut and mounted in such a way as to allow modifying its optical length and, therefore, the birefringence produced.

(3) Transformation of the birefringence into optical rotation. The elliptically polarized light emerging from the sample cell can be transformed back into plane polarized light by introducing a phase shift of $\lambda/4$ between its two components. The planes of polarization of the initial (before the sample cell) and final (after the $\lambda/4$ phase shift) beams make an angle α that is proportional to the birefringence produced by the sample (i.e., $\alpha = \delta/2$; see what follows). The measurement of birefringence is then converted into a simple polarimetry in which the angle α is determined either by rotating the analyzer until it gives complete extinction or by using a Faraday cell in which an optical rotation is produced by means of a crystal placed in a magnetic field.

The $\lambda/4$ phase shift is provided by a quarter-wave plate mounted in a rotatory disc. It is simply a sheet of birefringent material with a thickness such as to produce a phase shift of $\lambda/4$ for the wavelength of the He–Ne laser between the components that travel parallel and perpendicular to its optical axis. The correct orientation of the quarter-wave plate requires that its optical axis be parallel to that of the polarizer. This orientation is very easy to achieve, because when both optical axes are parallel, there is no perpendicular component of the polarized light and, therefore, the plate has no effect. Thus, the orientation procedure consists in placing the plate and rotating it between the crossed polarizers until the analyzer gives complete extinction.

A scheme of the Faraday cell is shown in Figure 7.6. It is formed by a $10 \times 10 \times 20$ mm prism of optical glass standing on a small platform placed between two pierced poles of a U-shaped electromagnet. The laser beam passes through the holes of the poles and through the glass, which is optically inactive, when the magnetic field is off and it becomes active by the effect of the field producing a rotation α of the plane of polarization of the light. The sign and absolute value of α are determined by the direction and intensity of the field, respectively. A power supply provides electrical current for the coils of the electromagnet; the sense of the current, and thus the direction of the magnetic field, can be easily changed while the intensity of the current is monitored with a digital voltmeter and can be adjusted up to 12 A.

In our opinion, either the Pockels or the Faraday cell is the best choice for compensators, since no mechanical manipulations of optical elements are required while the instrument is working. In particular, because the birefringences produced are small, any procedure

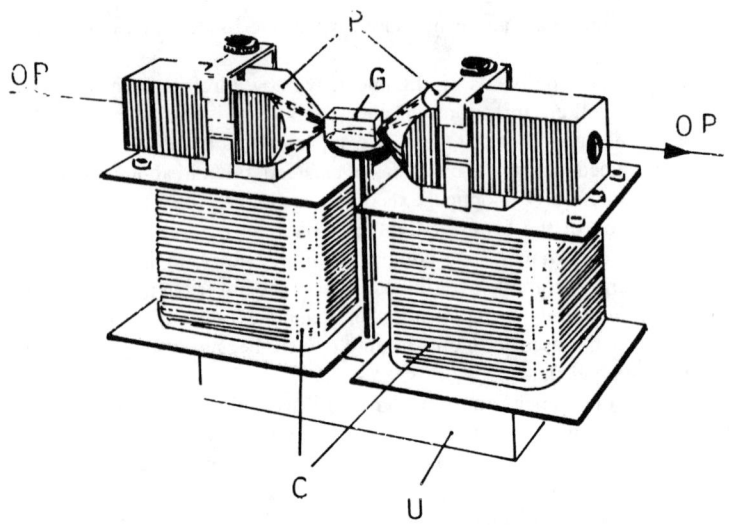

Figure 7.6   The Faraday cell: (C) coils, (U) core, and (P) poles of the electromagnet; (G) prism of optical glass; and (OP) path followed by the laser beam.

that requires rotating the analyzer is highly disadvantageous because it will have very little accuracy.

All the optical elements are mounted in a triangular steel optical bench and are placed inside a black-painted 220 × 60 × 50 cm wooden box divided in three compartments. The inside of the central compartment containing the sample cell is thermally isolated [123] by means of 40-mm-thick plates of expanded polystyrene with two small holes, covered with glass windows, allowing the passage of the light beam. The top and internal faces of the isolation plates are fitted with a coil made of 25 m of copper pipe through which a mixture of water and ethylene glycol is forced to circulate from a thermostat. The air close to the thermostatic coil is forced to circulate through the hollow electrodes of the cell by means of a small pump. The contact thermometer controlling the thermostat is placed close to the electrodes. Several tests proved that the whole system is able to keep a constant temperature in the range 0 to 80°C within ±0.2°C.

The best way for understanding the working of the whole apparatus is to perform a mathematical analysis of the amplitude of the radiation after passing through each of the optical elements. With this purpose, a Cartesian coordinate system for each of these elements [118, 119] (see Figure 7.7) is defined as follows: The x axis of the coordinate system for each element lies along its own optical axis. The z axis coincides with the direction of propagation of the radiation and, therefore, is common for all the systems (it is assumed that it goes up in Figure 7.7). Finally, the y axes complete right-handed systems. With these assumptions, the

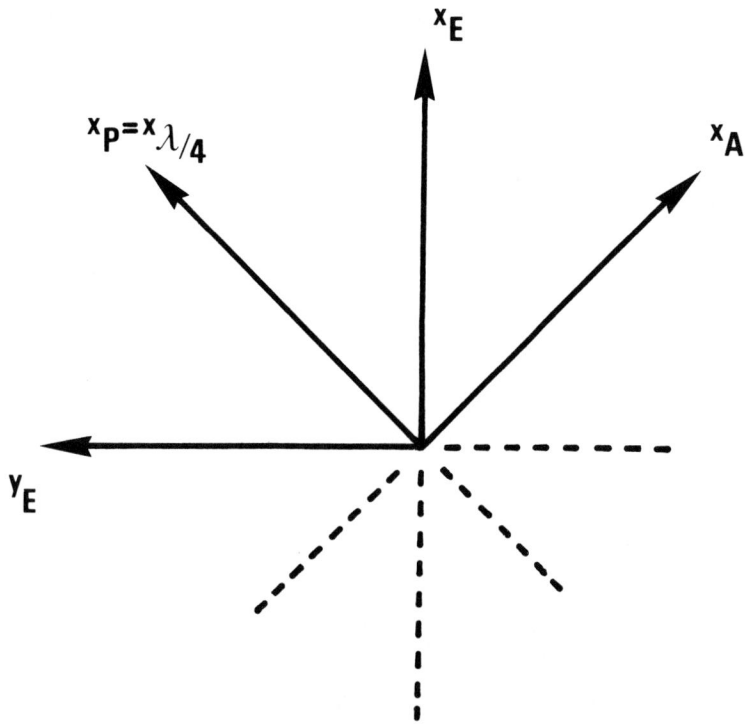

Figure 7.7 Cartesian coordinate systems used in the mathematical analysis of the amplitude of radiation.

radiation emerging from element j has two components in the xy plane that can be represented in matrix form by the real part of

$$\mathbf{A}^j = \begin{bmatrix} A_x^j \\ A_y^j \end{bmatrix} = \begin{bmatrix} A_x e^{i(\omega t + \phi_x)} \\ A_y e^{i(\omega t + \phi_y)} \end{bmatrix} \tag{7.32}$$

Assuming, for instance, that j is the polarizer, because it transmits only the component that vibrates along its optical axis (i.e., the x component), the perpendicular component will be zero and, therefore,

$$\mathbf{A}^p = \begin{bmatrix} A_x^p \\ 0 \end{bmatrix} = \begin{bmatrix} A_x e^{i(\omega t + \phi_x)} \\ 0 \end{bmatrix} \tag{7.33}$$

Rotations required to bring the coordinate system of element j into coincidence with that of element $j + i$ can be easily performed by means of a transformation matrix defined as

$$\mathbf{R}(\tau) = \begin{bmatrix} \cos \tau & \sin \tau \\ -\sin \tau & \cos \tau \end{bmatrix} \tag{7.34}$$

The effect of this matrix when it multiplies vector $\mathbf{A}$ is to rotate the x and y axes by an angle τ, leaving the z axis unchanged.

The effect of each optical element can be represented by a matrix operator. Thus, in the cases of the polarizer and analyzer, taking into account that each has its own coordinate system with the x axis lying in the direction of the optical axis, the operator should, in both cases, allow the transmission of an x component with the elimination of y. The operator for the sample cell should produce a phase shift of δ between the x and y components of the radiation, which are parallel and perpendicular to the direction of the applied electric field, respectively. In the case of the quarter-wave plate, the phase shift should be $\pi/2$, which is equivalent to $\lambda/4$. Finally, a rotation of the plane of polarization by an angle α should be produced by the operator of the Faraday cell. Therefore, the operators representing the action of each of these elements can be written as

$$\mathbf{O}_p = \mathbf{O}_a = \begin{bmatrix} 1 & 0 \\ 0 & 0 \end{bmatrix} \qquad \mathbf{O}_E = \begin{bmatrix} e^{i\delta} & 0 \\ 0 & 1 \end{bmatrix}$$

$$\mathbf{O}_{\lambda/4} = \begin{bmatrix} e^{i\pi/2} & 0 \\ 0 & 1 \end{bmatrix} \qquad \mathbf{O}_H = \begin{bmatrix} \cos\alpha & \sin\alpha \\ -\sin\alpha & \cos\alpha \end{bmatrix} \tag{7.35}$$

Application of these operators with the appropriate rotations gives the radiation emerging from the analyzer, and, therefore, reaching the photomultiplier as

$$\mathbf{A}^a = \mathbf{O}_a\mathbf{R}(-90)\mathbf{O}_H\mathbf{O}_{\lambda/4}\mathbf{R}(45)\mathbf{O}_E\mathbf{R}(-45)\mathbf{A}^P \tag{7.36}$$

This equation represents the following transformations:

1. Starting with the radiation emerging from the polarizer $\mathbf{A}^P$ given by Eq. (7.33), the x–y axes are rotated $-45°$ to bring them into coincidence with those of the sample cell; the operator of this element, $\mathbf{O}_E$, is then applied.

2. Next, the axes are rotated back $45°$ into coincidence with those of the quarter-wave plate (which are coincident with those of the polarizer); the operator $\mathbf{O}_{\lambda/4}$ is then applied.

3. The operator of the Faraday cell, $\mathbf{O}_H$, requires no previous transformation of the coordinate system because it always produces a rotation of the plane of polarization of the incoming radiation by an angle α.

4. After the radiation emerges from the Faraday cell, an axes rotation of $-90°$ is performed to bring the system into coincidence with that of the analyzer, whose operator $\mathbf{O}_a$ is the last one acting over the radiation.

A good exercise consists in checking that the radiation emerging from the sample cell has two components of the same amplitude, but with a phase shift of δ, and after passing through the quarter-wave plate, both components are back on phase.

Development of Eq. (7.36), followed by elimination of the time-dependent part and substitution of some simple trigonometric relationships gives the amplitude of the radiation reaching the photomultiplier as

$$A = K (\cos \alpha \sin \delta - 2 \sin \alpha \cos^2 \delta/2) \qquad (7.37)$$

where K represents a constant that is proportional to the initial amplitude of the laser beam and contains the attenuation factors produced by all the optical elements. The value of δ is normally very small; for instance, the cell described before filled up with benzene and with a voltage of 10^4 V (producing an electric field of 2.86×10^6 V/m) gives $\delta = 1.85°$; of course, many other compounds have much larger Kerr constants than that of benzene, but in these cases, much smaller electric fields are applied, so that values of δ are usually on the order of $1-2°$. Under these circumstances, the value of δ can be used with good approximation instead of $\sin \delta$ and Eq. (7.32) can be written as

$$A = K (\delta \cos \alpha - 2\alpha \cos^2 \delta/2) \qquad (7.38)$$

There are several conditions under which this equation is noticeably simplified. For instance:

(a) $\alpha = \delta = 0$. In this case, $A = 0$. This corresponds to the trivial situation in which both electric and magnetic fields are off so that sample and Faraday cells are inactive. Moreover, the quarter-wave plate is also inactive because the radiation passing through it is polarized along its optical axis. The whole setup reduces then to a pair of crossed polarizers that, of course, gives complete extinction.

(b) $\alpha \neq 0$ and $\delta = 0$. Then $A = -2K\alpha$. This situation arises when the electric field is off and, therefore, both the sample cell and the quarter-wave plate are inactive. However, the magnetic field is on and, consequently, the Faraday cell produces a rotation α of the plane of polarization of the light. The amplitude of the light reaching the photomultiplier is proportional to $\sin \alpha$ (intensity proportional to $\sin^2\alpha$) according to Malus's law. In this expression, the value of $\sin \alpha$ is approximated by the angle in radians.

Taking into account that the signal S produced by the photomultiplier (and amplified by the lock-in) is proportional to the intensity of the light emerging from the analyzer, and, therefore, to the square of its amplitude A, and that, on the other hand, the rotation α produced by the magnetic field is proportional to the strength of that field and, therefore, to the intensity I of the electric current passing through the coils of the electromagnet, one obtains

$$S^{1/2} = \beta I \qquad (7.39)$$

where β is a constant containing such things as the intensity of the initial beam of light, voltage applied to the photomultiplier, the number of turns in the electromagnet coils, the separation between the magnetic poles, etc. A representation of $S^{1/2}$ versus I should then give a straight line; the practical use of such a line will be shown in what follows.

(c) $\alpha = 0$ and $\delta \neq 0$. The amplitude is then $A = K\delta$. In this case, the magnetic field is off and, therefore, the Faraday cell is inactive. The electric field is on and the sample cell introduces a birefringence δ, producing elliptically polarized light. The quarter-wave plate transforms that light back into plane polarized light with a rotation with respect to the initial plane of polarization equivalent to $-\delta/2$. The amplitude of the radiation again follows Malus law with an angle of rotation of $-\delta/2$.

The value of δ is given by the product of the Kerr constant B of the sample times the optical length of the cell and times the square of the electric field E applied to it; see Eq. (7.1). Field E is proportional to applied voltage V; the factor between E and V is the gap between the electrodes, $E = V/g$. Therefore, in this case,

$$S^{1/2} = K\delta = K2\pi lB(V/g)^2 = \gamma BV^2 \tag{7.40}$$

where the constant γ contains, among other characteristics of the apparatus, the geometry of the sample cell. A representation of $S^{1/2}$ versus V^2 should give a straight line in this case that will be used in what follows.

(d) $\delta = 2\alpha$. In this case, by using Eq. (7.37) instead of (7.38), or in other words, by removing the approximation of using angles instead of sines, the value $A = 0$ is obtained. In this situation, the combined effect of the sample cell and the quarter-wave plate produces a rotation of the plane of polarization of the light, but the Faraday cell produces exactly the contrary rotation, thus restoring the original plane of polarization, and the radiation is extinguished by the analyzer.

These four particular situations provide two different methods for measuring, namely, the methods of *compensation* and *comparison*.

The compensation method is the most intuitive one. It consists in applying a given voltage V to the sample cell and increasing the electrical current I that flows through the electromagnetic coils until the analyzer gives complete extinction. This indicates that the Faraday cell compensates the combined effects of the sample cell and quarter-wave plate; or, in other words, the condition $\alpha = 2\delta$ is met. If the Faraday cell has been previously calibrated by measuring some liquids with known Kerr constants, the value of I can be easily converted into birefringence δ produced by the sample cell under the effect of the voltage V. In order to obtain the value of the Kerr constant B, the measurement is repeated at several values of V and the results of δ are plotted versus V^2. According to Eq. (7.1) and taking into account that $E = V/g$, the data should fit a straight line having a slope of $2\pi Bl/g^2$, where the length of the electrodes, l, and the gap between them, g, can be obtained either by measuring the geometry of the cell or by a previous calibration.

The biggest difficulty of this method is the determination of the exact value of I for which the condition $\alpha = 2\delta$ is met. Although complete extinction should be obtained under this condition, in practice, there is always some light reaching the photomultiplier and, on the other hand, this element gives a dark signal even without any light. This means that the situation of minimum intensity, and not that of complete extinction, should be found and, therefore, the measurement loses accuracy.

The comparison method seems to be more accurate. It consists in performing two sets of measurements: One set applies different voltages V to the sample cell while having the magnetic field off (this amounts to work under condition (c) of those explained before, having $\alpha = 0$ and $\delta \neq O$). The second set of measurements is carried out under condition (b) with $\alpha \neq 0$, $\delta = 0$, which is obtained by switching the electric field off and allowing different intensities of electric current I to circulate through the coils of the electromagnet, thus changing the intensity of the magnetic field.

By representing the signal from the photomultiplier by S, data of the first set should fit a straight line given by Eq. (7.40) and the second set will be represented by Eq. (7.39). The ratio between the slopes of these two lines is $(\gamma/\beta)B = \eta B$. If the measurement is performed with a liquid of known Kerr constant, the calibration constant η for the whole apparatus can be calculated, and, because it does not depend on the measured sample, can be used for all measurements.

Figure 7.8 shows the two kinds of representation for three liquids frequently

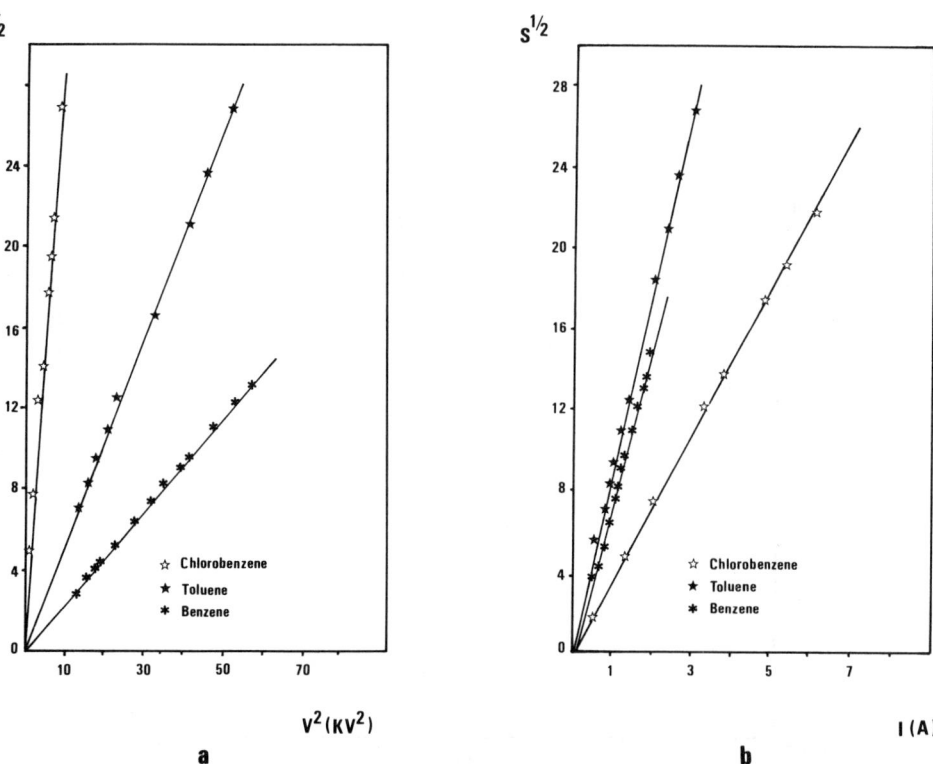

Figure 7.8 Square root of the signal received by the photomultiplier as function of (a) the square of the voltage applied to the sample cell, V^2, keeping the Faraday cell inactive ($I = 0$); (b) the intensity of the electric current I flowing through the electromagnet while no field is applied to the sample ($V = 0$). (Reprinted from reference [118] by courtesy of the Real Sociedad Española de Química.)

used in calibration, namely, benzene, toluene, and chlorobenzene, whose values of B are 0.416, 0.81, and 12.0 in units of 10^{-14} m V^{-2}, respectively [124]. Values for the calibration constant of our apparatus obtained with these three solvents are η = 0.0704, 0.0705, and 0.0703, respectively, when the voltage is measured in kilovolts, the intensity in A, and the B constant in 10^{-14} m V^{-2}.

The weakest point of the comparison method is the stability of the light intensity. If the power delivered by the light source changes with time, the method will not work. In this situation, the compensation method is recommended.

However, provided that the light source is very stable, the existence of a "dark signal" produced either by the dark current of the photomultiplier or by some light passing through the analyzer is not a serious problem for the comparison method unless it fluctuates with time. The reason is that a constant signal changes the intercepts of plots, but leaves unaffected the slopes, which are the only magnitudes used in the evaluation of B.

If the measured sample is a gas, the value of B is normally used as the Kerr constant and then no further manipulations are required. In the case of pure liquids, the value of B is converted into the molar Kerr constant by using Eq. (7.22). Finally,

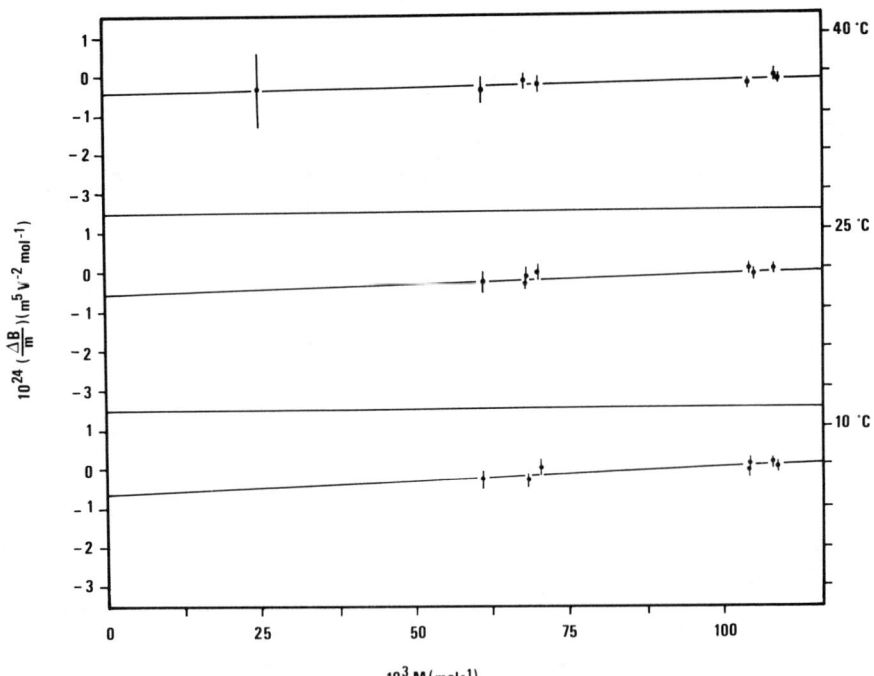

Figure 7.9 Extrapolation of $\Delta B/m$ versus m for a sample of poly(diethylene glycol terephthalate) (PDET) at the three temperatures indicated. (Reprinted from reference [123] by courtesy of Hüthig & Wepf, Basel.)

if the molecule of interest is the solute in a solution, as is always the case with polymers, several measurements are performed on solutions having different concentrations and the results are extrapolated to infinite dilution, according to Eq. (7.25) for molar Kerr constants or with Eq. (7.29) for specific Kerr constants at infinite dilution.

Figure 7.9 shows an example of the extrapolation of $\Delta B/m$ to $m \rightarrow 0$ as required in Eq. (7.25).

THEORETICAL CALCULATION OF THE KERR CONSTANT

As was indicated before, the only magnitude that is required for the calculation of Kerr constants, B with Eq. (7.15), $_mK$ with Eq. (7.18), or K_{sp} with Eq. (7.31), is the difference $\Delta\alpha$ between the polarizability of the molecules on the sample along directions parallel and perpendicular to the electric field. The only difficulty of the whole procedure is that this magnitude should be averaged over all possible orientations of the macromolecule inside the field and over all allowed conformations of the chain. Therefore, the average obtained will depend on both, the electric field and the characteristics of the macromolecule.

It is quite simple to seek out, at a qualitative level, the factors that determine the value of the average of $\Delta\alpha$. The physical situation is as follows: In the absence of the electric field, the macromolecules have a Brownian motion governed by the thermal energy kT that tends to randomize both position and orientation of the macromolecules inside the solution. In this situation, all the directions are equivalent and the sample is isotropic because any arbitrarily chosen direction will have, on average, the same number of macromolecules in every possible orientation. When the electric field is applied in a given direction, it interacts with the permanent (if any) and induced dipole moments of the macromolecules. The energy of this interaction depends on the orientation of those dipoles relative to the field, being minimum when they are in the direction of the field. Thus, the macromolecules tend to adopt the orientations of minimum energy and the sample becomes anisotropic because there are, on average, more molecules oriented with their dipole moments in the direction of the field than in the direction perpendicular to it.

The degree of orientation depends on the balance between the thermal energy and the interaction between the electric field and molecule. The energy of this interaction depends on the strength of the field and on the magnitude of both permanent and induced dipole moments. For instance, if the field is strong enough as to orientate all the macromolecules of the sample, there will be saturation and the birefringence will not increase any more with increasing field. Therefore, even if for weak fields the birefringence increases with E^2, it should reach an asymptotic limit for strong enough fields.

Thus, according to this intuitive analysis, the magnitudes that determine the average of $\Delta\alpha$ are thermal energy, strength of the electric field, permanent dipole moments, electrostatic polarizabilities α' (that determine the dipole moment induced by the field), and anisotropy of the optical polarizabilities $\Delta\alpha$, because if the molecule is isotropic ($\Delta\alpha = 0$), the average will be zero under any circumstance.

However, there are two points that we have not mentioned in this analysis, namely, the molecular structure of the polymer and the dynamic aspects of the whole process.

Most polymers may adopt an extremely large number of conformations when they are in solution or in an amorphous state above T_g. Therefore, values of many of their properties (dipole moments and polarizabilities among many others) are in fact averages over all the allowed conformations. The interaction with the electric field changes the relative energy of these conformations and, therefore, it could change the dipole moment and polarizability of the polymer. We will come back to this aspect later; at this point, we assume that the macromolecule is a rigid particle having cylindrical symmetry, for instance, a rigid rod.

As for the dynamic aspects, the preceding analysis was performed thinking only in the states of equilibrium, either with or without the electric field. However, when the field is switched on, the macromolecules have to rotate in order to align with the field; conversely, when the field is switched off, the macromolecules are forced to rotate again by the Brownian motion until their orientation is randomized. Two obvious characteristics of these rotations are that they require a certain time and that there is a force opposing them, namely, the friction of the polymer with the surrounding molecules of solvent. Obviously, the time required for a molecule to rotate depends on its size and shape; thus, small molecules in ordinary liquids rotate with lifetimes of about 10^{-9} to 10^{-10} seconds. However, in the case of polymers, these lifetimes are much larger and their measurement by the analysis of the variation of birefringence with time provides an important method of studying the size and shape of the polymer. The friction between macromolecule and solvent depends also on the size and shape of the polymer and it is quite easy to foresee that the rotational diffusion coefficient $\mathcal{D}$ will be an important magnitude for the variation of birefringence with time. Strictly speaking, the inertial effect opposing any change in the speed of the macromolecules should also been taken into account. However, O'Konski and Krause [125] have proven that inertial effects are negligible for ordinary solutions of macromolecules because the inertial energy is very quickly dissipated on viscous drag in the surrounding medium.

Next, let us try to put these qualitative ideas into quantitative form and calculate $_mK$ by evaluation of the average of $\Delta\alpha$. For this purpose, a laboratory reference frame is defined by the following conventions: The x axis has the direction of the electric field $\mathbf{E}$, the z axis coincides with the direction of propagation of the beam of light used for the measurement, and the y axis completes a right-handed coordinate system. In this frame, the difference $\Delta\alpha$ can be written as

$$\Delta\alpha = \alpha_{xx} - \alpha_{yy} \tag{7.41}$$

Let us assume that a rigid-rod macromolecule is placed in this coordinate system. By symmetry, the main axes of the α tensor of this particle lie along the longitudinal and transverse directions of the rod, so that written in these axes, the optical polarizability of the rod is a diagonal tensor whose nonzero elements are α_L, α_T, and α_T, with the L and T subscripts representing the longitudinal and transverse directions, respectively. The orientation of the rod is fully described by the angles θ and ψ defined, respectively, as the angle between the long axis of the rod and the x axis and the angle that the projection of the rod on the yz plane makes with the y axis (see Figure 7.10).

Performing the rotation of coordinates required to bring the axes of the rod, L, T, and T, into coincidence with xyz, one obtains the xx and yy components of α:

$$\begin{aligned}
\alpha_{xx} &= \alpha_L \cos^2\theta + \alpha_T \sin^2\theta \\
\alpha_{yy} &= \alpha_T \sin^2\psi + \alpha_T \cos^2\theta \sin^2\psi + \alpha_L \sin^2\theta \cos^2\psi
\end{aligned} \tag{7.42}$$

When there is no field applied to the sample, all the orientations (i.e., any set of values θ,ψ) are equivalent and one obtains $\langle \Delta\alpha \rangle = 0$. When the field is applied in the x direction, the equivalence of the orientations is destroyed and the energy depends on the orientation of the particle relative to x (i.e., on the θ angle);

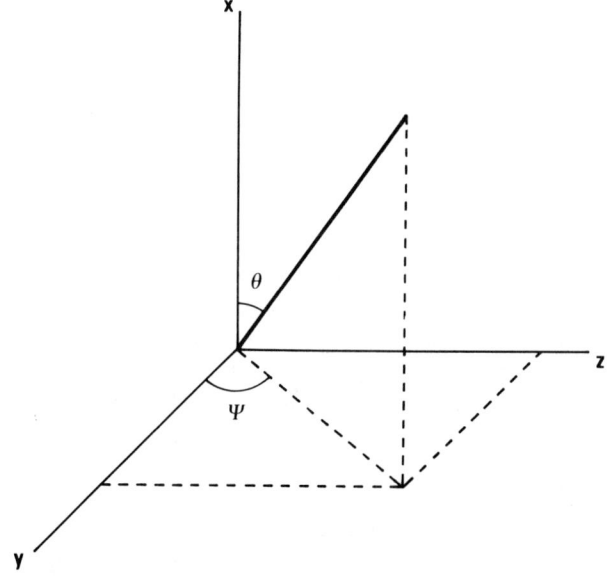

Figure 7.10 Orientation of a rigid-rod macromolecule in a laboratory coordinates system having x in the direction of the electric field, z in the direction of the beam of light, and y completing a right-handed frame; (θ) angle between the long axis of the rod and x; (ψ) angle between y and the projection of the rod in the yz plane.

however, it will be independent of the ψ angle. Thus, when performing the ensemble average over all the particles of the sample, all the values of ψ have the same weight and one obtains

$$\langle \Delta \alpha \rangle = (\alpha_L - \alpha_T) \langle (3 \cos^2\theta - 1)/2 \rangle = \Lambda \langle P_2(\cos \theta) \rangle = \Lambda \Phi \qquad (7.43)$$

where $\Lambda = \alpha_L - \alpha_T$ is the optical anisotropy of the particle, $P_2(\cos \theta) = (3 \cos^2\theta - 1)/2$ is the Legendre polynomial of degree 2, and the ensemble average of this polynomial over all possible orientations is the so-called orientation factor Φ. This ensemble average can be calculated as

$$\Phi(t) = \frac{\displaystyle\int_0^\pi P_2(\cos \theta) f(\theta,t) \sin \theta \, d\theta}{\displaystyle\int_0^\pi P_2(\cos \theta) \sin \theta \, d\theta} \qquad (7.44)$$

where $f(\theta,t)$ is the angular distribution function for the orientation of the rod with respect to the x axis. It is indicated explicitly that the function f may depend on time, and in that case, Φ will also be time-dependent.

The angular distribution function can be obtained by solving the rotational diffusion equation:

$$\frac{\delta f(\theta,t^*)}{\delta t^*} = \frac{1}{\sin \theta} \frac{\delta}{\delta \theta} \left\{ \sin \theta \left[\frac{\delta f(\theta,t^*)}{\delta \theta} + \frac{1}{kT} \frac{\delta V}{\delta \theta} f(\theta,t^*) \right] \right\} \qquad (7.45)$$

where t^* is a reduced time defined as $t^* = \mathscr{D}t$, with $\mathscr{D}$ being the rotational diffusion coefficient, and V is the energy of interaction between the electric field and the particle, which can be written as

$$V(\theta,t) = -\mathbf{E}_{eff}(t) \cdot (\mathbf{\mu} + \mathbf{m}) \qquad (7.46)$$

where $\mathbf{\mu}$ and $\mathbf{m}$ represent the permanent and induced dipole moments of the molecule, respectively. Assuming that the permanent dipole moment lies along the axis of the rod so that it makes an angle θ with the electric field, and representing by $\Lambda' = \alpha'_L - \alpha'_T$ the electrostatic anisotropy of the molecule and neglecting contributions of hyperpolarizabilities to $\mathbf{\alpha}'$, Eq. (7.46) can be rewritten as

$$V(\theta,t) = E_{eff}(t) \mu \cos \theta - \tfrac{1}{2}E_{eff}^2 \Lambda' \cos^2\theta \qquad (7.47)$$

Substitution of Eq. (7.47) into (7.45) followed by solution of the differential equation to obtain $f(\theta,t^*)$ and subsequent substitution of this function into Eq. (7.44) allows the calculation of the orientation factor Φ, which then permits the evaluation of the averaged anisotropy $\langle \Delta \alpha \rangle$ according to Eq. (7.43).

However, in many cases, this whole process is not an easy task. Readers interested in mathematical details are referred to the excellent revision published by Watanabe and Morita [87].

The solutions obtained in some particular cases for which either the procedure can be simplified or the results are specially interesting are presented in the next sections.

Static Fields

If the electric field does not change with time, $\delta f / \delta t^* = 0$ and then the solution of Eq. (7.45) gives a Boltzmann distribution:

$$f(\theta) = \frac{\exp(-V/kT)}{\displaystyle\int_0^\pi \exp(-V/kT) \sin \theta \, d\theta} \tag{7.48}$$

with V given by Eq. (7.47). Then, substituting into Eq. (7.44) and assuming that the time elapsed since the application of the field is long enough to allow equilibrium of rotational processes to be reached gives the orientation factor of the sample as

$$\Phi = \frac{3 \displaystyle\int_{-1}^{1} u^2 \exp(\beta u + \gamma u^2) \, du}{2 \displaystyle\int_{-1}^{1} \exp(\beta u + \gamma u^2) \, du} - \frac{1}{2} \tag{7.49}$$

where $u = \cos \theta$, and β and γ represent the strength of interactions produced by the electric field with permanent and induced dipole moments, respectively, as compared with the thermal energy, that is,

$$\beta = \mu E_{\text{eff}} / kT \quad \text{and} \quad \gamma = \Lambda' E_{\text{eff}}^2 / 2kT \tag{7.50}$$

The value of Φ can be obtained from Eq. (7.49) by any numerical integration procedure for a given combination of values of β and γ. Analytical expressions for Φ have also been obtained [50, 87, 126], although in many cases they are quite complicated. Three cases in which the integration of Eq. (7.49) is relatively simple are:

(a) *Low fields* ($\beta \ll 1$ *and* $\gamma \ll 1$). In this case, the exponentials of Eq. (7.49) can be written as a series expansion neglecting terms higher than E_{eff}^2 and the orientation factor becomes

$$\Phi = \frac{\beta^2 + 2\gamma}{15} = \frac{1}{15kT} \left(\frac{\mu^2}{kT} + \Lambda' \right) E_{\text{eff}}^2 \tag{7.51}$$

Thus, the orientation factor Φ, the anisotropy $\langle \Delta\alpha \rangle$, and the birefringence $\langle \Delta n \rangle$, which are all proportional to Φ, increase with the square of the field as indicated by Kerr's law, and $_mK$ is a constant independent of the field.

(b) *Orientation produced only by the permanent dipole moment* ($\gamma = 0$). If the contribution of the induced dipole moment to the orientation of the molecule is much smaller than that of the permanent dipole so that it can be disregarded ($\beta \gg \gamma \approx 0$), the integration of Eq. (7.49) gives

$$\Phi = 1 - \frac{3\mathscr{L}(\beta)}{\beta} = 1 - \frac{3(\coth \beta - 1/\beta)}{\beta} \tag{7.52}$$

where $\mathscr{L}(\beta)$ is the Langevin function.

(c) *Orientation produced only by the induced dipole moment* ($\beta = 0$). In the case that the term of the permanent dipole moment was negligible ($\gamma \gg \beta \approx 0$), the orientation factor becomes

$$\Phi = \frac{3}{4}\left[\frac{\exp(g^2)}{g\displaystyle\int_0^g \exp(x^2)\,dx} - \frac{1}{g^2}\right] - \frac{1}{2} \tag{7.53}$$

with $g = \gamma^{1/2}$.

Figure 7.11 shows a plot of Φ computed according to Eqs. (7.51), (7.52), and (7.53) versus the combination $\beta^2 + 2\gamma$, which is proportional to the square of the field. In the case of a low field, the plot is just a straight line with a slope of $1/15$, and in the other two cases, Φ tends asymptotically to unity for high enough fields, producing the so-called saturation effect. It is interesting to notice that even if there is saturation in both cases (b) and (c), the curves of Φ are quite different. Thus, permanent and induced dipoles produce different behavior in the region of high

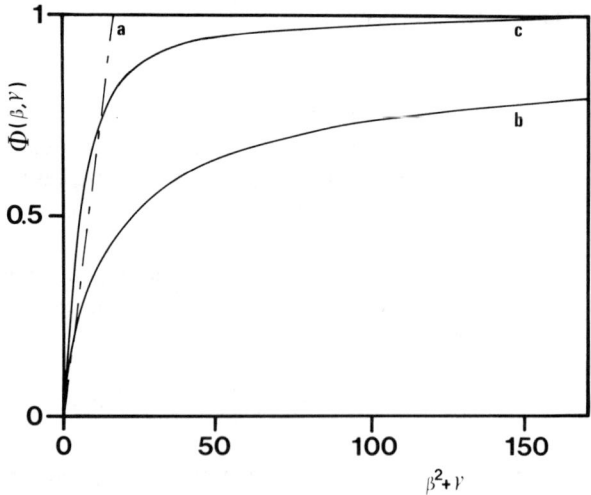

Figure 7.11 Variation of the orientation factor Φ versus the combination $\beta^2 + 2\gamma$ (proportional to E_{eff}^2) for (a) low fields, Eq. (7.51); (b) orientation produced by the permanent dipole moment, Eq. (7.52); and (c) orientation due to the induced dipole moment, Eq. (7.53).

fields, so that measurements performed in this region can give information about the electric properties of the macromolecule.

A much more detailed analysis of Φ as function of both the field strength represented by $\beta^2 + 2\gamma$ and the relative contribution of permanent versus induced dipole moments characterized by the ratio $\beta^2/2\gamma$ is given in reference [87].

Build-Up and Decay Processes

As indicated before, the rotational rearrangement of the macromolecules in a solution takes finite time. Therefore, if the electric field is applied instantaneously, the birefringence induced will increase with time as the molecules rotate to align with the field, reaching a constant value at equilibrium. This increase of birefringence with time after application of the field is the so-called build-up process. In a similar way, if an electric field that has been applied for a time long enough allowing equilibrium to be reached is instantaneously removed, the birefringence will decrease with time as the Brownian motion randomizes the orientation of the macromolecules; this is the decay process that eventually brings the birefringence to zero.

Both build-up and decay processes were observed by Kerr in his experiments with glass in which the reorientation time is long (of the order of seconds). On the contrary, small molecules in the liquid state have reorientation times too small to be detected because the time required for switching the field either on or off is larger than the reorientation time. Macromolecules in solution, depending on their size and shape, are somewhere between these two limits, and in many cases it is possible to take experimental measurements of these processes.

The first quantitative analysis of build-up and decay processes was performed by Benoit [45] both from experimental and theoretical points of view.

It is quite intuitive to foresee that the theoretical analysis of the decay process should be much easier than that of the build-up process because in decay, there is no field and therefore no interactions either with permanent or induced dipoles. There is only a Brownian motion that depends only on the size and shape of the molecule but not on its electric properties.

Let us assume that the electric field, which has been acting over a system for a long time, is suddenly removed at a given time that we take as $t = 0$. Then, according to Eq. (7.47), $V = 0$ and, therefore, $\delta V/\delta\theta = 0$ and Eq. (7.45) becomes

$$\frac{\delta f(\theta,t^*)}{\delta t^*} = \frac{1}{\sin\theta} \frac{\delta}{\delta\theta} \left[\sin\theta \left(\frac{\delta f}{\delta\theta} \right) \right] \tag{7.54}$$

The orientation factor Φ can then be easily obtained by differentiating Eq. (7.44) with respect to t^*. Taking into account that only $f(\theta,t^*)$ is a function of t^* and using Eq. (7.54), one obtains

$$d\Phi/dt^* = -6\Phi \tag{7.55}$$

which is immediately integrated to

$$\frac{\Phi(t^*)}{\Phi(0)} = \exp(-6t^*) \tag{7.56}$$

Taking into account the relationships between orientation factor Φ and anisotropy $\langle\Delta\alpha\rangle$, Eq. (7.43), anisotropy with birefringence Δn, Eq. (7.14), and the definition of the reduced time, $t^* = \mathcal{D}t$, Eq. (7.56) can be written as

$$\frac{\Phi(t^*)}{\Phi(0)} = \frac{\langle\Delta\alpha\rangle(t^*)}{\langle\Delta\alpha\rangle(0)} = \frac{\langle\Delta n\rangle(t^*)}{\langle\Delta n\rangle(0)} = \exp(-6t^*) = \exp(-6\mathcal{D}t) \tag{7.57}$$

Thus, the decay function of the orientation factor, the anisotropy, or the birefringence is a single exponential with a relaxation time of $\tau = 1/6\mathcal{D}$.

For the build-up process, the electric field is instantaneously applied at $t = 0$ and kept constant for a time long enough to allow the system to reach equilibrium. Thus, the orientation factor increases with time from $\Phi(0) = 0$ to a equilibrium value $\Phi(\infty)$. A complete analysis of $\Phi(t)$ for fields of any strength has been made by Watanabe and Morita [87]. However, most studies are still performed with the expressions first obtained by Benoit [45] for weak fields (the region where Kerr's law holds true), using serial expansion of the exponentials on Eq. (7.49) followed by truncation of terms depending on powers higher than E^2. The result thus obtained is

$$\frac{\Phi(t)}{\Phi(\infty)} = 1 - \frac{3\Gamma}{2\Gamma + 4} \exp(-2\mathcal{D}t) + \frac{\Gamma - 4}{2\Gamma + 4} \exp(-6\mathcal{D}t) \tag{7.58}$$

where the ratio $\Gamma = \beta^2/\gamma = 2\mu^2/kt\Lambda'$ represents the relative strength of permanent versus induced dipoles. Thus, the build-up process is a combination of two exponentials, one with a relaxation time $\tau = 1/6\mathcal{D}$ (the same as for the decay process) and a slower one with $\tau = 1/2\mathcal{D}$.

If the molecule has no dipole moment ($\mu = 0$), then $\beta = 0$, $\Gamma = 0$, and Eq. (7.58) reduces to

$$\frac{\Phi(t)}{\Phi(\infty)} = 1 - \exp(-6\mathcal{D}t) \tag{7.59}$$

In this case, the curve of the build-up process is symmetrical with that of decay, Eq. (7.56).

On the contrary, for very polar molecules having strong dipole moments, $\beta^2 \gg \gamma$ and $\Gamma \gg 1$, so that Eq. (7.58) can be approximated to

$$\frac{\Phi(t)}{\Phi(\infty)} = 1 - \frac{3}{2} \exp(-2\mathcal{D}t) + \frac{1}{2} \exp(-6\mathcal{D}t) \tag{7.60}$$

which gives a slower increase of $\Phi(t)$ than in the case of Eq. (7.59).

Figure 7.12 shows a representation of $\Phi(t)/\Phi(eq)$ versus time for the decay, Eq. (7.56), and build-up functions, both in the case of pure induced, Eq. (7.59), and permanent, Eq. (7.60), dipoles. In all cases, $\Phi(eq)$ represents the maximum value of this magnitude that is reached when the system obtains equilibrium with the field. Thus, $\Phi(eq) = \Phi(0)$ for the decay process and $\Phi(eq) = \Phi(\infty)$ for the build-up process. In the most general case, Eq. (7.58), provided that $\Gamma > 0$, the curve for the build-up process lies somewhere between the two extreme cases indicated in this figure, so that the analysis of the build-up process provides a way of determining the relative importance of permanent versus induced dipoles, although the use of pulsed fields simplifies this analysis, as is shown in the next section.

The value of Γ can be negative if the permanent dipole moment makes a large angle with the axis of maximum polarizability of the molecule, and, in this case, the build-up curves may differ substantially from those of Figure 7.12.

Because the decay is a pure diffusion process, it is not affected by the strength of the field. However, the preceding equations for the build-up process are not applicable for strong fields producing orientations energies of the order of kT; in this case, the variation of $\Phi(t)$ with t is much faster than indicated by Eq. (7.58) because then the effect of the field overcomes the thermal motions of the molecules.

It is quite simple to explain, at a qualitative level, the different shapes of the build-up curves for both kinds of orientations. If a rigid-rod molecule has a permanent dipole moment in the direction of the largest axis, it will be asymmetric with respect to a rotation of π over the transverse axis; thus, if the interaction with the field is V for a given orientation θ (see Figure 7.10), it will change to $-V$ for $\pi - \theta$.

By taking the most favorable orientation (the one giving the strongest negative value of V) as $\theta = 0$, then $\theta = \pi$ will be the most unfavorable one. Therefore, all the

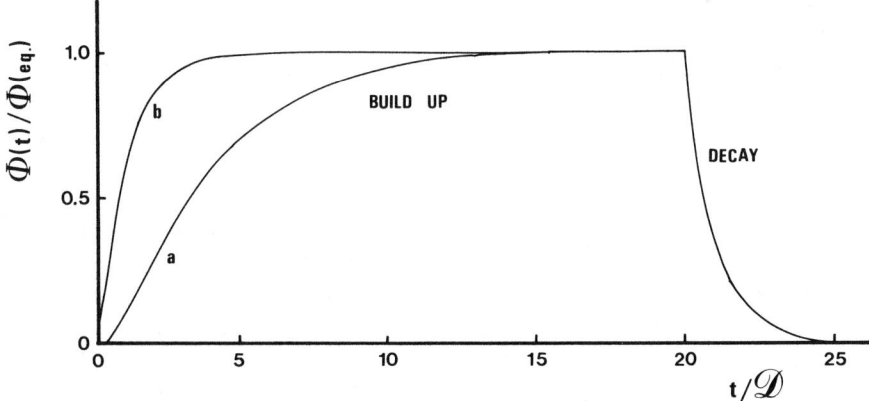

Figure 7.12 The orientation factor as a function of time for build-up and decay processes. Orientation produced by (a) the pure permanent dipole moment, and (b) the pure induced dipole moment.

molecules tend to orientate in the direction $\theta = 0$, so that molecules starting with orientations close to $\theta = \pi$ have to rotate almost π in order to align with the field. However, the polarizability tensors are symmetric with respect to a π rotation over the transverse axis of the molecule and, therefore, $\theta = 0$ and $\theta = \pi$ are equivalent. The anisotropy will be the same in both cases, although the induced dipole moment will have an opposite direction relative to the axis of the particle, but in both cases, it will be correctly aligned with the field. Thus, the molecules can "choose" to rotate toward $\theta = 0$ or $\theta = \pi$ and the orientation is faster.

In the decay process, because the birefringence is produced by the polarizability and not by the dipole moment, the orientation $\pi - \theta$ is as disordered (or as ordered) as θ regardless of the polarity of the molecule. The process has the same time dependence as the build-up process in the case of pure induced dipoles.

Pulsed Fields

The technique of pulsed fields is a good procedure for studying the rotational dynamics of polymers. Moreover, it also offers many advantages for the measurement of conducting samples because it minimizes the effects of electric conduction and heating of the sample. The combination of these factors makes pulsed fields the most widely used technique for the measurement of Kerr constants, especially in cases such as biopolymers or polyelectrolytes. Maybe its only disadvantage as compared with static fields is the higher complexity of the required equipment, so that many measurements on nonconducting and low-birefringence systems are performed with static fields.

A wide variety of shapes of pulses can be employed; however, the most frequently used are the square wave and the reverse-polarity square wave.

For the square-wave method, the field is applied almost instantaneously, kept constant for a given time, removed as quickly as possible, and the whole sequence is repeated with an adequate frequency. If the length of the pulse is large enough for the system to reach equilibrium under the field and between the pulses, a sequence of (1) build up, (2) steady state, (3) decay, and (4) isotropic state is repeatedly obtained.

For the technique of the reverse-polarity square wave, a field $\mathbf{E}$ is applied and kept constant for a time t. Then the polarity is reversed so that the field becomes $-\mathbf{E}$ almost instantaneously. It is kept at this value for the same time t. Then the polarity is reversed again, etc. The variation of Φ with time for the field reversal in the case of weak fields and particles having cylindrical symmetry was obtained by Tinoco and Yamaoka [51] as

$$\frac{\Phi(t)}{\Phi(\text{eq})} = 1 + \frac{3\Gamma}{\Gamma + 2}\left[\exp(-6\mathcal{D}t) - \exp(-2\mathcal{D}t)\right] \qquad (7.61)$$

where $\Phi(\text{eq})$ is the equilibrium value for a field of constant strength $|E|$ or $|-E|$.

As Eq. (7.61) indicates, if a molecule has no permanent dipole moment, $\boldsymbol{\mu} =$

0 and, therefore, $\beta = 0$, and taking into account that $\gamma \neq 0$, then $\Gamma = \beta^2/\gamma = 0$, which gives $\Phi(t) = \Phi(\text{eq})$ regardless of the time. Thus, the birefringence produced by a system in which the orientation of the molecules is only due to the interaction with the induced dipole does not change during the field reversal. This result is due to the symmetry of the polarizability, which makes molecules orientate in the same direction for both $+\mathbf{E}$ and $-\mathbf{E}$ fields.

On the contrary, because the permanent dipole moment in the case of polar molecules is asymmetric, the orientation produced by $+\mathbf{E}$ and $-\mathbf{E}$ differs in π. Thus, molecules orientated under $+\mathbf{E}$ are forced to rotate when the field is switched to $-\mathbf{E}$. Consequently, after the field reversal, the birefringence should decrease as the molecules start to rotate, and therefore to disorder, and later should increase again while the molecules orientate with the new polarity. In other words, the birefringence passes through a minimum before again reaching equilibrium. It is easy to show that if $\Gamma \neq 0$, then Eq. (7.61) shows a minimum $\Phi_m = \Phi(t_m)$, with t_m given by

$$\exp[-4\mathcal{D}t_m] = 1/3 \tag{7.62}$$

and the value of Γ can be obtained from the minimum value Φ_m through the relationship

$$\Gamma = \frac{2[1 - \Phi_m/\Phi(\text{eq})]}{0.1547 + \Phi_m/\Phi(\text{eq})} \tag{7.63}$$

Thus, by using reverse-field square waves, it is possible to obtain information about the rotational diffusion coefficient $\mathcal{D}$ and the ratio Γ between contributions of permanent and induced dipole moments.

Sinusoidal Fields

When the electric field applied to the system follows a sine-wave function like $E(t) = E_0 \sin \omega t$, the birefringence produced depends, as in all the cases shown before, on the strength of the field and on the optical, electrical, and rotational characteristics of the sample. But it also depends on the angular frequency ω of the field, and the study of this last dependence can give information about the structure of the molecules on the sample.

The first observation on the variation of the birefringence with ω was reported by Raman and Sirkar [127] in 1928 and since then it has been treated by many using different approaches [45, 87, 128–130]. The treatment developed by Thurston and Bowling [130] for weak fields and cylindrically symmetric particles is presented here because it gives a good compromise between relatively simple equations and possibilities of application. A more general treatment can be found in the revision published by Watanabe and Morita [87].

All the equations are presented as the ratio between the orientation factor at a given frequency ω and time, $\Phi(\omega,t)$, to the value obtained by extrapolation to $\omega \rightarrow$

0, which, of course, becomes independent of time. This ratio is represented by Φ/Φ_0.

The extrapolation to $\omega \to 0$ gives a time-independent field and, therefore, assuming that the field is weak, the limiting value of the orientation factor should be given by a relationship formally identical to Eq. (7.51). Thus,

$$\Phi_0 = (\beta_0^2 + \gamma_0)/15 \tag{7.64}$$

with the only difference that here the terms β_0 and γ_0 representing the contributions from permanent and induced dipole moments are defined as before, Eq. (7.50), but with the maximum strength of the field obtained when $\sin \omega t = 1$. Then

$$\beta_0 = \mu(E_0)_{\text{eff}}/kT \quad \text{and} \quad \gamma_0 = \Lambda'(E_0)^2_{\text{eff}}/2kT \tag{7.65}$$

and, as before, $\Gamma = \beta_2^0/\gamma_0$.

The ratio Φ/Φ_0 is written as the sum of two contributions, one stationary and the second one depending on time:

$$\Phi/\Phi_0 = \Phi_{\text{st}} + \Phi_{\text{alt}} \cos(2\omega_t - \delta) \tag{7.66}$$

where δ is the phase difference between the applied field and the response of the sample. The stationary component is given by

$$\Phi_{\text{st}} = \frac{1}{\Gamma + 1}\left(1 + \frac{4\mathscr{D}^2\Gamma}{\omega^2 + 4\mathscr{D}^2}\right) \tag{7.67}$$

so that, when $\omega \to 0$, $\Phi_{\text{st}} \to 1$.

The time-dependent component is calculated as

$$\Phi_{\text{alt}} = (\Phi_1^2 + \Phi_2^2)^{1/2} \tag{7.68}$$

and the phase difference is given by

$$\tan \delta = \Phi_2/\Phi_1 \tag{7.69}$$

The expressions for Φ_1 and Φ_2 are

$$\Phi_1 = \frac{6\mathscr{D}^2(6\mathscr{D}^2 - \omega^2)\Gamma + 9\mathscr{D}^2(4\mathscr{D} + \omega^2)}{(\Gamma + 1)(4\mathscr{D}^2 + \omega^2)(9\mathscr{D}^2 + \omega^2)}$$

$$\Phi_2 = \frac{30\mathscr{D}^3\omega\Gamma + 3\mathscr{D}\omega(4\mathscr{D}^2 + \omega^2)}{(\Gamma + 1)(4\mathscr{D}^2 + \omega^2)(9\mathscr{D}^2 + \omega^2)} \tag{7.70}$$

The behavior of Φ as a function of ω depends on the electric structure of the molecule as represented by parameter Γ. Let us examine the two extreme cases of $\Gamma = 0$ and $\Gamma \to \infty$.

If the molecule has no permanent dipole moment so that its orientation with the field is produced only by interaction with the induced dipole, then $\mu = 0$, $\beta_0 = 0$, and $\Gamma = 0$. Substituting these values in the preceding equations gives

$$\frac{\Phi}{\Phi_0} = 1 + \frac{3\mathscr{D}}{(9\mathscr{D}^2 + \omega^2)^{1/2}} \cos{(2\omega t - \delta)} \tag{7.71}$$

with

$$\tan{\delta} = \omega/3\mathscr{D}$$

$$\Phi_0 = \frac{\gamma_0}{15} = \frac{\Phi'}{30kT}(E_0)^2_{\text{eff}} = \frac{\Lambda'}{15\,kT}\left(\frac{E_0}{2^{1/2}}\right)^2_{\text{eff}} \tag{7.72}$$

The physical situation of the sample is as follows: The polarization of the molecules by effect of the field can be considered as instantaneous. This may not be true in the case of fields oscillating at optic frequencies obtained by using powerful lasers as orientating fields, but it holds for the most commonly used fields oscillating in the range of acoustic frequencies. Thus, the induced dipole, the orientation of the molecules, and, therefore, the birefringence produced reaches a maximum value twice each cycle of the electric field, one with $+E_0$ and a second time with $-E_0$. Consequently, all of these magnitudes oscillate with twice the angular frequency of the field, as indicated by Eq. (7.71).

The orientation of the molecules is the same for $+E_0$ and $-E_0$, however, the molecules orientated under the maximum strength of the field (either positive or negative) tend to disorient when the field decreases toward zero and to orientate again when the strength of the field increases. Thus, there is a rotational motion of the molecules following the oscillations of the field. But, because the movement of the molecules takes a finite time, the rotations lag behind the field with a phase difference δ, which, as indicated in Eq. (7.72), increases with increasing ω and with decreasing $\mathscr{D}$ (i.e., with decreasing rotational mobility); therefore, the coefficient $\mathscr{D}$ can be obtained by measuring δ as a function of ω.

At high frequencies, $\omega \to \infty$, the rotation of the molecules cannot follow the oscillations of the field so that they stay orientated during the whole cycle. In these conditions, the fraction on the right-hand side of Eq. (7.71) tends to zero and Φ tends to the time-independent value Φ_0, which, as a comparison between Eqs. (7.72) and (7.51) indicates, is the orientation that a weak static field of $E_0/2^{1/2}$ or E_{rms} would produce. At lower frequencies, the mean value of Φ is still Φ_0, which is reached every time that $\cos{(2\omega t - \delta)} = 0$, but it oscillates between a maximum and a minimum value obtained, respectively, when the value of the cosine is $+1$ and -1. As the frequency decreases, the difference between Φ_{max} and Φ_{min} increases, and in the limit $\omega \to 0$, the value of Φ oscillates between $2\Phi_0$ and 0.

Figure 7.13 shows the variation of Φ_{max} and Φ_{min} with ω according to Eq. (7.71). The decrease of Φ with increasing ω for very high frequencies, shown as a dashed line in this figure, is not predicted by the theory; it is experimentally found in solutions of nonpolar polymers such as the tobacco mosaic virus and it is attributed to the relaxation of the ionic atmosphere surrounding the macromolecules, although no theoretical treatment has been developed yet for this region.

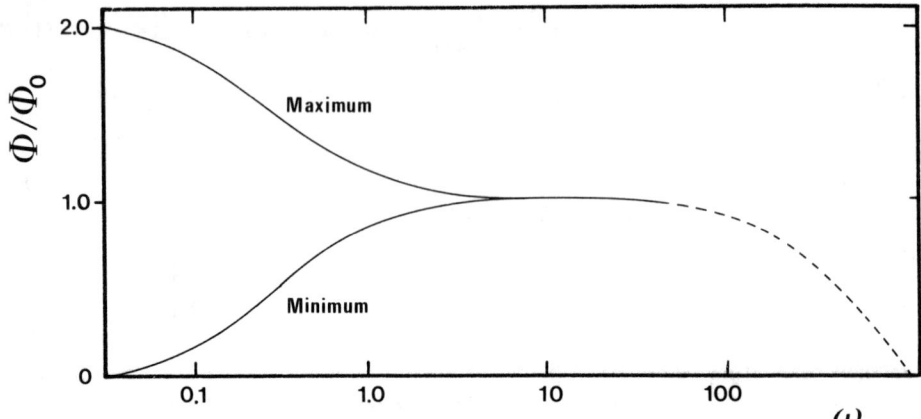

Figure 7.13   Maximum and minimum values of the orientation factor Φ for a nonpolar polymer under a sine-wave electric field of angular frequency ω. The decrease of Φ with increasing ω at high frequencies, shown as a dashed line, is not predicted by theory, although it is experimentally found and attributed to the relaxation of the ionic atmosphere.

For very polar molecules, $\beta_0 \gg \gamma_0$ and $\Gamma \to \infty$. Taking this limit in Eqs. (7.66) to (7.70), one obtains

$$\frac{\Phi}{\Phi_0} = \frac{4\mathcal{D}^2}{\omega^2 + 4\mathcal{D}^2} + \frac{6\mathcal{D}^2}{[(4\mathcal{D}^2 + \omega^2)(9\mathcal{D}^2 + \omega^2)]^{1/2}} \cos(2\omega t - \delta) \qquad (7.73)$$

with

$$\tan \delta = \frac{5\omega\mathcal{D}}{6\mathcal{D}^2 - \omega^2}$$

$$\Phi_0 = \frac{\beta_0^2}{15} = \frac{\mu^2 (E_0)_{\text{eff}}^2}{15 k^2 T^2} \qquad (7.74)$$

In this case, as in the previous one, the molecules are oriented in the direction of the field twice each cycle and, consequently, the response of the sample oscillates with twice the frequency of the field. Again, rotational friction is responsible for the phase difference δ between the field and the birefringence, which increases with increasing frequency, although this time it is not a linear function of ω as Eq. (7.74) indicates. However, due to the asymmetry of the dipole moment, the direction of the orientations with $+E_0$ and $-E_0$ differ in π; therefore, the molecules are forced to perform a full rotation (or a $+\pi$ rotation followed by a $-\pi$ rotation) during each cycle. Thus, when the frequency increases and the molecules cannot follow the field, they do not keep their orientation as in the case of the induced dipole. On the contrary, they stay disoriented as a response to the two opposite pulls of the field.

Consequently, at $\omega \to \infty$, both the static and the time-dependent components of Φ tend to zero, as Eq. (7.73) indicates.

When the frequency is low enough for the molecules to follow the oscillations of the field, they cross twice in each cycle in the direction in which they are aligned perpendicular to the field so that the birefringence produced in those moments has the opposite sign to that induced by a static field. This behavior is shown in Eq. (7.73) by the maximum and minimum values of the time-dependent component, which are larger in absolute value than the static one, and therefore dominates the total response of the sample.

In the limit $\omega \to 0$, the static component tends to unity while the time-dependent component oscillates between 1 and -1. Therefore, Φ oscillates between 0 and $2\Phi_0$ as in the previous case, with a mean value of Φ_0, which is given by Eq. (7.74).

Figure 7.14 shows the stationary and time-dependent components and the

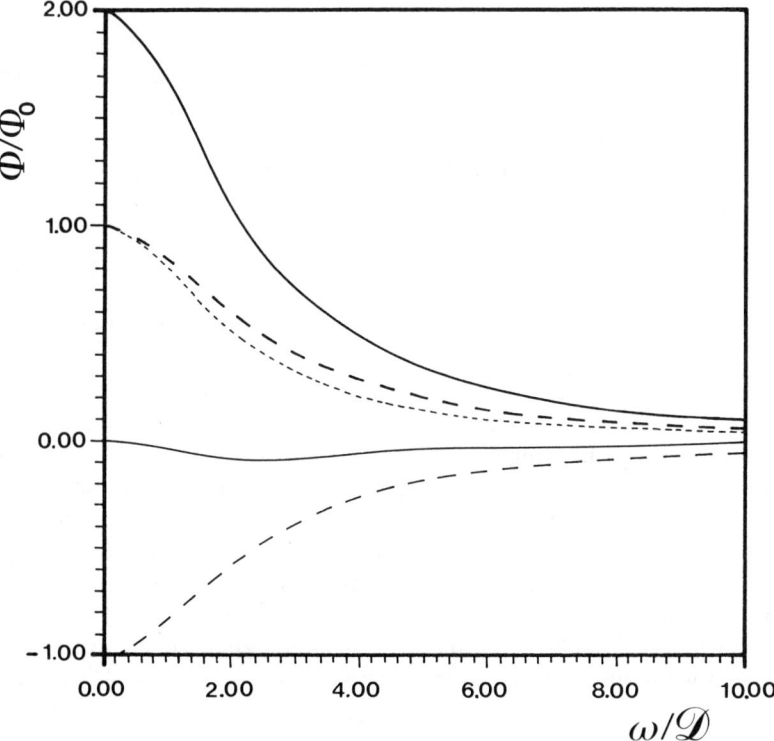

Figure 7.14 Maximum and minimum values of the total orientation factor (solid lines) and its time-dependent (dashed lines) and stationary (dotted line) components as functions of the angular frequency of the electric field for a molecule whose orientation is produced only by its permanent dipole moment.

total value of Φ as functions of angular frequency of the field ω computed according to Eq. (7.73).

Flexible Polymers

When the electric field is applied to a solution containing macromolecules, each group within the chain interacts with the field and tends to orientate in the direction of minimum energy according to its own electrical properties. However, because the groups are linked together forming a chain, they cannot move and orientate independently; the whole polymer, or at least long fragments of the chain in coupled motions, has to rotate in the direction of the field. If the polymer is flexible, the conformations giving stronger interactions with the field will be favored. Thus, the field not only orientates the macromolecules of the sample, it also perturbs their conformational statistics. This is not exclusive of macromolecules. On the contrary, any molecule having more than one possible conformation originated by rotations over single bonds (1,2-dichloroethane for example), and having different values of permanent dipole moment or polarizability, will be perturbed by the effect of the field since the energy of interaction will depend on the conformation of the molecule. Thus, the relative energy of the possible conformations will be changed when the field is applied.

Therefore, the calculation of the Kerr constant of a flexible polymer requires a detailed knowledge of the following characteristics:

(a) electric and optical properties of all the groups of the chain
(b) relative energy of the conformations available to the polymer
(c) effect of the electric field over the different conformations of the chain.

Consequently, the calculation of $_mK$ is, in general, more complicated than for many other conformation-dependent properties; however, it is also very sensitive to the microstructure of the sample. Therefore, it gives a potentially useful criterion of molecular structure and spatial configuration. This behavior is based on the fact that the molar Kerr constant is a molecular property whose range of values for low molecular-weight substances (about the size of a monomer unit) varies over four orders of magnitude and can be either positive or negative in sign [62]; therefore, the reinforcement or cancellation between contributions of different units becomes very strongly dependent on geometry, conformation, sequence, etc., of those units.

Three approximations are usually employed in order to simplify the calculations:

(a) *The applied field is weak and static.* Thus, the energy of interaction is much smaller than kT and does not change with time. Moreover, such a weak interaction is unable to modify either bond lengths or bond angles, although it is strong enough to change the distribution of rotational states over single bonds of the polymers.

(b) *Neglecting interactions between macromolecules.* The effective electric field act-ing over each macromolecule is taken to be different from the externally applied field E, so that the macromolecule may be immersed in a material medium rather than completely isolated (it may be in solution rather than in a gaseous state). However, every macromolecule acts independently of every other macromolecule present in the sample. Thus, the results of the calculations are compared with experimental values measured in solution and extrapo-lated to infinite dilution according to Eq. (7.25). Strictly speaking, the measurements should be made in theta conditions because neither polymer-solvent nor intramolecular long-range interactions are taken into account; however, it is usually assumed that Kerr constants are unaffected by excluded volume effects and measurements are carried out in nontheta condi-tions.

(c) *Neglecting hyperpolarizability contributions.* Thus, the dipole moment induced in the macromolecule by the effect of the electric field is taken to be given by Eq. (7.3) instead of (7.11). Molecules of polymers frequently have permanent dipole moments, or at least are highly polarizable. Under these circumstances, extremely high electric fields are required for the hyperpolarizability terms to be significant. Therefore, for most experimental determina-tions, the linear effect of the electric field is a good approximation.

The first calculations of Kerr constants of flexible macromolecules were per-formed assuming that polymeric chains were formed by a number of segments, each having cylindrical symmetry and linked according to a very simplified conforma-tional model. Thus, Stuart and Peterlin [131] treated a freely jointed chain, Dows [113] assumed a freely rotating chain, and Gotlib [132] treated the polymer chain whose bond rotations are hindered but remain independent of neighboring bond rotational states.

The calculation of $_mK$ with these models is rather simple. Thus, using Eq. (7.51) for the orientation factor Φ under weak static fields, and substituting it into Eq. (7.43) to obtain $\Delta\alpha$ and then into Eq. (7.18), one obtains the contribution of each segment of the polymeric chain as

$$_mK_i = \frac{4\pi N}{45kT} \left(\frac{\mu_i^2}{kT} + \Lambda_i' \right) \Lambda_i \tag{7.75}$$

where μ_i, Λ_i, and Λ_i' are, respectively, the permanent dipole moment, the elec-trooptic and the electrostatic anisotropies of the ith segment whose contribution to the Kerr constant of the polymer is represented by $_mK_i$.

If the chain contains N_s identical segments freely jointed, the average of all the magnitudes appearing in Eq. (7.75) will increase with N_s and, therefore, the Kerr constant of the whole chain is given simply by

$$\langle _mK \rangle = \frac{4\pi N N_s}{45kT} \left(\frac{\mu_i^2}{kT} + \Lambda_i' \right) \Lambda_i \tag{7.76}$$

If the segments form a rigid rod, the averages depend on N_s^2, and, therefore,

$$\langle _mK \rangle = \frac{4\pi N N_s^2}{45kT} \left(\frac{\mu_i^2}{kT} + \Lambda_i' \right) \Lambda_i \tag{7.77}$$

The first procedure of the calculation allowing the use of realistic models for the conformational characteristics of the polymer was developed by Nagai and Ishikawa [93] and later improved by Flory and co-workers [95, 96, 133]. It consists in the evaluation of the optical anisotropy $\Delta\alpha$ (i.e., $\Delta\alpha = \alpha_{xx} - \alpha_{yy}$ in the laboratory reference frame) averaged over all possible orientations of the macromolecule inside the field and over all possible conformations of the chain.

The energy V_E of the macromolecule in a given conformation, determined by a set of internal rotational angles ϕ_{int}, and a fixed orientation with respect to the field, determined by a set of Euler angles Ω_{ext}, can be calculated as the sum of the energy V that the chain, in the same conformation, would have in the absence of the field, and two perturbation terms representing the interaction of the electric field with both permanent and induced dipole moments. Thus, neglecting hyperpolarizability terms,

$$V_E = V - \mu_x E_{eff} - \tfrac{1}{2}\alpha'_{xx}E^2_{eff} \tag{7.78}$$

where μ_x and α'_{xx} represent the components of the dipole moment and the electrostatic polarizability tensor in the direction of the field (the direction of the x axis).

The average of any magnitude can be obtained by integration over both internal (ϕ_{int}) and external (Ω_{ext}) coordinates. Thus, for instance, the partition function Z_E is given by the integral of a Boltzmann function of V_E as

$$Z_E = \frac{1}{8\pi^2} \int_{ext} \cdots \int_{int} \exp\left(\frac{-V_E}{kT}\right) d\phi_{int}\, d\Omega_{ext} \tag{7.79}$$

where the constant $1/8\pi^2$ is a normalization factor for the external coordinates. By taking into account that the interactions of the field with either permanent or induced dipole moments are much smaller than kT, the Boltzmann function can be represented by a series expansion neglecting terms higher than E^2_{eff}:

$$\exp\left(\frac{-V_E}{kt}\right) = \exp\left(\frac{-V}{kt}\right)\left[1 + \mu_x\frac{E_{eff}}{kT} + \frac{1}{2}\left(\frac{\mu_x E_{eff}}{kT}\right)^2 + \frac{1}{2}\alpha'_{xx}\frac{E^2_{eff}}{kT}\right] \tag{7.80}$$

Substituting Eq. (7.80) into (7.79) and integrating gives the value of Z_E. The term on μ_x vanishes upon integration:

$$\int_{int} \mu_x \frac{E_{eff}}{kT}\exp\left(\frac{-V}{kT}\right)d\phi_{int} = \frac{E_{eff}}{k_T}\int_{int}\mu_x\exp\left(\frac{-V}{kT}\right)d\phi_{int} \tag{7.81}$$

$$= Z\frac{E_{eff}}{kT}\langle\mu_x\rangle = 0$$

because the integral in Eq. (7.81) divided by the partition function in the unperturbed (i.e., in absence of field) state Z represents the average of μ_x over all possible conformations of the chain, each of them weighted with a Boltzmann factor of the energy V that the polymer has in the unperturbed state. Thus, it is the unperturbed average $\langle \mu_x \rangle$ that is zero because in the absence of field, the three directions of space are equivalent, and for each conformation having a given value μ_x, there will be another one with the same energy and $-\mu_x$.

Integrating the other terms of Eq. (7.80) gives

$$\int_{int} \exp\left(\frac{-V}{kT} \right) d\phi_{int} = Z$$

$$\int_{int} \frac{1}{2} \left(\frac{\mu_x E_{eff}}{kT} \right)^2 \exp\left(\frac{-V}{kT} \right) d\phi_{int} = \frac{Z}{2} \left(\frac{E_{eff}}{kT} \right)^2 \langle \mu_x^2 \rangle_0 = \frac{Z}{6} \left(\frac{E_{eff}}{kT} \right)^2 \langle \mu^2 \rangle$$

$$\int_{int} \frac{1}{2} \alpha'_{xx} \left(\frac{E_{eff}^2}{kT} \right) \exp\left(\frac{-V}{kT} \right) d\phi_{int} = \frac{Z}{2} \langle \alpha'_{xx} \rangle \frac{E_{eff}^2}{kT} = \frac{ZE_{eff}^2}{6kT} \langle \text{Trace}\,(\alpha') \rangle$$

where the following relationships, validated by the symmetry of space in the absence of an electric field, have been used:

$$\langle \mu_x^2 \rangle = \langle \mu_y^2 \rangle = \langle \mu_z^2 \rangle = \tfrac{1}{3} \langle \mu^2 \rangle$$
$$\langle \alpha'_{xx} \rangle = \langle \alpha'_{yy} \rangle = \langle \alpha'_{zz} \rangle = \tfrac{1}{3} \langle \text{Trace}\,(\alpha') \rangle$$

Substituting these results into Eq. (7.79) and integrating over the external coordinates Ω_{ext}, taking into account that nothing on the integral depends on Ω_{ext} and that the expression is normalized with respect to these coordinates, gives finally the partition function in presence of the field:

$$Z_E = Z \left[1 + \frac{1}{6} \left(\frac{E_{eff}}{kT} \right)^2 \langle \mu^2 \rangle + \frac{E_{eff}^2}{6kT} \langle \text{Trace}(\alpha') \rangle \right] \tag{7.82}$$

This expression reduces to $Z_E = Z$ in the trivial case ($E_{eff} = 0$) or in the hypothetical case of a nonpolar ($\mu = 0$) and nonpolarizable ($\alpha' = 0$) molecule.

The required average of $\Delta\alpha$ can be calculated as

$$\langle \Delta\alpha \rangle_E = \langle \alpha_{xx} - \alpha_{yy} \rangle_E = \frac{1}{8\pi^2 Z_E} \int_{ext} \cdots \int_{int} (\alpha_{xx} - \alpha_{yy}) \exp\left(\frac{-V_E}{kT} \right) d\phi_{int}\, d\Omega_{ext}$$

$$\tag{7.83}$$

By using the series expansion for V_E, Eq. (7.80), and taking into account that the average $\langle \alpha_{xx} - \alpha_{yy} \rangle$ and $\langle (\alpha_{xx} - \alpha_{yy})\mu_x \rangle$ will vanish by the symmetry of space, Eq. (7.83) can be rewritten as

$$\langle \Delta\alpha \rangle_E = \frac{E_{eff}^2}{16\pi^2 Z_E} \int_{ext} \cdots \int_{int} (\alpha_{xx} - \alpha_{yy}) \left(\frac{\mu_x^2}{k^2 T^2} + \frac{\alpha'_{xx}}{kT} \right) \exp\left(\frac{-V}{kT} \right) d\phi_{int}\, d\Omega_{ext}$$

$$\tag{7.84}$$

The component μ_x^2 can be written as a function of the module of the total dipole moment of the chain μ and the angle τ_x that the direction of μ makes with the

x axis:

$$\mu_x^2 = (\mu \cos \tau_x)^2 \tag{7.85}$$

In a similar way, the components of the $\boldsymbol{\alpha}$ tensor in the laboratory coordinate system xyz can be written as functions of its components in the molecular main frame XYZ, in which the tensor is diagonal with nonzero elements α_X, α_Y, and α_Z and the angles τ_X^x, τ_X^y, τ_X^z, τ_Y^x . . . that the axes of the main frame make with those of the laboratory system:

$$\alpha_{xx} - \alpha_{yy} = \alpha_X(\cos^2\tau_X^x - \cos^2\tau_X^y) + \alpha_Y(\cos^2\tau_Y^x -$$
$$\cos^2\tau_Y^y) + \alpha_Z(\cos^2\tau_Z^x - \cos^2\tau_Z^y) \tag{7.86}$$

Thus, representing the three axes of the molecular main frame by $i = X, Y, Z$, the term $(\alpha_{xx} - \alpha_{yy})\mu_x^2$ appearing in Eq. (7.84) can be written as

$$(\alpha_{xx} - \alpha_{yy}) \, \mu_x^2 \sum_i \alpha_i \cos^2\tau_x \, (\cos^2\tau_i^x - \cos^2\tau_i^y) \tag{7.87}$$

Performing the averages over the external coordinates Ω_{ext} (i.e., over all possible orientations of $\boldsymbol{\mu}$ and XYZ relative to the laboratory coordinate system xyz) for a fixed conformation of the chain (i.e., for fixed values of the ϕ_{int} coordinates) gives

$$\langle(\alpha_{xx} - \alpha_{yy}) \, \mu_x^2\rangle_\phi = \frac{\mu^2}{15} \sum_i \alpha_i[(2 \cos^2\tau_i^\mu + 1) - (2 - \cos^2\tau_i^\mu)]$$

$$= \frac{\mu^2}{15} \sum_i (3\alpha_i \cos^2\tau_i^\mu - \alpha_i) \tag{7.88}$$

where τ_i^μ represents the angle that vector $\boldsymbol{\mu}$ makes with the main axis $i(X, Y, \text{or } Z)$ of the molecular reference frame. Thus, if the dipole moment of the molecule is written in the XYZ system, $\tau_i^\mu = 0$, and one can write

$$\mu^2 \sum_i \alpha_i \cos^2\tau_i^\mu = \sum_i \mu_i\alpha_i\mu_i = [\mu_X\mu_Y\mu_Z] \begin{bmatrix} \alpha_X & 0 & 0 \\ 0 & \alpha_Y & 0 \\ 0 & 0 & \alpha_Z \end{bmatrix} \begin{bmatrix} \mu_X \\ \mu_Y \\ \mu_Z \end{bmatrix}$$

$$= (\boldsymbol{\mu}^T\boldsymbol{\alpha}\boldsymbol{\mu})_{XYZ} \tag{7.89}$$

where the subscripts X, Y, and Z indicate the reference frame in which both $\boldsymbol{\alpha}$ and $\boldsymbol{\mu}$ are written. However, the value of the product does not change with the coordinate system used to define $\boldsymbol{\alpha}$ and $\boldsymbol{\mu}$ (as far as the same system is used for both magnitudes). Therefore, the subscript can be neglected and the product can be computed in any suitable coordinates system. On the other hand, the second term of Eq. (7.88) can be written as

$$\mu^2 \sum_i \alpha_i = \mu^2(\alpha_X + \alpha_Y + \alpha_Z) = \mu^2 \text{ Trace } (\boldsymbol{\alpha}) \tag{7.90}$$

which is also invariant with respect to the coordinates system.

Thus, substituting Eqs. (7.89) and (7.90) into (7.88) gives

$$\langle(\alpha_{xx} - \alpha_{yy})\,\mu_x^2\rangle_\phi = \tfrac{1}{15}[3(\boldsymbol{\mu}^T\boldsymbol{\alpha}\boldsymbol{\mu}) - \mu^2 \text{ Trace }(\boldsymbol{\alpha})] \qquad (7.91)$$

The integration of Eq. (7.91) over the internal coordinates ϕ_{int}, as required by Eq. (7.84), can be performed exactly in the same way as if there were no field because neither Eq. (7.91) nor the Boltzmann factor used in the integration depend on the electric field. The procedure for this integration is shown in what follows. At this point, we will represent the resulting average by the symbols $\langle\ \rangle$, indicating that the averaging is performed in absence of the field. Thus,

$$\langle(\alpha_{xx} - \alpha_{yy})\mu_x^2\rangle = \tfrac{1}{15}[3\langle\boldsymbol{\mu}^T\boldsymbol{\alpha}\boldsymbol{\mu}\rangle - \langle\mu^2 \text{ Trace }(\boldsymbol{\alpha})\rangle] \qquad (7.92)$$

Using the same procedure for the second term in the second set of parentheses in Eq. (7.84), one obtains the average over the external coordinates Ω_{ext} at a fixed conformation:

$$\langle(\alpha_{xx} - \alpha_{yy})\,\alpha'_{xx}\rangle_\phi = \frac{1}{15} \sum_i \sum_j \alpha_i\alpha'_j(3\cos^2\tau_j^i - 1) \qquad (7.93)$$

τ_j^i is the angle between axis i of tensor $\boldsymbol{\alpha}_i$ and axis j of $\boldsymbol{\alpha}'_j$, but since both tensors have the same main axes, $\tau_j^i = 0$ for $i = j$ and $\tau_j^i = 90°$ for $i \neq j$. Thus, Eq. (7.93) can be rewritten as

$$\begin{aligned}
\langle(\alpha_{xx} - \alpha_{yy})\alpha'_{xx}\rangle_\phi &= \tfrac{1}{15}[3(\alpha_X\alpha'_X + \alpha_y\alpha'_Y + \alpha_Z\alpha'_Z) - (\alpha_X + \alpha_Y + \alpha_Z) \\
&\quad (\alpha'_X + \alpha'_Y + \alpha'_Z)] \\
&= \tfrac{1}{15}[3\text{ Trace }(\boldsymbol{\alpha}\boldsymbol{\alpha}') - \text{Trace }(\boldsymbol{\alpha})\text{ Trace }(\boldsymbol{\alpha}')] \qquad (7.94)
\end{aligned}$$

and the result is independent of the coordinate system used to write $\boldsymbol{\alpha}$ and $\boldsymbol{\alpha}'$. Averaging over all the possible conformations of the polymer, one obtains

$$\langle(\alpha_{xx} - \alpha'_{yy})\alpha'_{xx}\rangle = \tfrac{1}{15}[3\,\langle\text{Trace }(\boldsymbol{\alpha}\boldsymbol{\alpha}')\rangle - \langle\text{Trace }(\boldsymbol{\alpha})\rangle\langle\text{Trace }(\boldsymbol{\alpha}')\rangle] \qquad (7.95)$$

Substituting Eqs. (7.92) and (7.95) into (7.84), taking into account that the normalization factors for the integrals over external, Ω_{ext}, and internal, ϕ_{int}, coordinates are, respectively, $8\pi^2$ and Z, one obtains

$$\begin{aligned}
\langle\Delta\alpha\rangle_E = \frac{E_{eff}^2}{30}\frac{Z}{Z_E}\Bigg\{ &\frac{1}{k^2T^2}[3\,\langle\boldsymbol{\mu}^T\boldsymbol{\alpha}\boldsymbol{\mu}\rangle - \langle\mu^2 \text{ Trace }(\boldsymbol{\alpha})\rangle] \\
&+ \frac{1}{kT}[3\,\langle\text{Trace }(\boldsymbol{\alpha}\boldsymbol{\alpha}')\rangle - \langle\text{Trace }(\boldsymbol{\alpha})\rangle\langle\text{Trace }(\boldsymbol{\alpha}')\rangle]\Bigg\}
\end{aligned} \qquad (7.96)$$

The ratio Z/Z_E in Eq. (7.96) usually is taken as unity. This approximation is reasonable when kT is much larger than the energy of interaction between the electric field with both permanent and induced dipole moments. Under these conditions, terms different from unity can be neglected in Eq. (7.82). On the other hand, the assumption that kT is much larger than the interaction energies has already been used when the Boltzmann factor of V_E was written as a series expansion neglecting terms higher than E_{eff}^2; see Eq. (7.80).

Equation (7.96) can be simplified through the definition of the anisotropic parts of the polarizability tensors $\boldsymbol{\alpha}$ and $\boldsymbol{\alpha}'$ as

$$\hat{\boldsymbol{\alpha}} = \boldsymbol{\alpha} - \tfrac{1}{3}\, \text{Trace}\,(\boldsymbol{\alpha})\, \mathbf{I}_3 \qquad (7.97)$$

and

$$\hat{\boldsymbol{\alpha}} = \boldsymbol{\alpha}' - \tfrac{1}{3}\, \text{Trace}\,(\boldsymbol{\alpha}')\, \mathbf{I}_3 \qquad (7.98)$$

where $\mathbf{I}_3$ represents a 3×3 identity matrix.

Then, representing $\tfrac{1}{3}\,\text{Trace}\,(\boldsymbol{\alpha})$ by a and $\tfrac{1}{3}\,\text{Trace}\,(\boldsymbol{\alpha}')$ by a', one obtains

$$
\begin{aligned}
\langle \boldsymbol{\mu}^T \boldsymbol{\alpha} \boldsymbol{\mu} \rangle &= \langle \boldsymbol{\mu}^{\mathrm{T}}(\hat{\boldsymbol{\alpha}} + a\mathbf{I}_3)\boldsymbol{\mu} \rangle = \langle \boldsymbol{\mu}^T \hat{\boldsymbol{\alpha}} \boldsymbol{\mu} \rangle + \langle a\boldsymbol{\mu}^T \mathbf{I}_3 \boldsymbol{\mu} \rangle \\
&= \langle \boldsymbol{\alpha}^T \hat{\boldsymbol{\alpha}} \boldsymbol{\mu} \rangle + \langle a(\mu_x^2 + \mu_y^2 + \mu_z^2) \rangle \\
&= \langle \boldsymbol{\mu}^T \hat{\boldsymbol{\alpha}} \boldsymbol{\mu} \rangle + \tfrac{1}{3}\langle \mu^2\, \text{Trace}\,(\boldsymbol{\alpha}) \rangle
\end{aligned} \qquad (7.99)
$$

$$
\begin{aligned}
\langle \text{Trace}\,(\boldsymbol{\alpha}\boldsymbol{\alpha}') \rangle &= \langle \text{Trace}\,[(\hat{\boldsymbol{\alpha}} + a\mathbf{I}_3)(\hat{\boldsymbol{\alpha}}' + a'\mathbf{I}_3)] \rangle \\
&= \langle \text{Trace}\,(\hat{\boldsymbol{\alpha}}\hat{\boldsymbol{\alpha}}') \rangle + \langle a\, \text{Trace}\,(\hat{\boldsymbol{\alpha}}'\mathbf{I}_3) + a'\, \text{Trace}\,(\hat{\boldsymbol{\alpha}}\mathbf{I}_3) \\
&\quad + aa'\, \text{Trace}\,(\mathbf{I}_3\mathbf{I}_3) \rangle \\
&= \langle \text{Trace}\,(\hat{\boldsymbol{\alpha}}\hat{\boldsymbol{\alpha}}') \rangle + \langle a\, \text{Trace}\,(\hat{\boldsymbol{\alpha}}') + a'\, \text{Trace}\,(\hat{\boldsymbol{\alpha}}) + 3aa' \rangle \quad (7.10) \\
&= \langle \text{Trace}\,(\hat{\boldsymbol{\alpha}}\hat{\boldsymbol{\alpha}}') \rangle + \langle 3aa' \rangle \\
&= \langle \text{Trace}\,(\hat{\boldsymbol{\alpha}}\hat{\boldsymbol{\alpha}}') \rangle + \tfrac{1}{3}\langle \text{Trace}\,(\hat{\boldsymbol{\alpha}}') \rangle\langle \text{Trace}\,(\hat{\boldsymbol{\alpha}}) \rangle
\end{aligned}
$$

because, as defined by Eqs. (7.97) and (7.98), both $\hat{\boldsymbol{\alpha}}$ and $\hat{\boldsymbol{\alpha}}'$ are traceless tensors.

Substituting Eqs. (7.99) and (7.100) together with the approximation $Z/Z_E \approx 1$ into Eq. (7.96) and then into Eq. (7.18) gives the average of the Kerr constant as

$$\langle {}_mK \rangle = \frac{2\pi N}{15kT}\left[\, \frac{1}{kT}\langle \boldsymbol{\mu}^T \hat{\boldsymbol{\alpha}} \boldsymbol{\mu} \rangle + \langle \text{Trace}\,(\hat{\boldsymbol{\alpha}}\hat{\boldsymbol{\alpha}}') \rangle \right] \qquad (7.101)$$

As Eq. (7.101) indicates, the averages required to calculate $\langle {}_mK \rangle$ can be obtained with the conformational statistics of the polymer in the absence of the electric field despite the fact that the field perturbs the energy of those conformations.

The electrostatic polarizability $\boldsymbol{\alpha}'$ is eliminated by assuming that the atomic contribution amounts to about 10% of the electronic polarizability and, therefore, $\boldsymbol{\alpha}' \approx 1.10\boldsymbol{\alpha}$. With all these convections, Eq. (7.101) is customarily written as

$$\langle {}_mK \rangle = \frac{2\pi N}{15kT}\left[\, \frac{1}{kT}\langle \boldsymbol{\mu}^T \hat{\boldsymbol{\alpha}} \boldsymbol{\mu} \rangle + 1.10\,\langle \text{Trace}\,(\hat{\boldsymbol{\alpha}}\hat{\boldsymbol{\alpha}}) \rangle \right] \qquad (7.102)$$

Most of the published values of dipole moments $\boldsymbol{\mu}$ and polarizabilities $\boldsymbol{\alpha}$ are given in Debyes and Å^3, respectively; thus, writing the values of the numerical constants of Eq. (7.102) and incorporating the conversion factors to SI units, one obtains

$$\langle_m K\rangle = \frac{2033.1244}{T}\left[\frac{7243.7512}{T}\langle\boldsymbol{\mu}^T\hat{\boldsymbol{\alpha}}\boldsymbol{\mu}\rangle + 1.10\langle\text{Trace }(\hat{\boldsymbol{\alpha}}\hat{\boldsymbol{\alpha}})\rangle\right]$$

$$\times\ 10^{-27}\ m^5\ V^{-2}\ mol^{-1}$$

(7.103)

when $\boldsymbol{\mu}$ is given in Debyes and $\hat{\boldsymbol{\alpha}}$ in $\mathring{A}^3$.

The magnitudes appearing on Eq. (7.103) are calculated using Flory's matrix-multiplication scheme [95, 96] together with the valence optical scheme (VOS), whose main assumption is that the $\boldsymbol{\alpha}$ tensor for the whole chain can be computed by addition of contributions $\boldsymbol{\alpha}_i$ from each skeletal bond of the polymer. The method of calculation is detailed in the Appendix. In the next section, the limitations of the VOS are shown.

Additivity of Bond Contributions: Valence Optical Scheme (VOS)

The whole procedure for the calculation of Kerr constants (and many other properties such as dimensions or dipole moments) of polymers rests upon the additivity of the contributions assigned to each unit of the polymeric chain. Let us examine the three kinds of contributions most commonly used in these calculations.

The additivity of bond lengths l_i required to compute values of the end-to-end vector **r** for the polymer is usually accepted without discussion. Bond lengths are obtained from crystallographic data of model compounds having structures similar to the repeating unit of the polymer to be studied and they are assumed to be independent of the molecular weight and the conformation of the chain.

The use of bond dipole moments $\boldsymbol{\mu}_i$ is more troublesome. In fact, bond dipoles *are not* additive. For instance, it is impossible to assign dipole moments to C=O and C—O bonds that could simultaneously reproduce the experimental dipoles of ketones, esters, and ether molecules. The reason is that when two polar bonds are close in a molecule, for instance, O=C—O in an ester, each of them modifies the polarity of the other. This modification of polarity is known as the *inductive effect* and prevents the possibility of using tables of bond dipoles similar to those of bond lengths. This difficulty is usually overcome using groups instead of bond dipole moments and assuming that the inductive effects do not propagate across a few nonpolar bonds. For instance, in the calculation of the dipole moment of poly(methyl acrylate) (PMA) chains, the dipole moment of each ester group is assumed to be unaffected by its neighbors. Then the experimental dipole moments of ester molecules such as methyl propionate or methyl isobutyrate are taken as contribution $\boldsymbol{\mu}_i$ for skeletal bond RCH—CH_2 and a value $\boldsymbol{\mu}_{i+1} = 0$ is used for the second bond of the repeating unit (skeletal bond CH_2—CHR).

A second difference between bond lengths and dipoles is that in the former, the directions of the l_i vectors are perfectly defined, whereas in the latter, it is relatively frequent that the asymmetry of the group precludes an unambiguous assignation of the direction of the dipole moment, for instance, in the case of ester groups. In those cases, the direction of the dipole should also be determined either by evaluation of the charge distribution on the group using any quantum mechanics procedure or through the analysis of experimental dipole moments of suitable model molecules [134, 135].

By bearing in mind that group contributions with well-determined directions should be used, the additivity of $\boldsymbol{\mu}_i$ contributions is widely accepted.

The case of $\hat{\boldsymbol{\alpha}}_i$ contributions to the anisotropic part of the optical polarizability tensor requires a more detailed analysis. In fact, the procedure for the calculation of optical properties based upon additivity of the $\hat{\boldsymbol{\alpha}}_i$ contributions is usually known as the *valence optical scheme* (VOS), even if it is conceptually identical to the case of the dipole moment or the end-to-end distance. The difference is that the inductive effects preventing the additivity of contributions are more important, and up to now less understood, in $\hat{\boldsymbol{\alpha}}_i$ than in $\boldsymbol{\mu}_i$ contributions. A great deal of effort is being spent to obtain group contributions $\hat{\boldsymbol{\alpha}}_i$ that take into account all the required inductive effects so that they could be added in a way similar to what is done for dipole moments.

The anisotropic part of the polarizability tensor $\hat{\boldsymbol{\alpha}}$ of a given molecule is defined, see Eq. (7.97), as

$$\hat{\boldsymbol{\alpha}} = \boldsymbol{\alpha} - \tfrac{1}{3} \text{Trace } (\boldsymbol{\alpha}) \, \mathbf{I}_3$$

Assuming that the polarizability tensor $\boldsymbol{\alpha}$ is written in its main coordinate system, so that it becomes diagonal,

$$\boldsymbol{\alpha} = \text{diag}(\alpha_x, \, \alpha_y, \, \alpha_z) \qquad (7.104)$$

the anisotropic part can be written as

$$\hat{\boldsymbol{\alpha}} = \text{diag}\left[\left(\alpha_x - \frac{\alpha_x + \alpha_y + \alpha_z}{3}\right), \left(\alpha_y - \frac{\alpha_x + \alpha_y + \alpha_z}{3}\right), \left(\alpha_z - \frac{\alpha_x + \alpha_y + \alpha_z}{3}\right) \right]$$

$$(7.105)$$

Equation (7.105) can be simplified by defining the following combinations of polarizabilities along the main axes:

$$\Delta\alpha = \alpha_x - \left[\frac{\alpha_y + \alpha_z}{2} \right] \quad \text{and} \quad \Delta\alpha^+ = \alpha_y - \alpha_z \qquad (7.106)$$

These two combinations can be regarded as *longitudinal* and *transverse* anisotropies. Thus, $\Delta\alpha$ represents the difference between the polarizability in the longitudinal (i.e., x) axis and the average of the two transverse axes, and $\Delta\alpha^+$ shows the difference between the two transverse directions. In the case of spherical symmetry, $\alpha_x = \alpha_y = \alpha_z$, and therefore, $\Delta\alpha = \Delta\alpha^+ = 0$. On the other hand, for cylindrical symmetry, $\alpha_y = \alpha_z$ and, consequently, $\Delta\alpha^+ = 0$.

Substituting Eq. (7.106) into (7.105) gives

$$\hat{\alpha} = \text{diag} \left[\tfrac{2}{3} \Delta\alpha, \ -\tfrac{1}{3} \Delta\alpha, \ -\tfrac{1}{3} \Delta\alpha \right]$$
$$+ \text{diag} \left[0, \ \tfrac{1}{2} \Delta\alpha^+, \ -\tfrac{1}{2} \Delta\alpha^+ \right] \tag{7.107}$$
$$= \Delta\alpha \ \text{diag} \left[\tfrac{2}{3}, \ -\tfrac{1}{3}, \ -\tfrac{1}{3} \right] + \Delta\alpha^+ \ \text{diag} \left[0, \ \tfrac{1}{2}, \ -\tfrac{1}{2} \right]$$

Therefore, assuming that the main axes of the α tensor are known, the two parameters $\Delta\alpha$ and $\Delta\alpha^+$ completely determine the anisotropic part $\hat{\alpha}$.

There are some cases in which the use of the main axes of α is not convenient. A good example is the ester group [136], for instance, in the molecule of methyl acetate, in which symmetry dictates that one of the main axes should be perpendicular to the plane of the molecule while the other two are contained in that plane. However, none of the axes lies along any bond of the molecule. In fact, one of the axes makes an angle ξ of about 20° with the C—C bond. Then the exact value of the ξ angle would be required in addition to the anisotropies $\Delta\alpha$ and $\Delta\alpha^+$ to define the $\hat{\alpha}$ tensor. Under these circumstances, it becomes easier to use the C—C bond as the x axis with z perpendicular to the plane of the molecule and write the $\hat{\alpha}$ tensor in nondiagonal form as a function of $\Delta\alpha$, $\Delta\alpha^+$, and the polarizability in the xy plane, $\alpha_{xy} = \alpha_{yx}$. Thus, the $\hat{\alpha}$ tensor of an ester group is customarily written as

$$\hat{\alpha} = \Delta\alpha \ \text{diag} \left[\tfrac{2}{3}, \ -\tfrac{1}{3}, \ -\tfrac{1}{3} \right] + \Delta\alpha^+ \ \text{diag} \left[0, \ \tfrac{1}{2}, \ -\tfrac{1}{2} \right] + \begin{bmatrix} 0 & \alpha_{xy} & 0 \\ \alpha_{yx} & 0 & 0 \\ 0 & 0 & 0 \end{bmatrix}$$

$$\tag{7.108}$$

Many attempts for obtaining either the α or $\hat{\alpha}$ tensor of a given molecule through addition of bond contributions have been made [137]. Maybe the most important one was performed by Denbigh [138], who tabulated polarizabilities of many chemical bonds using a coordinate system whose x axis lies in the direction of the bond and assuming that the polarizability of any bond should have cylindrical symmetry.

Despite all their inconveniences, Denbigh's values are quite useful and have been often employed in calculations of several optical properties. A reasonable application of these bond contributions is to obtain a first approximation of the desired magnitudes. Many times this approximation is compared with experimental results in order to correct the polarizabilities and obtain a consistent set of parameters [139]. They also can be used in calculations performed on molecules whose chemical structures are very similar so that possible differences in optical properties should be relatively independent of the polarizabilities assigned to each bond; good examples are homopolymer chains having different configurations and copolymers with different sequence distributions [140].

However, Denbigh's polarizabilities have raised a lot of criticism, not only because of the values assigned to each kind of bond, but also for the whole idea of computing polarizability tensors by addition of contributions. Besides the fact that the cylindrical symmetry is questionable in several cases, for instance, in double bonds, the most important point is that these contributions *are not additive*. Thus, for example, the tensor obtained by addition of the polarizabilities assigned to the bonds contained in a methyl acetate molecule is quite different from what is obtained through the analysis of experimental results of optical properties measured for the actual molecule. Moreover, it is quite intuitive to realize that these contributions should not be additive because the polarizability is governed by the ease with which the electron distribution can be distorted by the applied field. It is known, for instance, that the delocalization of electrons for two double bonds is greater when they are conjugated than when they are isolated. However, this situation cannot be regarded as a basic failure of the VOS scheme; it simply means that, as in the case of dipole moments, group contributions that are mutually independent should be used, and the way for obtaining these contributions is the same as was used for dipole moments: either a quantum mechanics calculation [141, 142] or the analysis of experimental results measured for suitable model compounds.

Unfortunately, the computational procedures required for the theoretical calculation of polarizabilities are not yet well developed. For instance, recently, Waite and Papadopoulos [142] calculated the polarizability and hyperpolarizabilities of a series of nitrogen heterocycles in order to correlate those properties with the molecular structure. The computational approach uses a Coupled Hartree–Fock Perturbation Theory (CHF-PT) for the calculation of the first- and second-order corrections to the density matrix and an extended-basis (EB) CNDO wave function to account for changes in the electron distribution due to the presence of an external electric field. They concluded that the theoretical results are in satisfactory agreement with the experimental data, but give more reliability to the trends shown by the calculated values for a series of compounds having similar structures than to absolute values obtained for a given molecule.

As for the analysis of model compounds, several examples are given in the next section. The general idea is to study small molecules, seeking values of the $\Delta\alpha$ and $\Delta\alpha^+$ parameters for some groups from which polymeric chains could be formulated. The determination of these two parameters for a given molecule requires the measurement of at least two macroscopic properties; with this purpose, electric and magnetic birefringences are often used together with the mean-squared optical anisotropy $\langle \gamma^2 \rangle$, defined as

$$\langle \gamma^2 \rangle = \tfrac{3}{2} \langle \text{Trace} (\hat{\alpha}\hat{\alpha}) \rangle \tag{7.109}$$

which can be obtained from measurements of depolarized Rayleigh scattering (DRS). The procedure consists in adjusting the values of the two parameters until they simultaneously reproduce the experimental results.

An illustrative example of application of this method is provided by Suter and Flory [143], who obtained mean-square optical anisotropies $\langle \gamma^2 \rangle$ of benzene, toluene, isopropyl benzene, t-butylbenzene, 2,4-diphenylpentane, 2,4,6-triphenyl heptane, and atactic polystyrene with average degrees of polymerization of 21, 38, and 96 from measurements of the depolarized Rayleigh scattering of their solutions in carbon tetrachloride at 25°C using $\lambda = 632.8$ nm. They analyzed their own results for the model compounds in addition to values of Kerr and Cotton–Mouton constants taken from the literature and found that in the case of the benzene molecule, the three experimental magnitudes could be simultaneously reproduced by using

$$\Delta\alpha_B = 1.83 \qquad \Delta_B^+ = 3.63 \qquad \text{both in Å}^3$$

written in a coordinate system having the z axis perpendicular to the molecular plane. In fact, the system that they used has the y axis perpendicular to the molecular plane, and, consequently, they give $\Delta_B^+ = -3.63$; see Eq. (7.106).

The toluene molecule can be schematically obtained by addition of one molecule of benzene and one of ethane with elimination of one molecule of methane, which, given its spherical symmetry, has $\hat{\alpha}_M = 0$. Consequently, assuming strict additivity of contributions,

$$\hat{\alpha}_T = \hat{\alpha}_B + \hat{\alpha}_E \tag{7.110}$$

The molecule of ethane has cylindrical symmetry [144] along the C—C bond and, therefore, taking the x axis along this bond and assuming additivity of the contributions from the C—C bond and the two methyl groups,

$$\hat{\alpha}_E = \Delta\alpha_E \, \text{diag}[\tfrac{2}{3}, -\tfrac{1}{3}, -\tfrac{2}{3}] = (\Delta\alpha_{C-C} + 2\Delta\alpha_{CH_3})$$
$$\text{diag}[\tfrac{2}{3}, -\tfrac{1}{3}, -\tfrac{1}{3}] \tag{7.111}$$

But, applying the additivity of contributions to the methane molecule,

$$\Delta\alpha_M = \Delta\alpha_{C-H} + \Delta\alpha_{CH_3} = 0 \;\Rightarrow\; \Delta\alpha_{C-H} = -\Delta\alpha_{CH_3} \tag{7.112}$$

and, therefore,

$$\Delta\alpha_E = (\Delta\alpha_{C-C} + 2\Delta\alpha_{CH_3}) = (\Delta\alpha_{C-C} - 2\,\Delta\alpha_{C-H}) = \Gamma_{cc} \tag{7.113}$$

This combination of contributions appears frequently in the formulation of polarizability tensors. It is customarily represented by Γ_{cc} and its value is 0.53 $\pm$ 0.05 Å^3 [144].

Substituting Eq. (7.113) into (7.111) and then into (7.110) gives

$$\Delta\alpha_T = \Delta\alpha_B + \Gamma_{cc} \qquad \Delta\alpha_T^+ = \Delta\alpha_B^+ \tag{7.114}$$

with the x axis in the direction of the C^{Ph}—CH_3 bond, and z perpendicular to the plane of the molecule.

However, the magnitude of the anisotropy attributable to the aromatic ring is affected markedly by substitutions, and cylindrical symmetry about the z axis is eliminated also. For these reasons, they introduce the tensor $\hat{\alpha}_{\phi H}$ to represent the phenyl group as it exists in a given derivative of benzene. Thus, the difference between the fictitious $\hat{\alpha}_{\phi H}$ tensor and $\hat{\alpha}_B$ of an actual molecule of benzene shows the inductive effects produced by the substitution. Consequently, Eq. (7.114) should be written as

$$\Delta\alpha_T = \Delta\alpha_{\phi H} + \Gamma_{cc} \qquad \Delta\alpha_T^+ = \Delta\alpha_{\phi}^+{}_H \qquad (7.115)$$

The experimental results of the molecule of toluene can now be examined to determine the two parameters $\Delta\alpha_{\phi H}$ and $\Delta\alpha_{\phi}^+{}_H$, as is shown in Figure 7.15, from which one obtains

$$\Delta\alpha_{\phi H} = 2.95 \qquad \Delta\alpha_{\phi}^+{}_H = 3.0 \qquad \text{both in Å}^3$$

and, therefore, $\Delta\alpha_T = 3.5$ and $\Delta\alpha_T^+ = 3.0$ Å^3.

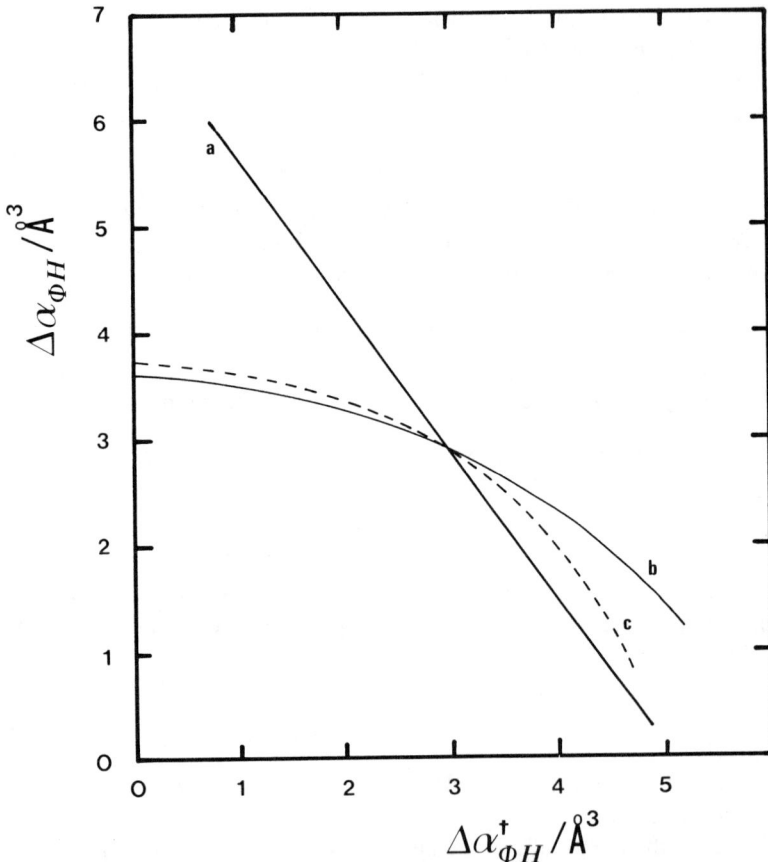

Figure 7.15 Determination of the $\Delta\alpha_{\phi H}$ and $\Delta\alpha_{\phi H}^+$ parameters for the toluene molecule. Lines represent the loci of values of these parameters that reproduce the experimental results of the following molar magnitudes: (a) Cotton–Mouton constant C_m, (b) mean-square optical anisotropy $\langle \gamma^2 \rangle$, and (c) Kerr constant $_mK$. (Reprinted from reference [143] by courtesy of The Royal Society of Chemistry.)

The same procedure is then applied to the molecule of isopropyl benzene from which

$$\Delta\alpha_{\phi H} = 3.85 \qquad \Delta\alpha^+_{\phi H} = 3.0 \qquad \text{both in } Å^3$$

Next, the molecule of t-butylbenzene is examined and the numerical values of the parameters that this molecule provides are identical to those obtained with t-butylbenzene. Consequently, they conclude that the inductive effects modify the contribution of the phenyl ring in passing from benzene to toluene and isopropyl benzene, but any further substitution of the alkyl group has only a minor effect. Thus, they calculate $\langle\gamma^2\rangle$ for several fractions of polystyrene and some of its oligomers using as additive contributions the values of $\Delta\alpha_{\phi H}$ and $\Delta\alpha^+_{\phi H}$ determined from the molecule of isopropylbenzene together with the parameter Γ_{cc} previously determined in the analysis of n-alkanes [144]. The excellent agreement between theory and experience that they obtain validates the assumption of additivity of these contributions.

Table 7.1 summarizes the values of $\Delta\alpha$ and $\Delta\alpha^+$ for some bonds or groups of bonds that are frequent constituents of polymeric chains. However, it is very important to realize that this kind of data should be used with great care and, whenever possible, a full analysis of some model compounds such as the one described before or a quantum mechanics calculation of the pertinent contributions should be performed. In particular, it should be pointed out that the values shown in Table 7.1 for individual bonds are additive only if they combine with other bonds having similar characteristics; they should never be used to formulate groups having large electronic mobility. For instance, the value assigned to the C—C single bond can be combined with that of C—H to study polyethylene or with C—O for the analysis of polyoxides. However, as the table indicates, the parameters for toluene cannot be obtained by adding those of benzene with one C—C and three C—H.

Thus, in brief, inductive effects between polarizabilities preclude a straightforward additivity of contributions. Moreover, these inductive effects are not only intramolecular (i.e., produced by different segments of a given molecule); on the contrary, they also arise from correlations with neighboring molecules (e.g., solvent) or segments of other molecules. This second aspect is especially important in the case of stress birefringence (see what follows) and renders it difficult to compare data obtained in solution (e.g., Kerr constants or mean-square optical anisotropies) with those determined in a solid network either unswollen or swollen with a given solvent. This is the worst problem that the calculation of optical properties has nowadays, and the only procedure available to overcome it consists in trying to obtain contributions in which the inductive effects are incorporated so that these contributions are additive. Unfortunately, neither quantum mechanics nor the analysis of model compounds is yet well developed as to give an easy and secure way for the evaluation of these additive contributions.

TABLE 7.1 Anisotropies (Å³) of Some Bonds and Groups of Bonds[†]

Bond or Group of Bonds	$\Delta\alpha$	$\Delta\alpha^+$	α_{xy}	Reference
C—C	0.95	0	0	[144]
C=C	2.05	0	0	[60]
C—H	0.21	0	0	[144]
CH$_3$—CH$_3$	0.53	0	0	[144]
C—N	0.12	0	0	[60]
C=N	1.71	0.44	0	[139]
C—O	0.58	0	0	[144]
C—S	0.19	0	0	[60]
C—Cl	1.75	0	0	[145]
C—Si	1.46	0	0	[146]
N—H	0.33	0	0	[60]
S—H	0.58	0	0	[138]
Si—CH$_3$	1.25	0	0	[146]
Si—O	1.30	0	0	[147]
Si—Cl	1.73	0	0	[146]
Si—H	1.02	0	0	[146]
Si—F	0.11	0	0	[146]
Benzene	1.83	3.63	0	[143]
Toluene[‡]	3.50	3.0	0	[143]
Fluorobenzene[‡]	2.93	5.67	0	[148]
Pentafluorobenzene[‡]	2.97	6.35	0	[148]
Chlorobenzene[‡]	4.7	3.3	0	[40]
Bromobenzene[‡]	5.4	3.9	0	[40]
ϕ—H	3.85	3.0	0	[143]
p—Cl—ϕ—H[‡]	6.0	3.3	0	[40]
p—Br—ϕ—H[‡]	6.9	3.9	0	[40]
ϕ—Si[‡]	5.1	3.0	0	[147]
CH$_3$—C*O*OCH$_3$[§‖]	1.55	1.21	−0.1	[136]
CH$_3$—C*O*OCH$_3$[§¶]	1.65	1.00	−0.28	[136]
H$_2$O[‡]	0.08	0.54	0	[149]
O—H	0.35	0	0	[150]

[†] In the case of individual bonds, the x axis is taken along the bond. In molecules, the z axis is perpendicular to the plane of the molecule. Although most results of Kerr constants are given in SI units, it is relatively common to write polarizabilities in cgs units as cm³ or Å³. To transform SI units of Cm² V^{-1} (equivalent to C² m² J^{-1}) to the volume units of cgs, take into account that the electrical permittivity of vacuum is $\epsilon_0 = 8.85419 \times 10^{-12}$ C² m^{-1} J^{-1} in SI and $\epsilon_0 = 1/4\pi$ in cgs. Then $\alpha/4\pi\epsilon_0 = $ C² m² J$^{-1}/4\pi 8.85419 \times 10^{-12}$ C² m^{-1} J^{-1}) = m³ = 10⁶ cm³. Therefore, 10^{-40} Cm²V^{-1} = 0.89875 Å³.

[‡] The x axis along the C$_2$ symmetry axis of the molecule.

[§] The x axis along the C—C*O* bond.

[‖] Linked by the acid residue (as in methyl isobutyrate).

[¶] Linked by the alcoholic residue (as in isopropyl acetate).

APPLICATIONS OF THE KERR CONSTANT TO CONFORMATIONAL ANALYSIS

As it was indicated before, the applicability of the Kerr constant to conformational analysis arises from the great sensitivity of this property to variations of molecular geometry and environment [151], including aspects such as geometry of solvation [152–155], aggregation of solutes [155], detection of critical points [156], etc.

The standard procedure used for these studies consists of a critical comparison between theoretical and experimental values of the molar Kerr constant $_mK$. In many cases, this comparison is done with the assistance of a similar analysis of some other conformation-dependent properties [157] such as dipole moments, mean-squared optical anisotropies, etc. Experimental values of $_mK$ are determined from measurements of the electrical birefringence of several solutions having different molar concentration of the molecule to be studied in a suitable solvent; see Eq. (7.25). The theoretical value of $_mK$ is calculated using Eq. (7.103), which requires the knowledge of the dipole moment μ and the optical anisotropy $\hat{\alpha}$ of the molecule together with a detailed model of its microscopic characteristics (including magnitudes such as geometry, conformational energies, stereochemical configuration of asymmetric centers, etc.) needed to perform the statistical average over all the allowed conformations of the molecule. The analysis may be aimed to determine any of these characteristics of the sample provided that all the others are known. Several examples are presented in the next paragraphs.

Although measurements performed on solutions are most suitable for this kind of analysis, determining B values of the Kerr constant, see Eq. (7.1), of pure substances either in the vapor or liquid states is also important in this context.

Measurements performed in the vapor state allow the evaluation of polarizabilities of the molecule. A good example is the analysis of fluorine-substituted benzenes (from C_6H_6 to C_6F_6) published by Gentle and co-workers [148, 158]. They measured Kerr constants and depolarization ratios in the vapor state at $\lambda = 632.8$ nm and analyzed their own results together with those previously reported for fluorobenzenes by others [152, 159–162]. For instance, using a coordinate system with the x axis along the C_2 symmetry axis of the molecule and the z axis perpendicular to the molecular plane, they obtained the following polarizabilities (in units of 10^{-40} Cm^2V^{-1}) for the fluorobenzene (FB) and pentafluorobenzene (PFB) molecules:

FB: $\alpha_{xx} = 13.54 \pm 0.31$ $\alpha_{yy} = 13.44 \pm 0.42$ $\alpha_{zz} = 7.13 \pm 0.16$
PFB: $\alpha_{xx} = 13.70 \pm 0.37$ $\alpha_{yy} = 13.93 \pm 0.47$ $\alpha_{zz} = 6.86 \pm 0.15$

They also examined the trends in the polarizability of the whole series from C_6H_6 to C_6F_6, finding that the main polarizability $\alpha_0 = (\alpha_{xx} + \alpha_{yy} + \alpha_{zz})/3$ is roughly independent of the number of fluorine atoms; however, the in-plane component α_{xx} increases and the out-plane component α_{zz} decreases with increasing number of fluorine atoms.

Measurements performed on pure liquids are needed to evaluate the B_1 Kerr constant of the solvent that is required to compute the molar Kerr constant of solute, $_mK$, according to Eq. (7.25). On the other hand, pure liquids of known Kerr constant are often used either for calibration or for testing the performance of experimental setups. The combination of these two factors makes it almost impossible to find an experimental determination of $_mK$ in which no measurements of B for some pure liquid is also performed. Some interesting publications containing values of B for several pure liquids are due to Aroney and co-workers [124], who measured B at 20°C using a He–Ne laser (λ = 632.8 nm), and Lewis and Orttung [163], who determined B as a function of λ, mostly in the ultraviolet region.

Determination of molar Kerr constants $_mK$ of small molecules allows the evaluation of optical properties (mainly polarizability tensors $\boldsymbol{\alpha}$) and energetic parameters of those molecules. This kind of analysis can be aimed to study the small molecules themselves, for instance, to determine the relative incidence of each allowed conformation. However, in most cases, the small molecules are oligomers or model compounds for the repeating unit of a polymer and the results of their analysis are used to study properties of the whole chain.

Conformational analyses of small molecules have been reported by Le Fevre and co-workers [164], who studied Kerr constants, dipole moments, and refractivities of cholesterol; cholest-5-ene; cholest-5-en-3-one; cholesteryl chloride, bromide, and iodide; and epicholesteryl chloride as solutes in carbon tetrachloride at 25°C. From the analysis of all these data, they concluded, for instance, that the halogen atoms at position 3 are attached equatorially in the three cholesteryl halides and axially in epicholesteryl chloride.

Simple carboxylic esters have also been studied with this method [165, 166]. Figure 7.16 shows, as an example, the variation of $_mK$ for methyl isobutyrate as a function of the rotation over the $(CH_3)_2CH—C^*O^*$ bond. Comparison of theoretical and experimental values of this magnitude confirms the small incidence of the conformation ϕ = 0, and the displacement of the other two conformations by about 20° from their perfectly staggered positions.

Another example was provided in the analysis of the unperturbed conformations of triacetin [167] by comparison of experimental and theoretical values of dipole moments, mean-square optical anisotropies, and Kerr constants as a means to determine conformational preferences of the glycerol moiety in unperturbed triglycerides, which are important components of biological membranes and also occur in complexes with specific proteins in solution.

Many more examples of conformational analysis of small molecules can be found in the revision published by Aroney [151].

Solvents having small dielectric permittivity and relatively low polarizability, such as carbon tetrachloride, p-dioxane, benzene, etc., are used for most measurements performed on small molecules. However, some results of experiments performed in water solutions, which are important for their application to biopolymers, have also been reported [57, 168, 169]. In particular, it is interesting to cite the work

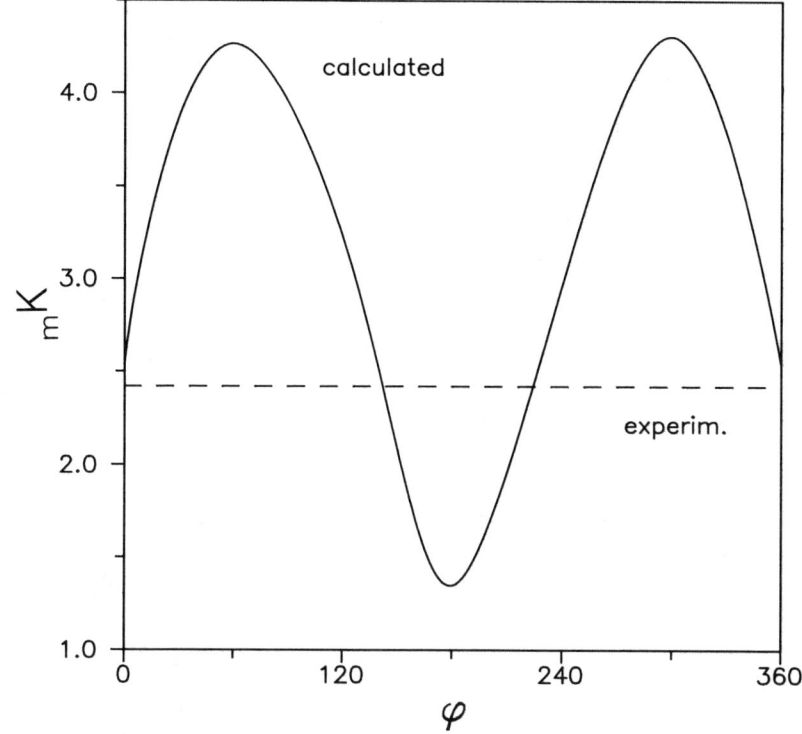

Figure 7.16 Variation of the molar Kerr constant $_mK$ (in units of 10^{-25} m^{5}V^{-2}mol^{-1}) for the molecule of methyl isobutyrate as a function of the rotation ϕ over the $(CH_3)_2CH—C*O*$ bond. The value $\phi = 0$ is assigned to the conformation in which the H of the acid residue is contained in the plane of the ester group. (Redrawn from reference [166] by courtesy of the Real Sociedad Española de Química.)

of Khanarian and Kent [149], who measured Kerr constants, densities, and refractive indices of water–dioxane mixtures over the entire range of compositions, finding the molar Kerr constant of water extrapolated to infinite dilution in dioxane to be 2.3×10^{-27} m^5 V^{-2} mol^{-1}. They also calculated the components of the polarizability tensor of water as $\alpha_{xx} = 1.69 \pm 0.05$, $\alpha_{yy} = 1.9 \pm 0.2$, and $\alpha_{zz} = 1.3 \pm 0.2$ in units of 10^{-40} C^2 m^2 J^{-1} with x along the C$_2$ symmetry axis of the molecule and z normal to the molecular plane.

Many small molecules have been studied in order to determine their polarizability, trying to obtain contributions of bonds or groups of bonds to the $\hat{\alpha}$ tensor that could be combined to evaluate the $\hat{\alpha}$ tensor of more complicated molecules. A good example is the work of Le Fevre and co-workers [170, 171], who studied dipole moments and Kerr constants of *para* disubstituted benzenes like *p*-XC$_6$H$_4$CH$_3$ (with X = CH$_3$, F, Cl, Br, I, CN, and NO$_2$), *p*-XC$_6$H$_4$t-Bu (X = H,

CH_3, Br, NO_2, and t-Bu), and p-$XC_6H_4CCl_3$ (X = Cl or CCl_3). Measurements were carried out in benzene and carbon tetrachloride solutions at 25°C. From their analysis, they concluded that in some cases the polarizability of the whole molecule can be predicted with reasonable accuracy from additivity of the appropriate component bond and group parameters. This behavior is exhibited, for instance, by the p-halogeno toluenes and the p-tButyl benzenes with X = CH_3, Br, and NO_2. However, in many other cases, deviations from additivity arise because of electromeric interactions, resulting in a modification of the electron mobility along the molecular 1,4-axis. This modification may be in the sense of either enhancing or decreasing it, as compared with what is calculated assuming additivity of contributions. Thus, for instance, in the cases of p-toluenes, with X = CN and NO_2, and p-chlorobenzotrichloride, there is an enhancement of the electron mobility and, consequently, the experimental result of $_mK$ is larger than the value calculated by addition of contributions. On the contrary, in the cases of p-xylene, p-diterbutylbenzene, and p-ditrichloridebenzene, the calculated values of Kerr constants are larger than the experimental results, which indicate a decrease of the electron mobility. Finally, they also conclude that stereospecific benzene-solute complex formation occurs in the aromatic solvent.

The two conclusions indicated in the preceding paragraph show the hardest problem that the calculation of Kerr constants, as well as other optical properties like stress birefringence or mean-square optical anisotropy, faces nowadays (see the preceding section), namely, is that, in many cases, bond or group contributions to the polarizability tensors are not additive, and even may change with the solvent used in the experimental determination. A typical example of determining group contributions to the polarizability tensor is the work of Suter and Flory [143] summarized in the preceding section.

Aliphatic esters have also been studied by Flory and co-workers [136]. They studied mean-square optical anisotropies and molar Kerr constants (in carbon tetrachloride solutions at 25°C and λ = 632.8 nm) of molecules having one or two ester groups, that is, methyl acetate (MA), methyl isobutyrate (MIB), dimethyl trans-1, 4-cyclohexanedicarboxylate (CDC), isopropyl acetate (IPA), and trans-1,4-cyclohexanediol diacetate (CDA). From this analysis, they were able of evaluating the $\hat{\alpha}$ tensor for an ester group; the result is slightly different, depending on whether the group is attached to the rest of the molecule through the acid residue (as in MIB and CDC) or by the alcoholic residue (as in IPA and CDA). Thus,

$$\hat{\alpha}_E(1) = \begin{bmatrix} 1.03 & -0.10 & 0 \\ -0.10 & 0.085 & 0 \\ 0 & 0 & -1.12 \end{bmatrix}$$

for groups with the first kind of substitutions, and

$$\hat{\alpha}_E(2) = \begin{bmatrix} 1.10 & -0.28 & 0 \\ -0.28 & -0.05 & 0 \\ 0 & 0 & -1.05 \end{bmatrix}$$

for the second. The units of both tensors are $\mathring{A}^3$ and the coordinate system used to write $\hat{\alpha}$ has the x axis along the C—C*O*O bond and the z axis perpendicular to the molecular plane.

However, when both ester and phenyl groups are linked together, as occurs, for instance, in benzoates, their contributions have to be adjusted. Thus, Flory and co-workers [172] studied several aromatic esters and oligomers of poly(p-oxy benzoate), which are very interesting compounds because of their nematogenic properties. They measured the mean-square optical anisotropy $\langle\gamma^2\rangle$ of phenyl acetate (PA), methyl benzoate (MB), phenyl benzoate (PB), and trimers and tetramers of the p-oxibenzoate series $C_6H_5CO(-OC_6H_4CO)_{n-2}-O_6C_5H$ from the DRS of their solutions in CCl_4. Molar Kerr constants of PA and trimer were evaluated from electric birefringence measurements in CCl_4 and dioxane, respectively, and values for MB and PB were taken from the work of Le Fevre and Sundaram [165]. They observed that the experimental magnitudes are well reproduced by calculations based on contributions to optical anisotropy tensors $\hat{\alpha}$. However, the $\Delta\alpha_{\phi\,H}^+$ contributions of the phenyl group had to be adjusted to $\Delta\alpha_{\phi\,H}^+ \approx 3.5 \mathring{A}^3$ in the case of phenyl acetate, which is not too different from the value of $\Delta\alpha_{\phi\,H}^+ \approx 3.85 \mathring{A}^3$ obtained for substituted benzenes; however, values of $\Delta\alpha_{\phi\,H}^+ \approx 5.4 \mathring{A}^3$ are required for methyl benzoate. Theoretical calculations performed with these adjusted optical parameters give a good account of the experimental results, including the most recent determinations of $\langle\gamma^2\rangle$ performed on oligomers of several p-oxybenzoates [173].

Something similar happens in the case of dibenzoates of some diols with different number of methylene units, that is, $C_6H_5COO(CH_2)_m OOCC_6H_5$ with $m =$ 2 to 6, reported by Mendicuti and co-workers [174]; the theoretical values of Kerr constants are very sensitive to the conformational energies of the glycol residue although the optical parameters have to be adjusted to achieve agreement between theory and experiment. However, once they are adjusted, the same set of values is able of reproducing the experimental results for the whole series of molecules.

Small modifications of the anisotropies are also required in the case of terephthalates [123]. Experimental results measured at 20°C for dimethyl terephthalate in p-dioxane solution can be used to optimize the optical parameters of the aromatic diester group, which can further reproduce the experimental values of both $_mK$ and its temperature coefficient for poly(diethylene glycol terephthalate) measured in the same conditions.

Another series of small molecules that has been extensively studied [175–180] contains a cyano group and either an alkyl or alkoxy group in positions 4,4' relative to a bicyclic residue; that is, molecules like NC—Cy—Cy—R, where Cy represents either a 1,4 disubstituted benzene residue or a trans-1,4 disubstituted cyclohexane ring, and R stands for either C_mH_{2m+1} or O—C_mH_{2m+1} with the number of carbon atoms, m, lying between 2 and 10. This kind of compound forms nematic liquid crystals whose stability is probably due to interactions between their large dipole moments.

We will cite, as a last example of the analysis performed on small molecules, the work of Armstrong and co-workers [146], who studied electric dipole moments and molar Kerr constants of nine organosilicon compounds to derive the polarizability of several silicon-containing bonds. Additivity of contributions appears valid for this series of compounds, and their results are often used in calculations performed on inorganic polymers. Table 7.2 summarizes the polarizabilities that were obtained.

Many flexible polymers have been analyzed, looking for different characteristics, using Kerr constant values. Reference to the work of Le Fevre's group is, once again, required, since they have studied polymers such as poly(ethylene glycol) [181], poly(methyl methacrylate) [182], and poly(methyl acrylate) [183]. For instance, in the case of PEG, they measured dipole moments and molar Kerr constants at 25°C and λ = 589.3 nm for benzene solutions of HO—$(CH_2$—CH_2—$O)_x$—H with $\langle x \rangle$ = 4.1, 6.4, 18, 34, 78, and 153. They concluded that both properties can be reconciliated with the supposition that gauche and trans conformations of the group O—CH_2—CH_2—O are equally probable for x up to about 5, but that thereafter gauche or cis arrangements occur more frequently than trans, explaining the tendency for the larger molecular chains to contain helical portions.

Most of the studies performed on biopolymers have been carried out under dynamic fields, assuming that the molecule is relatively rigid and looking for its hydrodynamic properties. However, flexible polypeptide chains have also been extensively studied. Thus, Ingwall and Flory [184] measured depolarized Rayleigh scattering of several amides, combining their results with those of Kerr constants determined by Aroney and co-workers [185] to evaluate the anisotropy tensors of several peptide units, which were then used to compute Kerr constants of several polypeptides and compare them with the results of experimental measurements [115]. Thus, they found that the values of $_mK$ calculated for various fixed conformations of polypeptides vary over a wide range, depending on the choice of conformation; hence, the configurational averages $\langle _mK \rangle$ are sensitive to small changes in conformational energy.

TABLE 7.2 Polarizability of Several Silicon-Containing Bonds in Units of 10^{-40} C m^2 V$^{-1\dagger}$

Bond	α_x	$\alpha_y = \alpha_z$
Si—Cl	4.35	2.43
Si—Me from Me$_3$SiCl	3.99	2.83
Si—Me from MeSiCl$_3$	4.12	2.75
Si—H from Me$_3$SiH	2.08	0.95
Si—H from HSiCl$_3$	1.88	1.01
Si-F from Me$_3$SiF	1.01	0.89

†The coordinate system has the x axis along the direction of the bond.

The sensitivity of $_mK$ to primary, secondary, and tertiary structures of simple model polypeptides has also been investigated [139, 186, 187]. Thus, for instance, López-Piñeiro and co-workers [139] carried out calculations of the Kerr constant for several polypeptides of the 20 natural α-aminoacids. Calculations were performed for both homopolymers and copolymers, including 10 actual proteins and 3 enzymes, assuming that all the residues were in their unperturbed random coil state. Their results, summarized in Tables 7.3 and 7.4, are very sensitive to the kind of aminoacid residues and to the sequence of those residues. Table 7.3 shows the calculated values of $_mK$ for three enzymes, each formed by nine aminoacid residues; according to the model used in the calculation, each of these residues is randomly coiled, and yet the values of $_mK$ are widely different; moreover, oxytocin and vasopressin differ only in the substitution of Ile by Phe in position 3 and Leu by Arg in position 8; however, their Kerr constants are as different as -316 versus 72.

Table 7.4 summarizes the results obtained for 10 proteins whose number of aminoacid residues is given in the second column. The third column indicates the result of $_mK$ calculated with the actual sequence of residues that the protein contains, although assuming that each residue adopts its random coil conformation; these results roughly cover the range ±400. The fourth and fifth columns of the table show the results of a calculation performed in the following way: The aminoacid residues of each protein were used, but they were linked to form a chain following a random sequence instead of using the actual order in which they are joined in the protein. The fourth column contains the average values computed for 10 such chains for each protein and the standard errors of these averages are given in parentheses in the fifth column. The differences between the values computed with actual and random sequences indicate the sensitivity of this magnitude to the sequence in which the residues appear in the chain; this sensitivity is even more confirmed by the large standard errors of the averages which indicate that the computed value strongly depends on the sequence of aminoacid residues that the random procedure generates.

The behavior of the Kerr constant in the region of the helix–coil transition on biopolymers and the possible conformational changes induced in the sample by the effect of the strong electric field applied to it have also been investigated [188–197]. The idea is that if the electric field induces a conformational change in the mac-

TABLE 7.3 Calculated Values of Kerr Constants for Some Enzymes in Units of 10^{-27} m^5 V^{-2} mol^{-1}[†]

Enzyme	Sequence	$_mK$
Bradykinin	Arg-Pro-Pro-Gly-Phe-Ser-Pro-Phe-Arg	629
Bovine oxytocin	Cys-Tyr-Ile-Gln-Asn-Cys-Pro-Leu-Gly	-316
Bovine vasopressin	Cys-Tyr-Phe-Gln-Asn-Cys-Pro-Arg-Gly	72

[†] Reprinted from reference [139] with permission of Springer-Verlag, Heidelberg.

TABLE 7.4 Calculated Values of Kerr Constants for Some Proteins in Units of 10^{-27} m^5 V^{-2} mol^{-1}[†]

Protein	Degree of Polymerization	$_mK$ for Sequences		
		Actual	Random	Standard Error
Myoglobin	151	−65	34	(17)
Lysozyme	129	−341	80	(47)
Chymotripsinogen	245	−155	173	(34)
Ribonuclease	124	39	−236	(83)
Human hemoglobin α	141	355	138	(73)
Human hemoglobin β	146	14	139	(44)
Human hemoglobin γ	146	−76	−31	(60)
Human hemoglobin δ	147	−278	11	(63)
Horse hemoglobin α	141	432	237	(39)
Horse hemoglobin β	146	−4	34	(45)

[†] Reprinted from reference [139] with permission of Springer-Verlag, Heidelberg.

romolecule in addition to preferential orientations, such a field effect will be detected as a change in electrooptical properties. This effect was reported especially for polynucleotides [188–194] and its biological implications were discussed [188, 192]. In the case of nonionic polypeptides such as poly(γ-benzyl L-glutamate), Watanabe and Yoshioka [195] obtained direct evidence of field-induced helix–coil transition. The situation is more complicated in the case of ionized polypeptides, which serve as the simplest model of ordered biopolyelectrolytes. Yoshioka and Kikuchi [196, 197] have performed some measurements on this kind of system, finding, for instance, that the specific Kerr constant and the molar ellipticity at 222 nm of poly(1-α,γ-diaminobutyric acid hydrochloride) in methanol–water mixtures underwent an abrupt change between 75 and 80% by volume of methanol at 25°C, corresponding to a solvent-induced helix–coil transition. On the helix side of the transition region, that is, between 78 and 80% by volume of methanol, anomalous birefringence transients indicative of field-induced helix–coil transition were observed at high fields.

The series of n-alkanes, as oligomers of polyethylene, has been studied by Champion and co-workers [198, 199], who performed measurements of the Kerr effect on liquid n-alkanes C_nH_{2n+2}, from $n = 5$ to 16, in the temperature range of 10 to 90°C and on carbon tetrachloride solutions for $n = 6$ to 12 at 23 ± 2°C. They also performed measurements of styrene and five polymer fractions [200], having M_w from 1 to 35 × 10^5 in carbon tetrachloride at 25°C. The calculations, performed using additivity of contributions, are in accordance with experiments provided that the vectors representing the segmental dipole and end-to-end distance are orthogonal.

A typical example of application of $_mK$ to conformational analysis is the work of Tonelli [140] who performed calculations of molar Kerr constants of five homopolymers: poly-(propylene) (PP), poly-(vinyl chloride) (PVC), poly-(styrene) (PS), poly-(p-methylstyrene) (PPMS) and poly-(p-chlorostyrene) (PPCS), and three copolymers (PP-PVC, PS-PPCS, and PPMS-PPCS) as functions of stereoregularities and sequence distribution. Among the five homopolymers studied, only the results obtained for the polar polymers PVC and PPCS exhibit marked sensitivities to stereosequence. The molar Kerr constants calculated for each of the three copolymer systems show high sensitivities to both stereosequence and monomer sequence distribution. Tonelli concluded that the molar Kerr constant may hold promise as an easily measured molecular property of a copolymer chain that is sensitive to its sequence distribution. The sensitivity of $_mK$ to the stereosequence in the case of polar vinyl polymers is confirmed by the analysis of the mean-square optical anisotropies and Kerr constants of p-chloro and p-bromo styrenes performed by Flory and co-workers [40].

Kelly and co-workers [201] have also investigated poly(oxyethylenes), $CH_3(-O-CH_2-CH_2)_x-O-CH_3$. They measured the Kerr constant and the dipole moments of the first four oligomers and of a high molecular weight ($x = 91$) in carbon tetrachloride solutions at 23°C. Calculated and measured values of dipole moments are in good agreement. However in the case of Kerr constants, the agreement is poor except in the sign of $_mK$ and its dependence upon chain length. In these compounds, the value of the dipolar contribution $\langle \mu^T \hat{\alpha} \mu \rangle$ to $_mK$ is negative and, therefore, opposes the positive term, depending upon the polarizability $\langle$Trace $(\hat{\alpha}\hat{\alpha})$. The value of the Kerr constant is then negative and very small; thus, they give the molar Kerr constant per repeating unit for the polymer with $x = 91$ as $_mK/x = -2.0 \times 10^{-12}$ cm^5 statvolt^{-2} mol^{-1} (equivalent to -2.2×10^{-27} m^5 V^{-2} mol^{-1}). Under these conditions, it seems likely that higher-order terms such as collision-induced anisotropy, hyperpolarizability, field-dependent polarizability, etc., which are usually neglected, become important. They conclude that their results would suggest that the utility of the Kerr-effect studies, as applied to polymers, may be restricted to those with substantial dipole moments, or at least to those in which $\langle \mu^T \hat{\alpha} \mu \rangle$ is positive and does not partially cancelates with the anisotropy term. Something similar happens in the case of poly(methyl phenyl siloxane), for which [202] $_mK/x = 1.9 \pm 0.5 \times 10^{-25}$ m^5 V^{-2} mol^{-1}, determined in p-dioxane solutions at 21°C. The dipole contribution is also negative although not large enough so as to overcome the positive contribution of polarizability. Agreement between theory and experience is very good for this polymer.

A good example of the limitations of the Kerr constant in the cases in which the two contributions have opposite signs is provided [203] in the evaluation of the unperturbed dimensions, dipole moments, mean-square optical anisotropies, and Kerr constants of 1,3-dioxolane polymers obtained by ring-opening polymerization of the cyclic monomer $\overline{CH_2-O-(CH_2)_2-O}$. Bond scissions occurring at the two different kinds of C—O bonds of the monomer produce structural irregularities in the polymers arising from the presence of the two different repeat units,

$CH_2OCH_2CH_2O$ and $CH_2CH_2OCH_2O$. Calculations performed as a function of the fraction ω_a of repeat units having one of these structures suggest that, due to the partial cancellation of contributions, an increase in ω_a from 0 to 0.5 causes a change in the Kerr constant of about 5%, while the variation in $\langle \mu^2 \rangle$ is about 60% and $\langle \gamma^2 \rangle$ (which depends only on the term $\langle \text{Trace } (\hat{\alpha}\hat{\alpha}) \rangle$) changes by 29%. The unperturbed dimensions are almost insensitive to these irregularities and the mentioned variation of ω_a only modifies $\langle r^2 \rangle$ by about 0.3%.

Flory and co-workers [204, 205] have also investigated polycarbonates and model analogues by analyzing data of depolarized light scattering and electric birefringence measurements for dimethyl carbonate, methyl phenyl carbonate, diphenyl carbonate, and 2,2-diphenyl propane. By combining the information on structure, anisotropy, and conformation of these model compounds, they estimated the optical anisotropy tensors for carbonate and phenyl groups occurring in the polycarbonate chain. Results of their calculations for these chains yield results in good agreement with experimental determinations.

We will cite, as a last example of the application of Kerr constants to conformational analysis, the study performed by Mumby and Beevers [206] on cyclic oligomers of dimethyl siloxane, having from $n = 8$ to 20 skeletal bonds, by calculating mean-squared optical anisotropies and molar Kerr constants of both cyclic and linear chains. They found substantial differences between calculated values for cyclic and linear oligomers and discussed them in terms of the restricted conformational population of the cyclic species. They also observed that, although there is good correlation between the trends calculated and those observed experimentally, discrepancies exist between the absolute values.

All the examples shown before seem to indicate that the Kerr constant is a powerful tool for the determination of many microscopic properties of a wide variety of samples. The exceptions may be those molecules in which both contributions to $_mK$ are of different sign and roughly cancel each other out. However, the worst limitation of the whole procedure arises from the apparent failure of the additivity of bond or group contributions to optical polarizabilities. Although considerable effort is been devoted to overcome, or at least minimize, this difficulty, it is obvious that there is still a long way to go and a very interesting challenge to answer.

8
Magnetic Birefringence

The first experimental observation of interaction between polarized light and a magnetic field was reported in 1846 by M. Faraday [207], who found that when a beam of polarized light passes through a sample that is subject to an externally applied magnetic field in the direction of propagation of the light, the plane of polarization of this radiation was rotated an angle θ that was proportional to the length of the sample l and to the strength of the magnetic field H:

$$\theta = V(\lambda) l H \qquad (8.1)$$

where $V(\lambda)$ is known as the Verdet constant [208], or the magnetooptical rotation constant, and depends on the characteristics of the measured sample, the temperature, and the wavelength of the radiation used in the measurement.

Equation (8.1) is the mathematical expression of the Faraday effect, which has many interesting applications, including the quantitative measurement of electric birefringence (see before).

At the beginning of the present century, it was observed [209] that it was possible to perform an experiment similar to that of Faraday, but applying the magnetic field in a direction perpendicular to the light beam. This arrangement of the magnetic field at right angle with the direction of propagation of light is known as the Voigt [210] configuration to distinguish it from the Faraday configuration in which the light beam and magnetic field are parallel.

At first sight, both Faraday and Voigt configurations give the same results. When an isotropic sample is observed between crossed polarizers, the analyzer produces complete extinction, but when the magnetic field is applied, part of the light passes through the analyzer. The intensity of the light emerging from the analyzer depends on the nature and length of the measured sample, the temperature, the wavelength of the radiation used, and the strength of the applied field. Upon removal of the magnetic field, the sample returns to its original isotropic behavior and the analyzer again gives complete extinction.

However, it is easy to prove that the effect produced in the sample by the two configurations is different, because in the case of the Faraday experiment, the extinction can be restored simply by rotating the analyzer, whereas with the Voigt configuration, the analyzer is unable of extinguishing the radiation in any orientation. Thus, the Faraday effect produces a rotation of the plane of polarization of the radiation, whereas the Voigt configuration produces birefringence in the sample so that the light, after traveling through the sample, emerges elliptically polarized and, therefore, cannot be extinguished by the analyzer.

The induction of optical birefringence in an isotropic sample by means of a magnetic field applied in a direction perpendicular to the propagation of the light is known as *magnetic birefringence*, or the Cotton–Mouton effect [211], and it is very similar to the electric birefringence (the Kerr effect). Thus, the magnetic field orientates the molecules of the sample. Consequently, the optical polarizability and the refractive index become different in directions parallel and perpendicular to the field, so that radiation traversing the sample will travel with different speed when polarized in a plane containing the direction of the field than when the polarization is perpendicular to that direction.

Therefore, the only difference between the Kerr and Cotton–Mouton effects lies in the driving force producing the orientation of the molecules of the sample, which is electric in the former and magnetic in the latter. Thus, most of the equations governing electric birefringence are also applicable to magnetic, by replacing electric field $\mathbf{E}$ and electrostatic polarizabilities α', respectively, by magnetic field $\mathbf{H}$ and magnetic susceptibilities χ. Thus, the difference in refractive indices between directions parallel and perpendicular (vertical) to the magnetic field produced by a field of strength $\mathbf{H}$ is

$$\Delta n = n_p - n_v = \lambda C H^2 \tag{8.2}$$

where C is the Cotton–Mouton constant (equivalent to the B Kerr constant of Eq. (7.1) of Chapter 7), which depends on the sample, the thermodynamical variables, and the light wavelength. The SI units of C are $m^{-1} T^{-2}$.

If a laboratory coordinate system is defined with the x axis in the direction of the magnetic field, the z axis in the direction of propagation of the light, and y completing a right-handed frame, a beam of light entering the sample polarized in a direction making an angle of 45° with the magnetic field can be resolved into two components of the same amplitude polarized in the xz and yz planes. These two components travel through the sample with different speeds and, therefore, emerge with a relative retardation R or a phase difference δ given, respectively, by

$$R = l\,\Delta n = l\lambda CH^2 \tag{8.3}$$

$$\delta = 2\pi R/\lambda = 2\pi lCH^2 \tag{8.4}$$

where l is the length of the sample.

The magnetic birefringence is measured with experimental equipment and techniques very similar to those employed with its electrical counterpart [212–217] with the only differences arising from the use of a magnetic instead of an electric field. Among these differences, it is interesting to note the following three:

1. Self-induction processes considerably increase the rise and decay times of magnetic fields. Therefore, it is impossible to obtain practically instantaneous switching as required for dynamic measurements. Consequently, all the magnetic birefringence experiments are performed with constant fields, and no information on the dynamics of the sample can be obtained with this technique.

2. However, on the other hand, the orientational force is not affected by complicated ionic polarization mechanisms such as those occurring in polyelectrolytic solutions and, therefore, the magnetic measurements are much easier in this kind of system.

3. A third important difference between electric and magnetic measurements is that in the case of magnetic, the interactions between molecules and field are, in general, weaker than in electric; furthermore, it is difficult to generate a very large homogeneous and stable magnetic field as required to produce strong interactions. The combination of these two effects dictates that the measurements are performed in conditions in which the thermal energy is much stronger than the orientational force and, therefore, no saturation effects similar to those produced under strong electric fields are usually observed in magnetic measurements.

As in the case of the Kerr effect, a molar Cotton–Mouton constant can be defined as the increment in molar refractions divided by the square of the applied field [212]:

$$_mC = (R_p - R_v)/H^2 \tag{8.5}$$

which, with the same approximations used before, can be transformed into

$$_mC = \frac{6\,\lambda M n}{(n^2 + 2)^2 d}\,C \tag{8.6}$$

where M is the molecular weight, n is the refractive index, and d is the density of the sample. Equation (8.6) is formally identical to Eq. (7.22) in Chapter 7 with the only difference being the factor $[(\epsilon + 2)/3]^2$ arising from the relationship between applied and effective electric fields.

In the case of a solution, the contributions of solute should be evaluated and extrapolated to zero concentration in order to eliminate the interactions between molecules of solute. This extrapolation can be performed with the same procedure explained for the Kerr constant. Thus, using subscripts 1 and 2 to represent the magnitudes of solvent and solute, respectively,

$$C_2 = \Delta C / \phi_2 + C_1 \tag{8.7}$$

which is identical to Eq. (7.2) in Chapter 7, where ϕ_2 is the volume fraction of solute, and ΔC is the difference between the values of C measured for the solution and for the pure solvent.

Substituting Eq. (8.7) into (8.6) followed by extrapolating to zero concentration gives the molar Cotton–Mouton constant at infinite dilution, which is equivalent to the Kerr constant defined in Eq. (7.25) in Chapter 7:

$$_m(C_2)_\infty = \frac{6\lambda n_1}{(n_1^2 + 2)^2} \left[\lim_{m \to 0} \left(\frac{\Delta C}{m} \right) + v_2 C_1 \right] \tag{8.8}$$

where m represents the molar concentration of solute.

The theoretical calculation of Cotton–Mouton constants requires the evaluation of the interaction between the magnetic field and the molecules of the sample. By representing the field by the vector $\mathbf{H}$ and assuming that the magnetic susceptibility tensor of one molecule is $\boldsymbol{\chi}$, the energy of interaction can be represented by the product

$$V = -\mathbf{H}^T \boldsymbol{\chi} \mathbf{H} \tag{8.9}$$

where $\mathbf{H}^T$ is the transpose of $\mathbf{H}$, both written in the same coordinate system used to define $\boldsymbol{\chi}$.

Equation (8.9) represents only the interaction of the field with the induced magnetic dipole, because the interaction with permanent magnetic dipoles is negligible even in the case of paramagnetic samples [218–220]. Thus, the interaction of a magnetic field with the sample is equivalent to the interaction of an electric field with a nonpolar molecule ($\boldsymbol{\mu} = 0$), represented by Eq. (7.46) in Chapter 7. Therefore, the equations of Cotton–Mouton constants can be directly obtained from those of Kerr constants of nonpolar molecules. For instance, if the molecule has cylindrical symmetry so that its magnetic susceptibility and optical polarizability tensors are diagonal when expressed in a coordinate system having the x axis along the symmetry axis of the molecule and can be represented, respectively, by

$$\boldsymbol{\chi} = \text{diag}(\chi_L, \chi_T, \chi_T) \tag{8.10}$$

$$\boldsymbol{\alpha} = \text{diag}(\alpha_L, \alpha_T, \alpha_T) \tag{8.11}$$

the orientation factor produced by a constant field weak enough as to give interactions smaller than the thermal energy kT is

$$\Phi = \frac{(\chi_L - \chi_T)H^2}{15kT} \tag{8.12}$$

which is equivalent to Eq. (7.51).

Substituting Eq. (8.12) into (7.43) gives the averaged anisotropy ($\Delta\alpha$) produced by the magnetic field as

$$\langle \Delta\alpha \rangle = (\alpha_L - \alpha_T)\Phi = \frac{(\alpha_L - \alpha_T)(\chi_L - \chi_T)\mathbf{H}^2}{15kT} \tag{8.13}$$

Equation (8.13) can now be substituted into (7.17) to calculate the difference in molar refractions, $R_p - R_v$. Substituting this difference into Eq. (8.5) gives the final expression of the molar Cotton–Mouton constant as

$$_mC = \frac{4\pi N}{45kT}(\alpha_L - \alpha_T)(\chi_L - \chi_T) = \frac{4\pi N}{45kT}\Lambda\Lambda_\chi \tag{8.14}$$

which is identical to Eq. (7.75) taking $\boldsymbol{\mu} = 0$ and replacing the electrostatic anisotropy Λ' by its magnetic counterpart Λ_χ.

If the molecules of the sample are polymers whose segments have cylindrical symmetry, the contribution of each segment to the molar Cotton–Mouton constant of the chain is given by Eq. (8.14) and the value for the whole polymer containing N_s segments becomes

$$_mC = \frac{4\pi N}{45kT}N_s\Lambda\Lambda_\chi \tag{8.15}$$

for a freely jointed chain, and

$$_mC = \frac{4\pi N}{45kT}N_s^2\Lambda\Lambda_\chi \tag{8.16}$$

for a rigid-rod polymer. Of course, Eqs. (8.15) and (8.16) are equivalent to (7.76) and (7.77), which give the Kerr constant for the same kind of polymers.

In the case of a flexible polymeric chain, the same procedure used to compute averaged magnitudes in the presence of an electric field can be applied for magnetic interactions. The final expression is equivalent to Eq. (7.101) and can be written as

$$_mC = \frac{2\pi N}{15kT}\langle \text{Trace}\,(\hat{\boldsymbol{\alpha}}\hat{\boldsymbol{\chi}}) \rangle \tag{8.17}$$

Provided that both the optical polarizabilities and magnetic susceptibilities of the repeating unit are known, the term $\langle \text{Trace}\,(\hat{\boldsymbol{\alpha}}\hat{\boldsymbol{\chi}}) \rangle$ appearing on Eq. (8.17) can be computed with the same matrix-multiplication scheme used to evaluate $\langle \text{Trace}\,(\hat{\boldsymbol{\alpha}}\hat{\boldsymbol{\alpha}}) \rangle$ of the Kerr constants with only a minor modification of the generator matrix (see the Appendix at the end of the book) consisting in replacing $\hat{\boldsymbol{\alpha}}^C$ by $\hat{\boldsymbol{\chi}}^C$.

If the experimental conditions are such that the contribution of the solvent can be neglected, so that the value of Δn measured for the solution is produced only by molecules of solute, then Eq. (7.14) can be used to write the birefringence produced by the sample as

$$\Delta n = \frac{2\pi}{9n} (n^2 + 2)^2 \mathcal{N} \langle \Delta \alpha \rangle = \frac{2\pi Nm}{9n} (n^2 + 2)^2 \langle \Delta \alpha \rangle \qquad (8.18)$$

where $\mathcal{N}$ is the number of molecules of solute per unit volume of solution, N is Avogadro's number, and m is the solute concentration in moles per unit of volume. Substituting the value $\langle \Delta \alpha \rangle$ given by Eq. (8.13) into (8.18) gives

$$\Delta n = \frac{2\pi Nm}{135nkT} (n^2 + 2)^2 \Lambda \Lambda_\chi H^2 \qquad (8.19)$$

Equation (8.19) is often used to define a specific Cotton–Mouton constant as [221]:

$$_sC = \frac{\Delta n}{\lambda g H^2} = \frac{2\pi N}{135 \lambda nkTM} (n^2 + 2)^2 \Lambda \Lambda_\chi \qquad (8.20)$$

where g is the concentration of solute in grams per unit volume, and M represents its molecular weight.

The practical applications of the Cotton–Mouton effect are very similar to those of the Kerr constant [212–241] and, in fact, it is quite common to measure both properties at the same time [223]. Values obtained for molecules of known geometry (including allowed conformations) give information about optical and magnetic anisotropies (i.e., values of Λ and Λ_χ). This kind of analysis is often performed on small molecules with structures similar to the repeating units of some polymers and the results are then used to compute values for the chain. On the other hand, if parameters Λ and Λ_χ are known, a comparison between experimental and theoretical values of Cotton–Mouton constants gives information on the conformations allowed to the molecule.

Many small molecules have been studied in order to determine their magnetic susceptibility tensor $\boldsymbol{\chi}$, and several reviews have been published [224–228]. A special mention should be made to the work of Flygare and co-workers [227–232], who have determined $\boldsymbol{\chi}$ for a large number of molecules (mostly by analysis of the molecular Zeeman effect), providing a theoretical framework for partitioning the individual elements in the total magnetic susceptibility tensor $\boldsymbol{\chi}$ into local contributions. Moreover, they have analyzed a considerable amount of molecular data in order to give local values of $\boldsymbol{\chi}$ for either atoms or groups [227, 232].

The family of aromatic compounds has been probably the series of molecules most extensively studied in this regard, both in crystalline state [224, 233, 234] and in solution [213, 220, 223, 235–239]. Just to give an example, Table 8.1 summarizes the values of Cotton–Mouton constants for a few aromatic compounds, determined from measurements of magnetic birefringences performed at 20°C in carbon tetrachloride solutions using $\lambda = 589$ nm. Columns 2 and 3 of this table give the parameters $\Delta \chi$ and $\Delta \chi^+$ (similar to $\Delta \alpha$ and $\Delta \alpha^+$ for optical anisotropy), which define the anisotropy of magnetic susceptibilities of these molecules, together with the values assigned to C—C and C—H bonds. The coordinate system used for this

TABLE 8.1 Some Examples of Molar Cotton–Mouton Constants at Infinite Dilution (in Units of 10^{-15} cm^3 G^{-2} mol^{-1}) and Anisotropies of Magnetic Susceptibilities (in Units of 10^{-13} J G^{-2} mol^{-1})

Bond or Molecule	$\Delta\chi$	$\Delta\chi^+$	$_mC$	Reference
C—C	7.71			[232]
C—H	2.53			[232]
Benzene	29.81	59.63	1.49	[220]
Toluene	32.16	59.63	1.70	[213]
Chlorobenzene	18.37	65.05	1.71	[213]
Bromobenzene	18.19	62.94	1.93	[220]

table in the case of molecules has the x axis along the C$_2$ symmetry axis and the z axis perpendicular to the molecular plane; in the case of the two bonds indicated, x lies along the bond.

In the case of polymers, if a detailed conformational model is available. Eq. (8.17) can be used to compute $_mC$, which is then compared with the experimental result to check the validity of the model used in the calculation. A typical example is provided by the work of Rosenblatt and Griffin [240], who measured magnetic birefringence of chains containing a rigid core formed by a phenyl benzoate residue and a flexible spacer so that they have the capability of forming liquid crystals. On the basis of the results, they concluded that the spacer exhibits a nearly completely random conformation in the isotropic phase and extends significantly in the nematic phase.

When a full theoretical calculation cannot be performed, a comparison between the experimental results with those indicated in Eqs. (8.15) and (8.16) for freely jointed chains and rigid rods gives an idea of the general shape of the polymer. In the cases of stiff chains that do not follow a straight line, for instance, in many biological samples, a correction factor is used to take into account the bending of the polymer; thus, the specific Cotton–Mouton constant is written as [221, 241]

$$_sC = \frac{2\pi N_s (n^2 + 2)^2}{135\lambda nkTM_0} \Lambda\Lambda_x \left(\frac{2q}{3l_0} \right) \left\{ 1 - \frac{q}{3L} \left[1 - \exp\left(-\frac{3L}{q} \right) \right] \right\}$$

$$(8.21)$$

where q and L are the persistence and contour lengths, respectively, of the polymeric chain containing N_s segments, each having molecular weight $M_0 = M/N_s$ and length $l_0 = L/N_0$. Comparison of the experimental results with those computed with Eq. (8.21) allows the evaluation of q. In particular, if some measurements are performed with samples of different L (i.e., with different molecular weight), the value of q can be obtained even if the parameters Λ and Λ_x are not accurately known. Some very interesting examples can be found in reference [221].

9
Stress Birefringence

INTRODUCTION

The first empirical observation of stress birefringence was reported by Sir David Brewster in 1816. He found that some optically isotropic materials, glass, for instance, become anisotropic when stressed. Under those conditions, the samples had optical properties similar, in some aspects, to those of crystals, one of which was birefringence. This anisotropic behavior lasted while the stress was kept, but disappeared upon removal of the stress.

The basic equations relating the stress applied to a sample [242] with induced birefringence were formulated by Maxwell in 1852. Let us assume that an external mechanical force is applied to an isotropic sample. The stress thus produced can be decomposed in three components, σ_x, σ_y, and σ_z, over orthogonal axes, and the differences between the refractive indices in those directions, n_x, n_y, and n_z, and the main value for the unstressed sample n_0 are given by

$$
\begin{aligned}
n_x - n_0 &= a\sigma_x + b(\sigma_y + \sigma_z) \\
n_y - n_0 &= a\sigma_y + b(\sigma_x + \sigma_z) \\
n_z - n_0 &= a\sigma_z + b(\sigma_y + \sigma_x)
\end{aligned}
\tag{9.1}
$$

where a and b are constants that depend on temperature and the characteristics of the sample, but are independent of stress.

Combining Eq. (9.1), one obtains the birefringence between two main directions of stress:

$$
\begin{aligned}
n_x - n_y &= C(\sigma_x - \sigma_y) \\
n_x - n_z &= C(\sigma_x - \sigma_z) \\
n_y - n_z &= C(\sigma_y - \sigma_z)
\end{aligned}
\tag{9.2}
$$

where $C = a - b$ is the so-called *stress optical coefficient*. Equation (9.2) indicates that, by effect of stress, the sample exhibits three values of refractive indices such that the difference between two of them is proportional to the difference in stress between two perpendicular directions. This linear relationship between birefringence and stress is known as Brewster's law.

If the stress applied to the sample is bidirectional (for instance, $\sigma_z = 0$), Eq. (9.2) reduces to

$$
\begin{aligned}
n_x - n_y &= C(\sigma_x - \sigma_y) \\
n_x - n_z &= C\sigma_x \\
n_y - n_z &= C\sigma_y
\end{aligned}
\tag{9.3}
$$

In most measurements performed to characterize polymeric samples, a uniaxial stress is applied by stretching the sample in one direction, for example, x. Therefore, $\sigma_y = \sigma_z = 0$ and Eq. (9.2) reduces to

$$
n_x - n_y = n_x - n_z = C\sigma_x
\tag{9.4}
$$

Assuming that the beam of light travels through the sample in the z direction, if it was polarized either in the xz or yz planes, it will find just one value of refractive index and, therefore, its polarization will not change, although the intensity may be smaller than the original due to absorption and scattering. Any other polarization of the incident radiation can be described by two components polarized in the xz and yz planes, which will travel at different speeds through the sample and emerge with an angular phase difference that, according to Eq. (9.4), will be given by

$$
\delta = 2\pi l \, \Delta n / \lambda = 2\pi l C \sigma_x / \lambda
\tag{9.5}
$$

Therefore, values of birefringence obtained at different stresses should fit a straight line from whose slope the stress-optical coefficient C can be obtained.

THEORETICAL CALCULATIONS

A sample of polymeric material should have some kind of network structure in order to exhibit stress birefringence. The macromolecules forming the sample should be cross-linked at occasional points and those links should be permanent, at least while the experiment is performed; otherwise, the imposed stress and, therefore, the induced birefringence would be dissipated simply by the sliding of the macromolecules. This condition introduces some differences between stress and the other kind of induced birefringences in the sense that the network structure restrains the orientation of the molecules by effect of the external perturbation that produces the birefringence. In the context of strain birefringence, the word "orientation" means a deformation of the molecules that, on average, is larger in the direction of stress. The word "chain" also has a restricted meaning in the case of networks; it designates a portion of the macromolecule extending between two consecutive links.

On the other hand, the molecules of the sample should be able of undergoing deformations produced by the stress and of returning to their unperturbed isotropic state upon removal of stress. This condition eliminates small, or, in general, nonchain, molecules. For instance, individual molecules of vinyl chloride, $CH_2{=}CHCl$, exhibit electric and magnetic birefringence, however, they do not show stress birefringence unless a network is formed by polymerizing the individual molecules and then cross-linking the macromolecules of the resulting PVC. Moreover, this condition also eliminates samples of glassy, rigid, or crystalline polymers that are anisotropic even in absence of an external stress. As indicated before, many times it is interesting to measure the anisotropy of this kind of sample, without applying any external perturbation, in order to analyze the internal stress or the degree of order of such a sample and its evolution with time; however, if an external force is superimposed, it would be difficult to separate internal and external effects on the resulting birefringence.

In brief, measurements and calculations of stress birefringence are performed on networks formed from flexible, amorphous, and noncrystalline macromolecules when they are to be used to correlate the macroscopic property with the microscopic characteristics of the polymeric chains.

The first theoretical treatment of stress birefringence in polymer networks was developed by Kuhn and Grün [243] for the simplest model of the freely jointed chain. Of course, the use of such an unrealistic model imposes severe restrictions to the practical application of this first theory. However, besides the doubtless importance of any pioneer work, Kuhn and Grün were able to prove two interesting points, namely, the observed linear relationship between stress and the increment in refractive index, as indicated by the constant value of the stress optical coefficient, see Eq. (9.4), and the intrinsic connection between the stress birefringence and the elasticity of the polymeric networks. This connection should be quite evident because both macroscopic properties have the same microscopic origin: the change in the averaged molecular dimensions produced by the applied stress and achieved through a modification of the conformational distribution of the polymer [244]; however, Kuhn and Grün proved that both properties depend on the same function of the strain and of the network structure, as indicated by the degree of cross-linking.

A more realistic treatment allowing for the use of microscopic properties such as chemical nature and conformational features of the sample was started by Gotlib [245, 246] and Volkenstein [94] who initiated the application of the rotational isomeric states model for the evaluation of this property. The work of Nagai [247] allowed the calculation of the stress optical coefficient of practically any Gaussian polymeric chain. Finally, the matrix-multiplication scheme developed by Flory and co-workers [95, 96, 284] simplifies the calculations, making them possible for practically any microscopic features of the sample provided that it meets two conditions:

(a) The chains are long enough as to allow the distribution function of unstressed end-to-end vectors **r** to be described by a Gaussian function having spherical symmetry.

(b) Both the **r** vector and the optical polarizability tensor of the chain **α** can be obtained by addition of contributions $\mathbf{r}_i$ and $\boldsymbol{\alpha}_i$ from each bond or groups of bonds of the polymer.

In the case of dimensions, the contributions $\mathbf{r}_i$ are simply the skeletal bond lengths and it is commonly accepted that they are additive. However, the obtainment of contributions $\boldsymbol{\alpha}_i$ that could be added to calculate **α** is troublesome (see the section on the additivity of bond contributions), and even more complicated in the case of stress birefringence due to intermolecular correlation either between different chains on the sample or between chains and molecules of solvent eventually used to swell the network (see what follows). At this point, we will assume that such contributions are known.

As in the case of electric birefringence, the stress optical coefficient is calculated by evaluation of the averaged difference between optical polarizabilities in the direction of the external perturbation (direction of the x axis) and a direction perpendicular to it (the y axis, assuming that the beam of light used in the measurement travels in the z direction). The Lorentz–Lorenz relationship between polarizabilities and refractive indices is then used to obtain the birefringence Δn; see Eq. 7.12 in Chapter 7.

Let us take a polymeric chain extending between two consecutive junctions of the network and assume that those links are fixed. Then, both extremes of the chain, and, therefore, its end-to-end vector **r**, would be fixed, too. However, despite this restriction, the chain may still adopt any conformation that gives that particular value of **r**; the situation is similar to the string of a guitar that can vibrate even if it has both extremes tied. But the constraint of having both extremes fixed imposes cylindrical symmetry to the average of any magnitude taken over all those allowed conformations. Therefore, the average of the optical polarizability tensor **α** of the chain should exhibit cylindrical symmetry over the direction of **r**; thus, when written in a coordinate system having one of the axes along **r** and the other two perpendicular to it, the tensor should be diagonal and the two transverse components should be equal. Therefore,

$$(\boldsymbol{\alpha})_r = \text{diag}(\alpha_r, \alpha_t, \alpha_t) \tag{9.6}$$

where the symbol $(\boldsymbol{\alpha})_r$ represents the average over all the conformations that are compatible with that particular value of the vector $\mathbf{r}$ written in its own molecular reference frame. The subscripts r and t on the elements of $(\boldsymbol{\alpha})_r$ designate longitudinal (along $\mathbf{r}$) and transverse components, respectively. This tensor can be written in a more convenient way by defining the main polarizability α_0 as

$$\alpha_0 = \tfrac{1}{3}\text{Trace}\,[(\boldsymbol{\alpha})_r] = \tfrac{1}{3}(\alpha_r + 2\alpha_t) \tag{9.7}$$

then

$$\alpha_t = (3\alpha_0 - \alpha_r)/2 \tag{9.8}$$

and

$$(\boldsymbol{\alpha})_r = \text{diag}[\alpha_r, (3\alpha_0 - \alpha_r)/2, (3\alpha_0 - \alpha_r)/2] \tag{9.9}$$

If the molecular reference frame is rotated into coincidence with any arbitrarily chosen xyz system, the components of the $(\boldsymbol{\alpha})_r$ tensor in the new system are given by

$$\begin{aligned}(\alpha_{xx})_r &= \alpha_r \cos^2\tau_r^x + \alpha_t \sin^2\tau_r^x \\ &= \tfrac{1}{2}(\alpha_r - \alpha_0)(3\cos^2\tau_r^x - 1) + \alpha_0\end{aligned} \tag{9.10}$$

where τ_r^x is the angle between the $\mathbf{r}$ vector and the x axis. This angle can be eliminated by using the relationship between the $\mathbf{r}$ vector and its projection over the x axis: $x = \mathbf{r}\cos\tau_r^x$. Then

$$x^2 = \mathbf{r}^2 \cos^2\tau_r^x \tag{9.11}$$

The term $\alpha_r - \alpha_0$ appearing in Eq. (9.10) can be related to the difference between longitudinal and transverse polarizabilities ($\Delta\alpha_r = \alpha_r - \alpha_t$) by using Eq. (9.8) to write

$$\Delta\alpha_r = \alpha_r - \alpha_t = \tfrac{3}{2}(\alpha_r - \alpha_0) \tag{9.12}$$

Substituting Eqs. (9.11) and (9.12) into (9.10) gives

$$(\alpha_{xx})_r = \tfrac{1}{3}\,\Delta\alpha_r(3x^2/r^2 - 1) + \alpha_0 \tag{9.13}$$

In a similar way, the other two diagonal elements can be written as

$$(\alpha_{yy})_r = \tfrac{1}{3}\,\Delta\alpha_r(3y^2/r^2 - 1) + \alpha_0 \tag{9.14}$$

$$(\alpha_{zz})_r = \tfrac{1}{3}\,\Delta\alpha_r(3z^2/r^2 - 1) + \alpha_0 \tag{9.15}$$

The anisotropy of the sample is produced by the difference between the xx and yy components, which according to Eqs. (9.13) and (9.14) is given by

$$(\alpha_{xx} - \alpha_{yy})_r = \Delta\alpha_r(x^2 - y^2)/r^2 \qquad (9.16)$$

However, Eq. (9.16) has been obtained for a fixed **r**, which means fixed molecular dimension and orientation of the chain. Therefore, it should be averaged over all possible orientations and lengths of the chain in the network before using it to evaluate a macroscopic property of the sample. A molecular model of both the network and the applied strain is required to perform this average and the difficulty of the whole procedure will be determined by the complexity of the model used.

Thus, it is very easy to average Eq. (9.16) for the unstressed sample. If the sample is in thermodynamical equilibrium and no external perturbation is applied to it, all the directions of space are equivalent. Therefore, any orientation of **r** has the same probability of occurrence, and the average of the components along three axes arbitrarily chosen should be the same; consequently,

$$\langle x^2 - y^2 \rangle = 0 \quad \text{and} \quad \langle \alpha_{xx} - \alpha_{yy} \rangle = 0 \qquad (9.17)$$

meaning that the sample is isotropic, as it was assumed from the beginning.

In order to perform the average of Eq. (9.16) for a stressed sample, three consecutive steps are taken:

(a) The chains are supposed to be formed by n_s segments, each having an optical polarizability tensor $\boldsymbol{\alpha}_i$.

(b) A fixed vector **r** is taken and the contributions $\boldsymbol{\alpha}_i$ of each segment of the chain are added and averaged to obtain the tensor $(\boldsymbol{\alpha})_r$ (given by Eq. (9.6)) and the parameter $\Delta\alpha_r$ (according to Eq. (9.12)) for all the conformations of the chain that give that particular **r**.

(c) The value of $\Delta\alpha_r$ obtained in the previous step is substituted into Eq. (9.16), which is then averaged over all the possible values of the **r** vector.

Let us perform these three steps for the simplest conformational model, which follows.

The Freely Jointed Chain

Let us assume that the segments of the chain have cylindrical symmetry so that the $\boldsymbol{\alpha}_i$ for each segment is diagonal when written in a molecular reference frame having one of the axes in the direction of the segment and the other two perpendicular to it. Then:

$$\boldsymbol{\alpha}_i = \text{diag}(\alpha_p, \alpha_n, \alpha_n)_i \qquad (9.18)$$

where subscripts p and n indicate components along directions parallel and perpendicular (normal) to the axis of the segment, respectively.

The orientation of the ith segment with respect to axes r, t, and t used to write the $\boldsymbol{\alpha}_r$ tensor for the whole chain can be determined by the angles θ and ϕ of a standard set of polar coordinates. Thus, θ is the angle between the directions of the segment (p) and the **r** vector (r), and ϕ is defined as the angle between the projection of the segment over the tt plane and one of the t axes (see Figure 9.1). Rotation of the p, n, and n axes into coincidence with r, t, and t gives the components of $\boldsymbol{\alpha}_i$ in the molecular frame used for the **r** vector as

$$(\alpha_r)_i = [\alpha_p \cos^2\theta + \alpha_n \sin^2\theta]_i \qquad (9.19)$$

$$(\alpha_t)_i = [(\alpha_p - \alpha_n) \sin^2\theta \cos^2\phi + \alpha_n]_i \qquad (9.20)$$

These contributions should be added for all the segments of the chain and the result averaged over all the conformations giving the fixed value of **r**. Let Θ represent

$$\Theta = \langle \sin^2\theta \rangle = 1 - \langle \cos^2\theta \rangle \qquad (9.21)$$

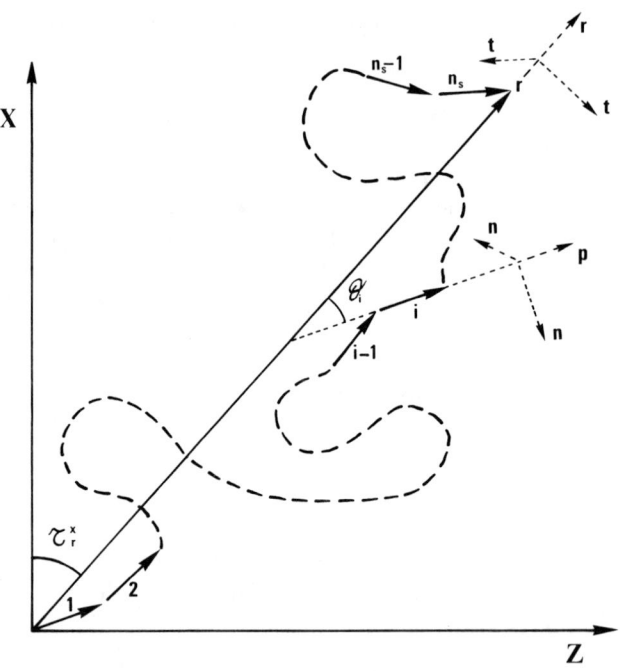

Figure 9.1 Laboratory-fixed coordinate axes x and z defined by the directions of stress and propagation, respectively, of the laser beam. The orientation of any particular end-to-end vector **r** is defined by the angle τ_r^x. The chain having this particular **r** may adopt different conformations, one of which is shown here. The orientation of a given segment i with respect to **r** is defined by the angle θ_i. Coordinate system rtt is used to write the $(\boldsymbol{\alpha})_r$ tensor for the whole chain (Eq. 9.6) and pnn is employed for the tensor $\boldsymbol{\alpha}_i$ of the ith segment (Eq. 9.18).

However, for a freely jointed chain, all the values of ϕ are equivalent and, therefore,

$$\langle \cos^2\phi \rangle = \tfrac{1}{2} \tag{9.22}$$

Then, assuming that all the segments are identical, the components of the tensor for the whole chain having a predetermined value of $\mathbf{r}$ are

$$\alpha_r = \langle \sum_{i=1}^{n_s} (\alpha_r)_i \rangle_r = n_s[\alpha_p(1 - \Theta) + \alpha_n\Theta)] \tag{9.23}$$

$$\alpha_t = \langle \sum_{i=1}^{n_s} (\alpha_t)_i \rangle_r = n_s\left[(\alpha_p - \alpha_n)\frac{\Theta}{2} + \alpha_n \right] \tag{9.24}$$

and the parameter $\Delta\alpha_r$ defined in Eq. (9.12) becomes

$$\begin{aligned}
\Delta\alpha_r &= \alpha_r - \alpha_t = n_s(\alpha_p - \alpha_n)(1 - 3\Theta/2) \\
&= n_s \, \Delta a(1 - 3\Theta/2)
\end{aligned} \tag{9.25}$$

where

$$\Delta a = \alpha_p - \alpha_n \tag{9.26}$$

is the optical anisotropy (difference between parallel and normal polarizabilities) of one segment of the chain.

The value of Θ is calculated using the following relationships:

$$\langle \cos^2\theta \rangle = 1 - \langle \sin^2\theta \rangle = 1 - \Theta = \langle (l_r/l)^2 \rangle \tag{9.27}$$

where l is the length of the segment, and l_r represents its projection in the direction of the end-to-end vector $\mathbf{r}$. By using the distribution function for the segments of a freely jointed chain, the last average appearing in Eq. (9.27) is obtained as

$$\begin{aligned}
\langle (l_r/l)^2 \rangle &= 1 - 2\beta^{-1}\mathcal{L}(\beta) \\
&\cong \frac{1}{3} + \frac{2}{5}\left(\frac{r}{n_s l} \right)^2 + \frac{24}{175}\left(\frac{r}{n_s l} \right)^4 + \cdots
\end{aligned} \tag{9.28}$$

where $r/n_s l$ is the ratio between the actual length of the chain r and the maximum length that would be obtained if all the segments were placed over a straight line. The symbol $\mathcal{L}(\beta)$ in Eq. (9.28) indicates the Langevin function and the coefficient β can be related to the ratio $r/n_s l$ through the equation

$$\left(\frac{r}{n_s l} \right) = \mathcal{L}(\beta)$$

and can be written as a power expansion:

$$\beta = 3\left(\frac{r}{n_s l} \right) + \frac{9}{5}\left(\frac{r}{n_s l} \right)^3 + \frac{297}{175}\left(\frac{r}{n_s l} \right)^5 + \cdots \tag{9.29}$$

If the number of segments n_s is large and the stress is not too high so that the chain is far from fully extended, $r/n_s l \ll 1$ and terms of powers higher than $(r/nsl)^2$ can be neglected in Eq. (9.28). Using this approximation to compute Θ according to Eq. (9.27) and substituting its value into Eq. (9.25) gives the parameter $\Delta\alpha_r$ as

$$\Delta\alpha_r = \frac{3}{5} n_s \Delta a \beta^2 = \frac{3}{5} \Delta a \frac{r^2}{n_s l^2} = \frac{3 \Delta a r^2}{5 \langle r^2 \rangle_0} \tag{9.30}$$

where $\langle r^2 \rangle_0 = n_s l^2$ is the mean-square end-to-end distance for an unstressed freely jointed chain.

Substituting Eq. (9.30) into Eq. (9.16) gives the optical anisotropy of the chains having the fixed value of $\mathbf{r}$ referred to the laboratory coordinate system xyz as

$$(\alpha_{xx} - \alpha_{yy})_r = \frac{3 \Delta a}{5} \frac{x^2 - y^2}{\langle r^2 \rangle_0} \tag{9.31}$$

The next step consists in averaging Eq. (9.31) over all possible values of the $\mathbf{r}$ vector (including both modulus and direction) for the stressed network, and a microscopic model of the deformation of the sample under stress is required in order to perform this average.

Let us assume that the network is unstressed and $\langle r^2 \rangle_u$ is the mean-square end-to-end distance of the chains on the network. The difference between $\langle r^2 \rangle_u$ and $\langle r^2 \rangle_0$ is that the former corresponds to chains of a network that therefore have both ends linked to other chains of the sample, and the second magnitude is the value for the same chains having both ends free.

In absence of external perturbation, the three directions of space are equivalent and, therefore, the projections over three arbitrary axes are

$$\langle x^2 \rangle_u = \langle y^2 \rangle_u = \langle z^2 \rangle_u = \langle r^2 \rangle_u / 3 \tag{9.32}$$

and, therefore, the average of Eq. (9.31) vanishes as it was indicated before.

During the experimental measurements, a mechanical force is applied in a given direction (the x axis in the laboratory reference frame). The deformation of the sample under a given force can be represented by three elongation ratios defined as the lengths of the sample along the three coordinate axes under a given force, divided by their corresponding values in the unperturbed (i.e., $F = 0$) state. Thus,

$$\begin{aligned} \lambda_x &= L_x^F / L_x^u \\ \lambda_y &= L_y^F / L_y^u \\ \lambda_z &= L_z^F / L_z^u \end{aligned} \tag{9.33}$$

The simplest model describing both the macroscopic and the microscopic behaviors of the network is defined by the three following assumptions:

(a) The volume of the sample is supposed to be independent of the elongation, that is,

$$\lambda_x \cdot \lambda_y \cdot \lambda_z = 1 \quad \text{or} \quad \lambda = \lambda_x = 1/(\lambda_y \cdot \lambda_z) \tag{9.34}$$

(b) The two transverse directions are supposed to undergo the same deformation, that is,

$$\lambda_y = \lambda_z = \lambda^{-1/2} \tag{9.35}$$

Therefore, a single elongation ratio λ represents the deformation of the sample.

(c) Chains of the network are supposed to suffer, on average, the same deformations than the macroscopic sample; thus,

$$(L_x^F)^2 = (L_x^u)^2\lambda_x^2 = (L_x^u)^2\lambda^2 \rightarrow \langle x^2\rangle_F = \langle x^2\rangle_u\lambda^2 = \lambda^2\langle r^2\rangle_u/3 \tag{9.36}$$

$$(L_y^F)^2 = (L_y^u)^2\lambda_y^2 = (L_x^u)^2/\lambda \rightarrow \langle y^2\rangle_F = \langle y^2\rangle_u/\lambda = \langle r^2\rangle_u/3\lambda \tag{9.37}$$

Assumptions (a) and (b), Eqs. (9.34) and (9.35) represent a *simple elongation* of the macroscopic sample, and the last assumption, Eqs. (9.36) and (9.37) and the equivalent one for z, is the so-called *phantom network model*.

Substituting Eqs. (9.36) and (9.37) into Eq. (9.31) gives

$$\langle\alpha_{xx} - \alpha_{yy}\rangle = \frac{\Delta a}{5}\frac{\langle r^2\rangle_u}{\langle r^2\rangle_0}(\lambda^2 - \lambda^{-1}) \tag{9.38}$$

The averaged optical anisotropy given in Eq. (9.38) can be related to the difference in refractive indices, Δn, by using the Lorentz–Lorenz expression that could be written as

$$\frac{n^2 - 1}{n^2 + 2} = \frac{4\pi}{3}\frac{dN}{M}\alpha = \frac{4\pi}{3}\frac{v}{V}\alpha \tag{9.39}$$

where v is the number of chains in volume V.

Applying Eq. (9.39) for both x and y components and assuming that n_x and n_y are not too different from the refractive index of the unstressed sample n, so that $(n_x^2 + 2)(n_y^2 + 2) \approx (n^2 + 2)^2$ and $n_x + n_y \approx 2n$, one obtains

$$\Delta n = n_x - n_y = \frac{2\pi}{9}\frac{v}{V}\frac{(n^2 + 2)^2}{n}\langle\alpha_{xx} - \alpha_{yy}\rangle$$

$$= \frac{2\pi\,\Delta a(n^2 + 2)^2}{45n}\frac{v\langle r^2\rangle_u}{V\langle r^2\rangle_0}(\lambda^2 - \lambda^{-1}) \tag{9.40}$$

According to the theory of rubber elasticity, in the case of a phantom network, the stress σ and the elongation ratio λ are related by the equation

$$\sigma = \frac{kTv\langle r^2\rangle_u}{V\langle r^2\rangle_0}(\lambda^2 - \lambda^{-1}) \tag{9.41}$$

Substituting Eq. (9.41) into (9.40) gives the birefringence induced in the sample by the stress σ as

$$\Delta n = \frac{2\pi\,\Delta a(n^2 + 2)^2}{45kTn}\sigma = C\sigma \tag{9.42}$$

Equation (9.42) indicates that birefringence Δn is proportional to the stress σ according to Brewster's law. The stress-optical coefficient C can be evaluated as

$$C = \frac{2\pi \, \Delta a (n^2 + 2)^2}{45kTn} \tag{9.43}$$

where the only microscopic property of the chain appearing on the final expression of C is the anisotropy of the segment Δa defined in Eq. (9.26). If this parameter is known, the value of C can be easily calculated. Conversely, experimental measurement of C allows the evaluation of Δa as

$$\Delta a = \frac{45kTnC}{2\pi(n^2 + 2)^2} \tag{9.44}$$

Real Chains

The calculation of C for real chains is almost identical to the one shown before for the freely jointed model although there are two noteworthy differences in:

(a) the treatment of the contributions α_i of each segment to the optical polarizability tensor of the chain α

(b) the conformational model used to perform the averages

Let us represent by α_i the actual contribution of skeletal bond i (or a group of bonds associated to it) to the polarizability tensor of the chain. By using 1, 2, and 3 as the main axes of α_i, this tensor adopts the diagonal form when written in that molecular frame. Thus,

$$\alpha_i = \mathrm{diag}(\alpha_1, \alpha_2, \alpha_3)_i \tag{9.45}$$

Equation (9.45) differs from its counterpart in the freely jointed model, Eq. (9.18), in two aspects. First, α_i represented in Eq. (9.45) may not be cylindrically symmetric, that is, α_2 may be different than α_3. Moreover, the principal axis 1 may not lie in the direction of the ith skeletal bond. Thus, the location of the main axes and the values of the three components of α_i depend on the chemical structure of the ith segment.

The component of α_i in the direction of the end-to-end vector of the chain $\mathbf{r}$ can be calculated with an equation similar to Eq. (9.19):

$$(\alpha_r)_i = (\alpha_1 \cos^2\tau_1^r + \alpha_2 \cos^2\tau_2^r + \alpha_3 \cos^2\tau_3^r)_i \tag{9.46}$$

where τ_j^r represents the angle between the j axis ($j = 1, 2, 3$) and the direction of r.

The total value of the α_r component is obtained by adding Eq. (9.46) over all the segments of the chain and averaging over all the conformations allowed for the fixed vector $\mathbf{r}$. The mathematical details of this average are given elsewhere [95].

The result for a Gaussian chain having a large number of segments under small extensions is

$$\langle \cos^2 \tau_j^r \rangle_r \cong \frac{1}{3} + \frac{1}{5} \left(\frac{3 \langle r^2 \cos^2 \tau_j^r \rangle_0}{\langle r^2 \rangle_0} - 1 \right) \frac{r^2}{\langle r^2 \rangle_0} \tag{9.47}$$

where the subscript r indicates the average over those conformations of the chain giving the particular value of $\mathbf{r}$, and subscript 0 represents averages for all the conformations allowed to a chain with the same number of segments but free of any constraint. Therefore, Eq. (9.47) relates an average over the conformations having a predetermined value of $\mathbf{r}$, which is difficult to evaluate, with some other averages for an unconstrained chain that can be performed using an adequate conformational model of the polymer, rotational isomeric states, for instance.

The actual form of Eq. (9.47) is a series expansion on terms $r^{2m}/\langle r^2 \rangle_0$. Terms with $m > 1$ are neglected under the approximation of small elongations assuming that, on average, r^2 is not much larger than the unperturbed value; therefore, $r^4/\langle r^2 \rangle_0 \sim 1/\langle r^2 \rangle_0 \ll 1$ and can be neglected.

Substituting Eq. (9.47) into (9.46) gives

$$(\alpha_r)_i = \frac{1}{3} (\alpha_1 + \alpha_2 + \alpha_3) + \frac{r^2}{5 \langle r^2 \rangle_0} \sum_{j=1}^{3} \left(\frac{3 \langle \alpha_j r^2 \cos^2 \tau_j^r \rangle_0}{\langle r^2 \rangle_0} - \alpha_j \right)$$

$$= \alpha_0 + \frac{3 r^2}{5 \langle r^2 \rangle_0} \left[\left(\sum_{j=1}^{3} \frac{\langle \alpha_j r^2 \cos^2 \tau_j^r \rangle_0}{\langle r^2 \rangle_0} \right) - \alpha_0 \right] \tag{9.48}$$

The sum appearing on Eq. (9.48) is very similar to that in Eq. (7.88) in Chapter 7 for the Kerr constant and can be handled with the same procedure used there. Thus, $\mathbf{r} \cos \tau_j^r$ represents the projection of the $\mathbf{r}$ vector over the axis j used to define tensor $\boldsymbol{\alpha}_i$, and, therefore, using the same coordinate system for both $\mathbf{r}$ and $\boldsymbol{\alpha}_i$,

$$\left(\sum_{j=1}^{3} \alpha_j r^2 \cos^2 \tau_j^r \right)_i = \mathbf{r}^T \boldsymbol{\alpha}_i \mathbf{r} \tag{9.49}$$

Substituting Eq. (9.49) into (9.48) and then into (9.12) gives the contribution of the ith segment to the anisotropy of the chain as

$$(\Delta \alpha_r)_i = \tfrac{3}{2} [(\alpha_r)_i - (\alpha_0)_i]$$

$$= \frac{9 r^2}{10 \langle r^2 \rangle_0} \left[\frac{\langle \mathbf{r}^T \boldsymbol{\alpha}_i \mathbf{r} \rangle_0}{\langle r^2 \rangle_0} - \alpha_0 \right] \tag{9.50}$$

Introducing the anisotropic part of the polarizability tensor defined in Eq. (7.97): $\hat{\boldsymbol{\alpha}}_i = \boldsymbol{\alpha} - \alpha_0 \mathbf{I}_3$, Eq. (9.50) reduces to

$$(\Delta \alpha_r)_i = \frac{9 r^2}{10 \langle r^2 \rangle_0} \left[\frac{\langle \mathbf{r}^T \hat{\boldsymbol{\alpha}}_i \mathbf{r} \rangle_0}{\langle r^2 \rangle_0} \right] \tag{9.51}$$

Addition over all the segments gives the anisotropy of the chain:

$$\Delta\alpha_r = \frac{9r^2}{10\langle r^2\rangle_0} \left[\frac{\langle \mathbf{r}^T\hat{\boldsymbol{\alpha}}\mathbf{r}\rangle_0}{\langle r^2\rangle_0} \right] \tag{9.52}$$

Equation (9.52) obtained for real chains becomes identical to its counterpart in the case of a freely jointed chain, Eq. (9.30), if a fictitious anisotropy of a freely jointed segment (or a Kuhn segment), Δa is defined as

$$\Delta a = \frac{3}{2} \left[\frac{\langle \mathbf{r}^T\hat{\boldsymbol{\alpha}}\mathbf{r}\rangle_0}{\langle r^2\rangle_0} \right] \tag{9.53}$$

Δa is the so-called *optical configuration parameter* and represents the anisotropy that would be required for a segment of a freely jointed chain in order to have the same stress birefringence than the actual chain under study.

Then, the stress-optical coefficient can be obtained by substituting Eq. (9.53) into (9.43) as

$$C = \frac{2\pi \,\Delta a(n^2 + 2)^2}{45kTn} = \frac{2\pi(n^2 + 2)^2}{45kTn} \left[\frac{3\langle \mathbf{r}^T\hat{\boldsymbol{\alpha}}\mathbf{r}\rangle_0}{2\langle r^2\rangle_0} \right] \tag{9.54}$$

The only difference between Eq. (9.43) for the freely jointed model and Eq. (9.54) for real chains is that in the former case, Δa represents the actual anisotropy of one of the segments on the chain, whereas in the real case, it should be obtained as the ratio of two averaged magnitudes that are calculated using a realistic conformational model.

The term $\langle \mathbf{r}^T\hat{\boldsymbol{\alpha}}\mathbf{r}\rangle$ appearing in Eq. (9.54) is formally identical to the $\langle \boldsymbol{\mu}^T\hat{\boldsymbol{\alpha}}\boldsymbol{\mu}\rangle$ of the Kerr constant, and the unperturbed value of the mean-square end-to-end distance $\langle r^2\rangle_0$ is similar to $\langle \mu^2\rangle$. The procedure used to compute all these averages is explained in the Appendix.

Thus, the theoretical calculation gives the value of Δa that can be transformed into stress-optical coefficient C according to Eq. (9.54) and compared with experimental results. Conversely, experimental values of C can be converted into optical configuration parameter Δa according to Eq. (9.44) and compared with the theoretical results. In practice, this second procedure is the most frequently used and most comparisons are performed on the basis of Δa.

It is interesting to note that the stress optical coefficient does not depend on the degree of cross-linking of the network. The reason is that this magnitude appears as the number of chains per unit volume (ν/V) in the relationship between birefringence Δn and elongation λ, Eq. (9.40), but it also appears in the equation relating λ and stress σ, Eq. (9.41). Consequently, the ratio $C = \Delta n/\sigma$ is independent of ν. This behavior is confirmed by experimental measurements provided that the cross-linking agent does not contribute to the birefringence of the sample and that the chains are long enough as to be Gaussian.

Something similar should happen if the network is swollen with an adequate solvent. Assume that the solvent itself is not affected by the stress so that its molecules are randomly oriented and therefore isotropic on average even when the sample is strained. Considering also that the optical polarizability of the network is not modified by the presence of solvent, the only effect of swelling should be to change the average of the unstressed dimensions of the chains, $\langle r^2 \rangle_u$. But, again, this magnitude appears both in Eqs. (9.40) and (9.41), so that any modification vanishes when both equations are divided to obtain C. However, this analysis is not complete, because the presence of solvent changes the main refractive index of the network and, therefore, the term $(n^2 + 2)^2/n$ appearing in C should depend on both the nature and quantity of the solvent used. Consequently, the theory predicts a small variation of C upon swelling produced by changes in the main refractive index of the network. Experimental measurements, however, show variations of C in swollen samples that cannot be explained by the change in refractive index. These changes are attributed to modifications on the polarizability of the segments of the chain produced by solvation.

A more recent theory of strain birefringence has been developed by Erman and Flory [249] by extension of their rubber elasticity theory. They introduce some constraints on the fluctuations of the molecular dimensions that impart a pattern of behavior intermediate between the extremes of phantom and affine networks. In contrast with those theories, they predict a nonlinear variation of Δn with σ, that is, the stress-optical coefficient $C = \Delta n/\sigma$ should decrease with elongation.

EXPERIMENTS

Preparation of the Samples

The samples used in stress optical measurements are strips (about $1 \times 10 \times 50$ mm) of a network formed by the polymer to be studied. Thus, a step previous to any kind of measurement is the preparation of a suitable sample. The procedure used to prepare the sample depends on the polymer under study, but, in general, it implies cross-linking the polymer, molding the network to obtain the desired geometry, and characterizing the resulting sample.

Sometimes, the network structure can be directly obtained in the polymerization reaction using an appropriate amount of polyfunctional monomers. However, it is more frequent to obtain linear chains of the polymer and cross-link them by using either high-energy radiation (with frequencies in the X or γ regions) or an adequate chemical cross-linking agent. A typical procedure for this second option consists in casting a solution containing the adequate proportions of both the polymer and the cross-linking agent into a mold, removing the solvent by evaporation, curing the homogeneous mixture several hours at moderate temperature (about 24 to 72 h at 70–90°C), and finally extracting all the nonnetwork material with a solvent.

Density ρ and refractive index n of the network are measured as functions of temperature within the range to be covered in the stress measurements. The physical dimensions and volume of the sample are carefully measured and two marks are drawn at a given separation; these marks are used to measure the deformation of the sample by determining their separation, with the aid of a cathetometer, as functions of the applied force and temperature.

Assuming that the total volume of the sample does not change with the applied stress, the cross-section S at a given force can be obtained from its unperturbed value as $S = S_0\lambda_y \cdot \lambda_z = S_0/\lambda$, and, therefore, the true stress σ appearing in the definition of the stress optical coefficient is given by $\sigma = \lambda F/S_0$. However, the volume of the sample changes with temperature, and, for this reason, either the thermal expansion coefficient or the density as a function of temperature is required.

Frequently, measurements are performed on samples swollen with an adequate solvent. In these cases, the network is allowed to reach equilibrium with the appropriate amount of swelling agent, and volume fractions of solvent and network are required in addition to the regular data on dimensions, refractive index, and density (or thermal expansion coefficient) of the swollen sample.

Apparatus

The experimental setup used in stress birefringence measurements [147, 250–254] is very similar to that employed for electric birefringence with only two main differences, namely, the sample cell, including the device used to produce and measure the mechanical force applied to the sample, which of course is different than in the case of an electric field, and the detection system, which usually is less sophisticated in the case of stress measurements.

Figure 9.2 is a block diagram of a standard setup for stress birefringence measurements. The light source is usually a low-power laser, for example, a He–Ne laser of about 2 mW. The polarizer is oriented at 45° with the direction of the stress

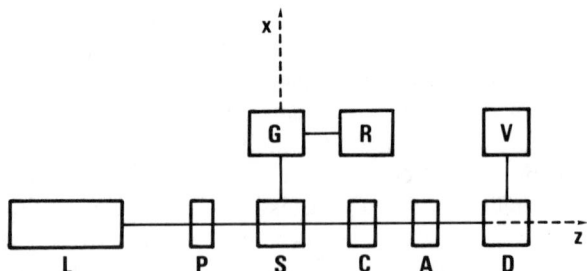

Figure 9.2 Block diagram of the experimental setup for stress birefringence measurements: (L) laser, (P) polarizer, (A) analyzer, (S) sample, (G) strain gauge, (R) register, (D) detector, (V) voltmeter, and (C) compensator or quarter-wave plate.

and the analyzer is crossed with it. The orientation of the polarizer can be easily determined, as in the case of electric birefringence, by locating the orientation in which the applied stress produces no birefringence, indicating that the plane of polarization of the incident light contains the direction in which the stress is applied (i.e., it is polarized in the zx plane), and rotating 45° from that position.

The birefringence produced during stress experiments is usually much larger that in the case of electric fields. Moreover, only equilibrium values are measured and, therefore, no variation of birefringence with time has to be recorded. These two characteristics allow the use of very simple detection systems, which, in general, are formed by a photodiode, many times covered by a narrow-band interference filter centered at the wavelength of the laser, whose signal is shown in a digital voltmeter. The measurements are performed using the compensation method with either a Babinet–Soleil compensator or a combination of a quarter-wave plate with an analyzer mounted on a precision rotary stage; in both cases, the manipulations required to obtain minimum light intensity are manually performed.

A diagram of a sample cell commonly used is shown in Figure 9.3. The sample strips are mounted between two clamps: the upper one hangs from a inextensible wire attached to a strain gauge and the lower one can be lowered by means of a screw. The force thus applied is measured by the strain gauge, whose signal is recorded by a register. The glass sample cell has two small windows of negligible birefringence in the path of the laser beam; the rest is surrounded by a double-walled jacket allowing the circulation of water or any other liquid from a thermostat; the temperature is measured by a digital thermometer whose probe is placed close to the sample.

Measuring Procedure

Usually, both stress–strain and stress–optical measurements are performed simultaneously by two sets of experiments, one at fixed temperature, changing the elongation of the sample, and another one at constant elongation with different temperatures.

Measurements at constant temperature are performed stretching the sample to the maximum desired elongation at a given temperature and allowing it to reach equilibrium, which is indicated by constant values of both force and light intensity reaching the detector. The birefringence is then measured by moving the compensator (or rotating the analyzer) until the minimum light intensity is obtained. Then, the elongation of the sample is decreased and a new measurement is performed after equilibrium has been reached. The procedure is repeated until the lowest desired elongation is reached. The reversibility of the process is checked by increasing the elongation again and performing some measurements in the opposite sequence (i.e., from low to high elongations). The whole process is repeated at different temperatures, covering the interval to be studied. Values of the measured birefringences are written as Δn and plotted versus strain; they fit a straight line whose slope is the

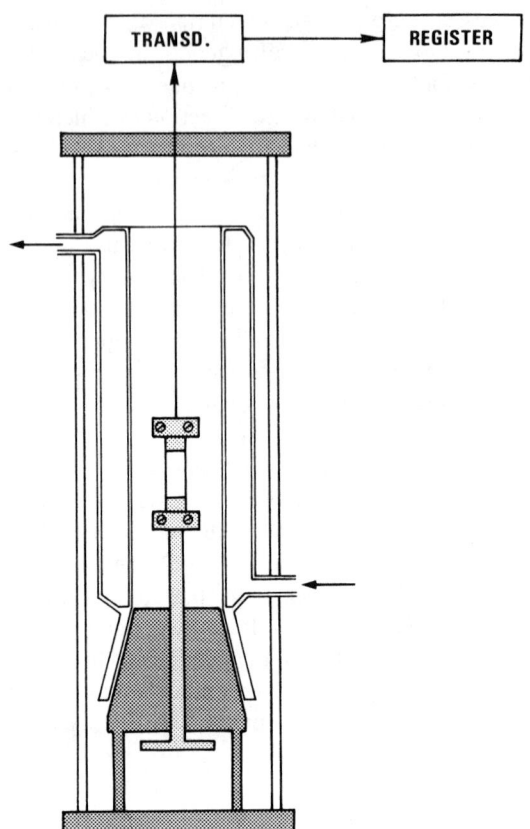

Figure 9.3 Diagram of the sample cell for stress optical measurements. The direction of the laser beam is perpendicular to the plane of the figure.

stress–optical coefficient C. Some typical results of these measurements are shown in Figure 9.4.

Measurements at constant elongation are performed by stretching the sample to the desired elongation at the highest temperature to be measured. Once the equilibrium has been reached and the measurement performed, the temperature is decreased to a new value and the process repeated until the lowest measuring temperature is reached. Some measurements are also performed in the opposite sequence to check the reversibility. After the whole measurement has been performed for a given elongation, it is repeated for a different value of that parameter until the desired measuring range has been covered. The results are shown by plotting Δn versus temperature at constant elongation ratios, as indicated in Figure 9.5.

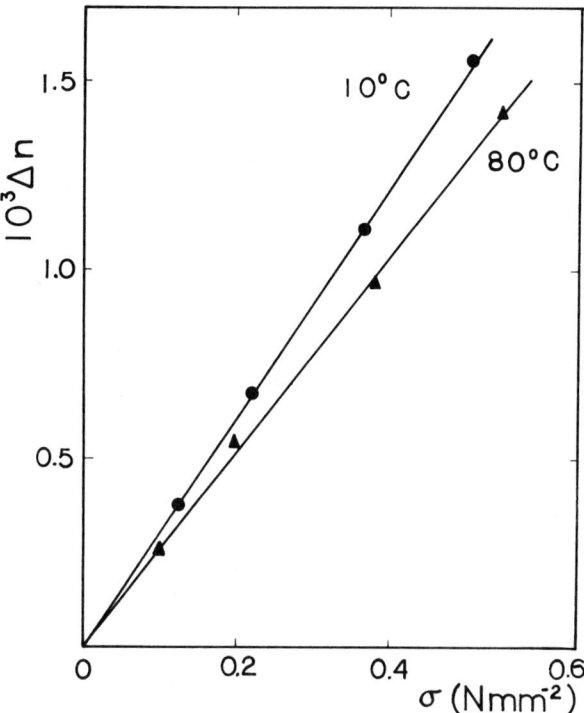

Figure 9.4 Typical values of birefringence versus stress. The results correspond to a sample of poly(cis/trans-1,4-cyclohexane dimethanol-alt-formaldehyde) network (PCDO). (Reprinted from reference [253] by courtesy of John Wiley & Sons, Inc.)

Values of stress-optical coefficient C are commonly converted into optical configuration parameter Δa using Eq. (9.44) and compared with the results of the theoretical calculation.

According to theory, the birefringence Δn should be a linear function of the applied stress σ. Good linearity is normally found in experimental measurements with two noteworthy exceptions:

(a) samples that can crystallize under stress
(b) very high elongations (close to the breaking point) in highly cross-linked samples

If the measured sample undergoes stress-induced crystallization, the results of Δn versus σ show a very sharp upturn over the linear behavior; in fact, this is one of the most sensitive ways of detecting this kind of process. The second nonlinear behavior is found with short chains under high elongations; in this situation, the chains are not Gaussian and departures from the theoretical predictions, obtained

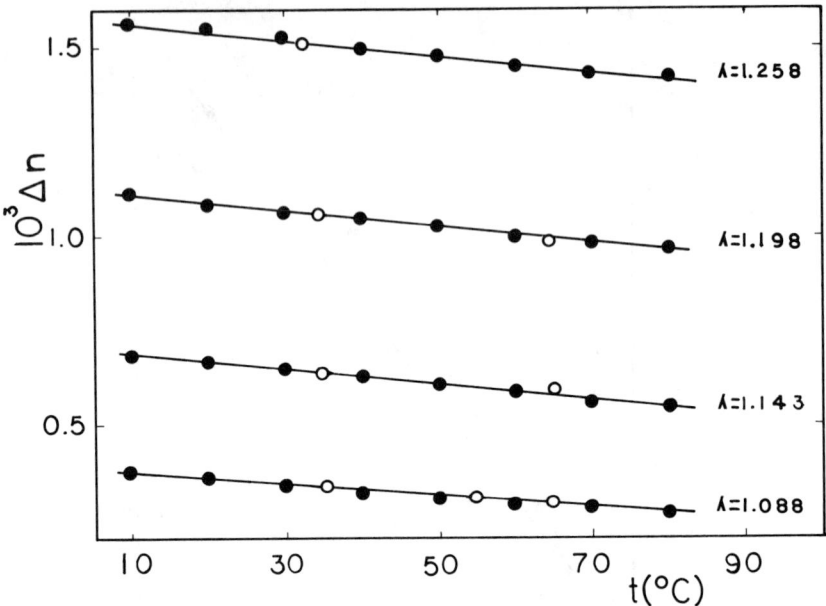

Figure 9.5 Typical values of birefringence versus temperature at constant elongation. Results from a sample of a PCDO network. (Reprinted from reference [253] by courtesy of John Wiley & Sons, Inc.)

assuming Gaussian behavior, occur. Therefore, if the experimental results show linear variation of Δn with σ, it can be asserted that there is no crystallization and that the chains of the network are Gaussian.

COMPARISON BETWEEN THEORETICAL AND EXPERIMENTAL RESULTS

The methods used for the analysis of the stress birefringence of a given sample and the practical applications of such an analysis are, in principle, identical to the case of electric birefringence explained before. Thus, the birefringence induced in the sample, Δn, can be measured as a function of the applied stress at a given temperature with the procedures explained in the preceding section, and the experimental results of the stress-optical coefficient C, Eq. (9.43), and the optical configuration parameter Δa, Eq. (9.44), are then computed. On the other hand, theoretical values of Δa are calculated with Eq. (9.53) and compared with the experiment.

This procedure has been applied to the study of many polymers [242–283]. However, the results obtained to date are quite disappointing since in most of the cases there are severe disagreements (sometimes larger than one order of magnitude) between theoretical and experimental values.

These discrepancies are attributed to a failure of the values assigned to the contributions $\hat{\alpha}_i$ used for the theoretical computation of Δa. As it was explained in the section on the valence optical scheme in Chapter 7, inductive effects are responsible of the nonstraight additivity of contributions assigned to bonds or groups. And, what is worse, those inductive effects are not only due to intramolecular correlations between different groups of the molecule, but are also produced by the neighboring molecules as well.

Most of the contributions $\hat{\alpha}_i$ are evaluated either in the gaseous state at low pressure or, more frequently, in solution in a nearly isotropic solvent (carbon tetrachloride or p-dioxane) extrapolated to zero concentration. Consequently, it is reasonable to assume that, with some limitations, those contributions may contain the intramolecular inductive effects and, at the most, intermolecular effects produced by a spherically symmetric surrounding.

However, the situation in the case of stress birefringence measurements is quite different because the macromolecules of the sample are closely packed together, and even when the network is swollen with a suitable solvent, the environment of any given segment is not the same as it was either in dilute solution or in the gaseous phase. Consequently, the inductive effects and, therefore, the values of the contributions should be different.

There are several experimental results that seem to corroborate this interpretation of the disagreement between theory and experiment because of the numerical values assigned to the parameters rather than to a failure on the general theory used for the calculation. Thus, for instance:

(a) Although the quantitative concordance between theoretical and experimental values is bad, the agreement is, in general, good at a qualitative level. Thus, general trends of variation of Δa with molecular weight, tacticity, composition of copolymers, size, and optical characteristics of the side groups, etc., are reasonably well predicted by theory.

(b) Swelling of the network with an adequate solvent produces changes in Δa that are much larger than those accounted for considering only the variation of the main refractive index of the network upon swelling. The variation of Δa depends strongly on the degree of swelling and on the nature of the solvent used; in general, the largest variations are obtained with solvents whose molecules have roughly spherical symmetry. In most cases, the agreement between theory and experiment is better when experimental results obtained on swollen samples, instead of those measured for the unswollen network, are used for the comparison. The idea of inductive effects that modify the values of $\hat{\alpha}_i$ gives a straightforward explanation to this behavior because they would depend on whether or not a solvent is used. In the former case, they would change with the amount of solvent and its optical characteristics, becoming more similar to those occurring in solution when a large amount of isotropic solvent is used.

(c) As a general rule, the agreement between theory and experience is better in the case of asymmetric chains than in symmetric polymers. For instance, vinyl polymers, CH_2—CHR—, having a bulky side group R give better concordance than many other more symmetric chains such as polyethylene, polyoxides, polydimethyl siloxane, etc. A tentative explanation of this behavior is that the molecular packing is less compact and, consequently, the intermolecular inductive effects would be smaller for asymmetric chains.

At first sight, the solution to this problem may appear to be very simple: Let us determine new values of $\hat{\alpha}_i$, including the intermolecular inductive effects, and use them to compute Δa. Unfortunately, according to Murphy's law, it is not as easy as it looks. The problem is that a theoretical treatment that could be used to evaluate those inductive effects is not yet available. Moreover, the analysis of low molecular-weight model compounds, which is frequently used to provide values of $\hat{\alpha}_i$ in solution, is precluded in the case of stress birefringence because it is impossible to prepare an elastic network with that kind of molecule. Of course, the comparison between theoretical and experimental values of Δa could be used to compute the $\hat{\alpha}$ tensor for the whole chain, which can then be decomposed into contributions $\hat{\alpha}_i$ for its bonds or groups of bonds. However, those contributions would reproduce the results of a given polymer under certain conditions but would fail for other polymers and even for the same polymer if the conditions of swelling are changed.

Sometimes the agreement between theory and experience can be improved by adjusting some of the geometrical parameters of the chain, for instance, the bond angles [274, 276, 280]. In other cases, the values assigned to the $\hat{\alpha}_i$ contributions can be optimized as to give a reasonable simultaneous agreement for several optical properties [284], for example, Δa, $\langle \gamma^2 \rangle$, and $_mK$. However, most of the times the parameters required to reproduce Δa do not work for any other property that depends on $\hat{\alpha}$.

Some suggestions for a procedure to obviate this problem have been made. For instance, it is quite reasonable to assume that the temperature coefficient of the optical configuration parameter (i.e, $A_T = 10^3[(\Delta a)^{-1} \, d(\Delta a)/dT] = 10^3 \, d[\ln(|\Delta a|)]/dt$) should be less sensitive than Δa to the values assigned to $\hat{\alpha}_i$; consequently, it would be a better magnitude for the comparison [273, 274]. Unfortunately, A_T is in general very small and, therefore, difficult to measure and very sensitive to experimental uncertainties. Moreover, the theoretical values of this magnitude are also little accurate, mainly in the cases of atactic samples or when the conformational energies are not extremely well known. Therefore, in most cases, the comparison between theoretical and experimental values of A_T are performed only at a qualitative level due to the uncertainty of both kind of results.

Another possibility would be to measure Δa as a function of the degree of swelling with an isotropic solvent and extrapolate to zero concentration of polymer before comparing with theoretical values. But, again, this procedure has difficulties because only relatively small degrees of swelling can be reached and, therefore, the extrapolation is not too reliable. Moreover, it is not clear that increasing the degree of swelling will make the surroundings of a given segment in a cross-linked sample more similar to those of the same segment appearing on an individual chain within a dilute solution.

Some examples illustrating all of these aspects are presented in this section. However, it is quite evident that a lot of work, both theoretical and experimental, will have to be done before the mechanisms governing intermolecular inductive effects are understood so that those effects could be incorporated to the theoretical calculations.

Abe and co-workers [272] calculated the optical configuration parameter of polypropylene (PP) and polystyrene (PS) using the bond polarizabilities tabulated by Denbigh [138] and compared with the results of experimental measurements carried out for both polymers [260, 261]. They checked the effect produced in Δa by variations on the statistical-weight parameters, placement of the rotational isomers, and stereoregularity of the chain (represented by the fraction of meso dyads, W_m). They obtained good concordance between theory and experience for both polymers. In the case of PP, the agreement with the experimental measurements performed on atactic samples [260] is obtained with a second-order statistical weight $\omega \approx 0.05$ in the range of tacticities $W_m = 0.2$ to 0.6. Values of $\omega \approx 0.5$ and $W_m = 0.5$ are assigned for PS and the agreement with experience [261] is improved by displacement of the rotational states by about $\Delta\phi \approx 10°$ from their perfectly staggered positions and by allowance for torsional oscillations of the phenyl group about the C^{α}—C^{Ph} bond.

Liberman and co-workers [273] carried out measurements and calculations of Δa of polymethylene (PM) and poly(dimethyl siloxane) (PDMS). Measurements were performed for networks both unswollen and swollen with diluents in the temperature ranges 115–220°C for PM and 15–90°C for PDMS. Their experimental results are summarized in Table 9.1 along with those of some others.

As Table 9.1 indicates, dilution of PM with decalin reduces both Δa and the absolute value of its temperature coefficient. Swelling with n-$C_{12}H_{26}$ and $C_{22}H_{46}$ effects smaller reductions in both magnitudes. For PDMS, the value of Δa is markedly reduced by swelling with decalin, cyclohexane, and especially with CCl_4.

They performed the theoretical analysis by evaluating the contributions to the optical polarizabilities that would be required to reproduce the experimental results of Δa and A_T and comparing those values with the results obtained from some other optical properties such as depolarized Raleigh scattering and electric birefringence. Thus, they concluded that the values of A_T measured on PM networks swollen with decalin are consistent with the results of previous analysis, whereas the agreement for all the other values is poor and suggested that residual intermolecular correlations are responsible for the discrepancy.

In the case of polyisobutylene networks, Liberman and co-workers [274] found that the largest reduction of Δa upon swelling is not obtained with the most symmetrical solvent, CCl_4, but with the one whose refractive index matches best that of the network ($n = 1.508$, 1.454, 1.421, and 1.501 at 30°C, respectively, for the PIB network, CCl_4, cyclohexane, and $CBrCl_3$). The temperature coefficient is $A_T \approx 0.2$ K^{-1} (see Table 9.1). Both the optical configuration parameter and its temperature coefficient are much greater than calculated from rotational isomeric

TABLE 9.1   Experimental Results of the Optical Configuration Parameter
Δa (in Å^3) and its Temperature Coefficient $A_T = 10^3 d[(\ln (\Delta a)]/dT$ (in K^{-1})
for Polymethylene (PM), Poly(Dimethyl Siloxane) (PDMS),
and Polyisobutylene (PIB) Networks

Polymer and Solvent	$v_2{}^\dagger$	Δa	A_T	Reference
PM at 150°C				
None	1.0	7.8	-4.1 ± 0.1	[273]
		8.4	5.0	[266]
		8.4	3.0	[263]
n-$C_{22}H_{46}$	0.41	6.5	-2.6 ± 0.1	[273]
n-$C_{12}H_{26}$	0.44	5.1	-2.2 ± 0.3	[273]
Decalin	0.33	4.0	-1.4 ± 0.1	[273]
		3.9	0 ± 1	[266]
	0.36	4.0		[268]
PDMS at 70°C				
None	1.0	0.81	8.5 ± 0.1	[273]
		0.68	8.2	[262]
		0.3–0.7	7.3	[264]
Decalin	0.25	0.51	4.9 ± 0.5	[273]
Cyclohexane	0.16	0.38	11.7 ± 0.7	[273]
CCl_4	0.16	0.18	18 ± 3	[273]
	0.35–0.71	0.15–0.20	> 0	[265]
PIB at 30°C				
None	1.0	4.10	0.16	[274]
		3.5		[256, 257]
CCl_4	0.16	3.36	0.27	[274]
Cyclohexane	0.17	3.79		[274]
$CBrCl_3$	0.19	2.84		[274]

$^\dagger v_2$ is the volume fraction of the polymer in the network.

state theory assuming additivity of bond polarizabilities ($\Delta a = 0.19\ \text{Å}^3$; $A_T = -0.35\ K^{-1}$). The disparity (more than tenfold for Δa) cannot be relieved by any rational adjustment of the structural parameters, although it is slightly improved by modifications of the valence-bond angle. In fact, assuming additivity of bond contributions and tetrahedral geometry, the four C—C bonds meeting at the quaternary C would cancel out their contributions; departure from tetrahedral geometry increases the value of Δa, although not in the amount required to give agreement with the experimental result.

Poly(dimethyl-sil-methylene) (PDMSM) has a repeating unit, $Si(CH_3)_2$—CH_2—, that is closely related to those of PDMS, $Si(CH_3)_2$—O—, and PIB, $C(CH_3)_2$—CH_2—. Llorente and co-workers [276] studied PDMSM networks and obtained results virtually identical to those reported for PIB [274] (see Table 9.2). Thus, the value of $\Delta a = 4.27\ \text{Å}^3$ measured for the unswollen network at 50°C is reduced to about 2.9 and 2.5, respectively, by swelling with decalin and the more nearly symmetrical solvent 1,3,5-triethylbenzene, even though decalin gave a higher degree of swelling. It is interesting to note that swelling with cyclic PDMS

TABLE 9.2 Some Examples of Comparing Theoretical and Experimental Values of the Optical Configuration Parameter Δa in Units of Å³

| Polymer | Δa (Experimental) | | Δa (Theoretical) | Reference |
	Unswollen	Swollen[+]		
PP		0.3	0.25–0.42	[272, 260]
PS		−1.7	−1.91−−2.25	[272, 261]
PM	7.8	4.0		[273]
PDMS	0.81	0.18		[273]
PIB	4.1	2.84	0.19	[274]
PDMSM	4.27	2.5	0.06	[276]
PVAc	−2.68	−1.90	−1.6	[278]
PMA[‡]	−0.84		−5.4	[277]
PMA[‡]	−0.34		−6.95	[275, 282]
PEA[‡]	−0.82		−7.20	[275, 282]
PBA[‡]	−0.95		−7.44	[275, 282]
POA[‡]	−1.10		−7.80	[275, 282]
PDET	19.8	26.5	4.27	[279]
PTET	13.3		4.1	[279]
PCCS	9.2	7.5	3.0	[279]
PDGC	7.4		2.3	[279]
PMTHF	2.56		1.5–2.0	[280]
PCDO[§]	9.99		1.38	[253]
PCDO[¶]	7.24		0.67	[253]
PMPS	−12.1	−8.50	−12.6	[147]
POP	4.33	3.05	1.2	[254]

[+] Values correspond to results obtained with the solvent that produces the smallest absolute value of Δa.

[‡] PMA from reference [277], $T = 50°C$, and fraction of meso dyads, $W_m = 0.5$. PMA, PEA, PBA, and POA from references [275] and [282], $T = 50°C$, and $W_m = 0.3$.

[§] Fraction of rings in trans configuration, $W_t = 0.7$.

[¶] Fraction of rings in trans configuration, $W_t = 0.1$.

pentamer increased the value of Δa to about 4.4 Å³, presumably because of inter-molecular correlation effects. The temperature coefficient A_T for the unswollen network is about 1.2 K⁻¹. Theoretical calculations gave values of $\Delta a = 0.064$ Å³ and $A_T = -0.28$ K⁻¹. Modification of the parameters used for the calculation has only relatively small effects on Δa with the noteworthy exception of the valence angle θ at the silicon atom. Changes in this angle remove the symmetry and associated cancellation of bond anisotropies, which are the origin of the very small theoretical value of Δa; thus, increasing θ from 109.5 to 130° raises Δa from 0.064 to 1.76 Å³. Therefore, departures of θ from the tetrahedral value improve the agreement between theory and experience; on the other hand, if the deformation of θ also depended significantly on temperature, the agreement for the temperature coefficient A_T might also be improved.

Poly(methyl acrylate) (PMA) [277] and poly(vinyl acetate) (PVAc) [278] exhibit negative values of stress birefringence, that is, the refractive index in a direction perpendicular to the applied force becomes larger than in the direction of stress. This behavior was reported before [255] in the case of poly(ethyl acrylate) (PEA) and can be explained by the presence of a highly polarizable side group such as the carbonyl. When the $\hat{\alpha}$ tensor of the repeating unit is written in a molecular reference frame with the x axis along any of the C—C skeletal bonds, the contribution of the side group produces large and negative α_{xx} and α_{xy} components that overcomes the positive contributions of the hydrocarbon skeleton. Thus, the product $\langle \mathbf{r}^T \hat{\alpha} \mathbf{r} \rangle$ determining the value of Δa for the polymer becomes negative. This interpretation is corroborated because the experimental values [275] of Δa in the series of poly n-alkyl acrylates (measured for methyl PMA, ethyl PEA, n-butyl PBA, and n-octyl POA) are negative and their absolute values increase with increasing length of the side group. Theoretical calculations [282] for this series show the same behavior although the agreement with experience is only qualitative. However, the discrepancy between theory and experience becomes smaller as the length of the side chain increases, suggesting that the flexible long side chains act as diluent, reducing the intermolecular interactions that enhance Δa.

In the case of PVAc [278], the value of Δa measured at 50°C is -2.68 Å³; swelling of the network with 1,3,5-triethylbenzene up to a volume fraction of polymer in the network, $v_2 = 0.615$, reduces the value of the stress-optical coefficient to about -1.90 (extrapolated from actual measurements performed at lower temperatures). The values of the temperature coefficient, evaluated as $A_T = 10^3[(\Delta a)^{-1} d(\Delta a)/dT]$, are -5.22 and -10.5 K^{-1} for unswollen and swollen networks, respectively. The theoretical values are very sensitive to the rotational angle χ over the bond C$^\alpha$—O joining the side group to the chain skeleton and to the tacticity of the polymer represented by the fraction of meso dyads, W_m. Thus, taking $\chi = \pm 45°$, the values calculated for Δa are -0.76, -1.55, and -2.43 Å³, respectively, for $W_m = 0$, 0.5, and 1, and the values of the temperature coefficient are, respectively, -3.44, -2.26, and -1.36 K^{-1} for the same tacticities.

Unswollen PMA [277] gives $\Delta a = -0.84$ Å³ at 50°C and $A_T = -12.5$ K^{-1}. Experiments were not performed on swollen samples because of difficulties in measuring the birefringence Δn for these networks. The theoretical values of Δa are very sensitive to the tacticity of the chain, mainly for predominantly syndiotactic polymers (i.e., $W_m < 0.5$). Values calculated at $W_m = 0$, 0.5, and 1 are, respectively, -13.2, -5.44, and -4.92 Å³ for Δa and -1.80, -1.64, and -1.84 K^{-1} for A_T.

In the experimental study of the stress birefringence of poly(diethylene glycol terephthalate) (PDET) [279] it was found that $\Delta a = 19.8$ Å³ at 70°C and $A_T = -1.1$ K^{-1}. Swelling of the network with tricresylphosphate up to a volume fraction of polymer of $v_2 = 0.58$ produces a notable increase of birefringence, giving $\Delta a = 26.5$ Å³ at 70°C. Theoretical calculations give poor agreement for both Δa and A_T, that is, $\Delta a \approx 4.3$ Å³ at 70°C (extrapolated from calculations performed at lower temperatures) and $A_T = 4.1$ K^{-1}. No reasonable modification of either conforma-

TABLE 9.2   Optical Configuration
Parameter Δa (in units of Å^3) and Its
Temperature Coefficient
$A_T = 10^3(\Delta a)^{-1}d(\Delta a)/DT$ (in units of
K^{-1}) for Unswollen and Swollen
Networks of Poly(Methyl Phenyl
Siloxane) at 25°C[†]

v^2	Δa	A_T
1.000	−12.1	2.02
0.673	−11.5	0.64
0.488	−9.12	1.29
0.350	−8.32	0.85
0.211	−8.10	1.38
0.204	−8.50	1.99

[†] Solvent is decalin.

tional energies or optical parameters would bring theory into agreement with experience; the disagreement can only be explained, at a qualitative level, by intermolecular interactions that increase upon swelling. In contrast with the results of Δa, theoretical and experimental values of the Kerr constant and its temperature coefficient are in good agreement for this polymer [123]. Some examples of other polyesters that have been studied [279] are poly(triethylene glycol terephthalate) (PTET), poly(cis-1,4-cyclohexane dimethanol sebacate) (PCCS), and poly(diethylene glycol trans-1,4-cyclohexane dicarboxylate) (PDGC); comparison between theoretical and experimental values of Δa for these polymers is shown in Table 9.3.

Saiz and co-workers [280] studied poly(3-methyl tetrahydrofuran) (PMTHF) obtained by ring-opening polymerization of the cyclic monomer 3-methyl tetrahydrofuran, $\overline{\text{O—CH}_2— \text{CH(CH}_3)\text{—CH}_2\text{—CH}_2}$. The ring opening takes place by one of the O—CH_2 bonds; however, simultaneous occurrence of both possible kinds of bonds scissions (i.e., β scissions (CH_3) $\text{CHCH}_2\text{—O}$ and δ scissions $\text{CH}_2\text{CH}_2\text{—O}$ produces structural irregularities on the chain. $^{13}\text{C—NMR}$ indicates that 70% of scissions take place by one of the two possibilities, but cannot distinguish which one, that is, it may be the fraction of β scissions, $W_\beta = 0.7$, and therefore $W_\delta = 0.3$, or the other way around, $W_\beta = 0.3$ and $W_\delta = 0.7$. Critical analysis of the dipole moments of the polymer indicates that $W_\delta = 0.7$. The strain birefringence experiments were performed on unswollen networks in the temperature range 20 to 60°C, giving $\Delta a = 2.56$ Å^3 at 30°C with a temperature coefficient $A_T = 3.1$ K^{-1}. No measurements were performed on swollen samples due to their poor mechanical properties. Theoretical calculations show a noticeable dependence of Δa on the skeletal bond angles and a small increase of that magnitude with increasing W_δ, although the results are always smaller than the experimental values. Thus, $\Delta a = 1.5$ and 1.7, respectively, for $W_\delta = 0.3$ and 0.7 when all the skeletal bond angles are taken to be 110°. Adjustment of those angles to 114, 112, and 110°, respectively, for CCC, CCO, and COC (i.e., departure from tetrahedral geometry) increases the values of Δa to about 1.8 and 2.0 Å^3 for $W_\delta = 0.3$ and 0.7, respectively. Thus, even if the theoretical results do not agree with the experimental values, they seem to support the value of $W_\delta = 0.7$ obtained by the analysis of the dipole moment. The agreement between theory and experience is worse for the temperature coefficient A_T, whose theoretical value is -0.3 K^{-1}. The concordance can be improved by assuming that the skeletal bond angles change with temperature; thus, assuming a variation of those angles of about $d\theta/dT \approx 0.025$ deg K^{-1}, the value of A_T increases to 1.3 without worsening the agreement for other conformation-dependent magnitudes.

A very similar analysis has been performed [253] on poly(cis/trans-1,4-cyclohexane dimethanol-alt-formaldehyde) (PCDO) networks. Two samples having different fractions of cyclohexane rings in the trans configuration W_t were prepared. Thus, sample A had $W_t = 0.7$ and sample B contained $W_t = 0.1$. Measurements performed between 10 and 70°C gave $\Delta a = 9.99$ and 7.24 Å^3, respectively, for samples A and B at 30°C with an undetectable temperature coefficient, that is, $A_T \approx 0$, for both samples. Again, the poor mechanical properties precluded measurements on swollen networks. Theoretical values of Δa are roughly one order of magnitude smaller than the experimental results; however, they reproduce satisfactorily both the variation with W_t (i.e., $\Delta a = 1.38$ and 0.67 Å^3, respectively, for samples A and B at 30°C) and the temperature coefficient, which falls below the estimated error of the calculation.

An interesting example is provided by the work of Llorente and co-workers [147] on poly(methyl phenyl siloxane) (PMPS) networks. They used a sample having a fraction of meso dyads of $W_m = 0.5$ (characterized by ^{1}H—NMR) and measured the stress birefringence of networks both unswollen and swollen with decalin up to different volume fractions of polymer v_2. Table 9.3 summarizes their results obtained at 25°C.

The optical configuration parameter Δa is negative as with the other polymers having large and anisotropic side groups explained before, that is, PS, PVAc, PMA, PEA, etc. The absolute value of Δa decreases with increasing degree of swelling. Theoretical calculations show a very strong dependence of Δa with the tacticity of the sample; thus, $\Delta a = -21$ for syndiotactic chains, that is, $W_m = 0$, and decreases to $\Delta a = -7$ for isotactic polymers, that is, $W_m = 1$, both at 25°C. The values calculated for atactic samples, $W_m = 0.5$, are $\Delta a = -12.6$ and $A_T = 1.67$ and are in excellent agreement with the experimental results measured on unswollen samples; however, swelling with decalin produces disagreement between theory and experience. Thus, it seems that the intermolecular correlations are negligible for unswollen networks of this polymer, but increase upon swelling with decalin.

The analysis of Kerr constants for this polymer [202] performed with the same set of parameters used in the calculation of Δa gives also a very good agreement between theory and experience. However, the values of mean-square optical anisotropy [284] $\langle \gamma^2 \rangle$ computed with the same parameters give only a qualitative account (i.e., variation with tacticity and degree of polymerization) of the main features of the experimental results although theoretical values are in general larger than experimental. Attempts to improve the agreement of $\langle \gamma^2 \rangle$ worsens that of the other two magnitudes. A compromise with a reasonable simultaneous agreement for the three magnitudes can be achieved through small adjustments of the polarizabilities of the bonds containing silicon.

Results have also been reported [281] on the stress birefringence of ethene-propene copolymers as a function of chemical composition, chemical-sequence distribution, and stereochemical structure of the propylene sequences. Values of Δa were generally found to decrease significantly with an increase in the fraction of propylene units, but to be relatively insensitive to chemical-sequence distribution and stereochemical structure. Theoretical values are roughly one-third of the experimental results; thus, agreement between theory and experience is intermediate between the very poor obtained for symmetric chains [253, 276, 279] and the relatively good found in the case of asymmetric polymers [147, 278, 280, 283].

The last example that will be presented is the work of Andrady and co-workers [254] on poly(oxypropylene) (POP) networks. They used several hydroxyl-terminated POP chains of different lengths and cross-linked them into trifunctional networks having average molecular weights between cross-links, M_c ranging from 725 up to 3000 (i.e., the number of repeating units x_c from 12 to 50) using tris(p-isocyanoatophenyl) thiophosphate. The use of this cross-linking agent allows the samples to be cured at room temperature, minimizing any side reactions so that model networks virtually free of defects can be prepared. The birefringence was found to decrease with increasing length of the chain (i.e., with increasing M_c). Thus, values of Δa, measured on unswollen samples at 25°C, are 4.66 ± 0.04 and 4.33 ± 0.09 Å^3, respectively, for M_c = 2000 and 3000. This behavior could be explained theoretically by the effect of the end groups introduced by the cross-linking agent. Thus, if the calculations are performed for POP chains alone, the results of Δa reach an asymptotic limit for $x_c \approx 10$. However, assuming that the chains have N-phenyl carbamates as end groups (approximately one-third of the cross-linking molecule), the calculated values of Δa show a steep decrease with increasing x_c and only reach the asymptotic limit for $x_c > 100$. Thus, values of Δa = 1.48 and 1.20 Å^3 are obtained at 25°C for M_c = 2000 and 3000, respectively. The reason is that N-phenyl carbamate is more anisotropic than the POP chain, but the effect of the end groups becomes smaller with increasing chain length. Swelling of the network having M_c = 3000 with POP oligomers and with decalin reduces the experimental values of Δa measured at 25°C to 3.87 and 3.05 Å^3, respectively, thus bringing them closer to the theoretical value. However, the main features of the whole analysis remain the same, namely, that the qualitative concordance between theory and experience is satisfactory although the quantitative agreement is poor.

Table 9.2 summarizes the experimental and theoretical values of Δa *for all polymers cited before.*

10
Flow Birefringence

The Canada balsam is a resin produced by a special kind of tree (*Abies balsamifera*); it is a viscous liquid with a refractive index very close to that of glass and for this reason it has many optical applications. In 1873, Maxwell [285] observed a sample of Canada balsam between crossed polarizers and found that it was isotropic (i.e., the analyzer gave complete extinction) as far as the sample was at rest; however, it became birefringent on flowing. It was soon discovered that this behavior is not exclusive of Canada balsam; on the contrary, many liquids such as resins and oils that are isotropic if the sample stands still exhibit birefringence when forced to flow [286–293]. The induction of optical birefringence in an isotropic sample by forcing it to flow is known as the Maxwell effect, or either flow or streaming birefringence.

The molecular origin of flow birefringence is, as in all the other cases of induced birefringence, the orientation and possible deformation of the molecules on the sample produced by external perturbation, which in this case is the flowing of the liquid or, more precisely, the difference in shear rate among neighboring layers of the sample originated in a laminar flow.

Let us assume a liquid moving along the direction of the x axis in a laminar flow, such as that shown in Figure 10.1(a), with a constant gradient of velocity Γ in the direction of the y axis:

$$\Gamma = dv_x/dy \qquad (10.1)$$

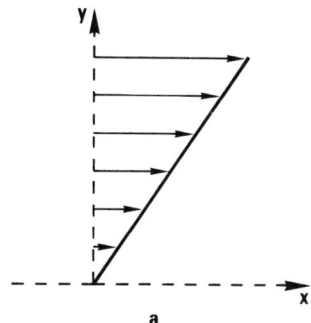

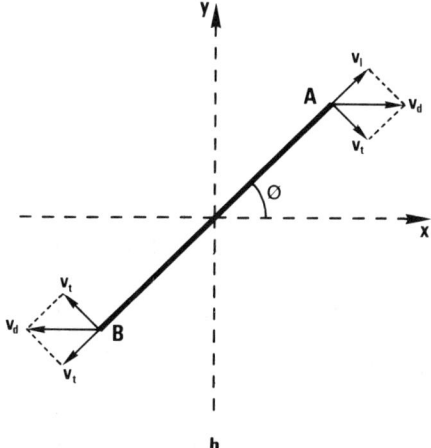

Figure 10.1 (a) Laminar flow with a constant-velocity gradient. (b) Orientation of a rigid-rod molecule in laminar flow.

Let us now represent the molecules of the liquid by rigid rods and place the center of mass of one of them at the origin of the coordinates, as shown in Figure 10.1(b). The orientation of the rod with respect to the direction of flow (direction x) is determined by the angle ϕ between x and the axis of the rod. The velocity of the center of mass is v_x and it is determined by the location of this particular molecule within the flowing liquid. If the length of the rod is l, the y coordinates of the two extremes will be $\pm(l/2)\sin\phi$ and, therefore, the velocities of the liquid at those positions are

$$v_A = v_x + v_d \quad \text{and} \quad v_B = v_x - v_d \tag{10.2}$$

where

$$v_d = \Gamma(l/2) \sin \phi \qquad (10.3)$$

represents the difference in velocity between the center of mass and each of the extremes of the rod.

As Figure 10.1(b) indicates, v_d can be written as the sum of two contributions, one along the axis of the molecule and another one normal to that direction. The values of those components are

$$v_1 = v_d \cos \phi = \Gamma(l/2) \sin \phi \cos \phi \qquad (10.4)$$

$$v_t = v_d \sin \phi = \Gamma(l/2) \sin^2 \phi \qquad (10.5)$$

The longitudinal component v_1 tries to stretch the molecule (and we assume that this hypothetical molecule is undeformable), whereas tangential v_t produces a rotation of the rod in the sense of aligning its axis with the direction of the flow, that is, in the sense of making $\phi \rightarrow 0$. The angular velocity can be easily calculated taking into account that the extreme of the rod rotates over the center of mass with a radius $l/2$ and a linear velocity v_t; therefore,

$$\omega = d\phi/dt = v_t/(l/2) = \Gamma \sin^2 \phi \qquad (10.6)$$

According to Eq. (10.6), the rotational motion of the molecules is non-uniform; on the contrary, ω is maximum when $\phi = 90°$ (i.e., the molecules are perpendicular to the direction of flow, which defines the x axis of the laboratory coordinate system), and becomes zero when the molecular axis is aligned in the direction of flow ($\phi = 0$). Consequently, all the molecules should orientate in the direction of the x axis (i.e., with $\phi = 0$) and, therefore, a beam of light traveling through the sample in the z direction finds a different refractive index when polarized in the zx plane than if the polarization were in xy. Thus, the sample becomes birefringent on flowing.

Up to now we have purposely neglected the thermal motion of the molecules. We will take care of this motion in what follows, but thinking qualitatively, it is easy to see that thermal motion will have two effects:

(a) Randomize the orientation of the molecules when the flow stops, thus restoring the isotropy of the sample.

(b) Prevent obtaining a "perfect orientation" of the sample in which all the molecules would have $\phi = 0$. Instead, a steady-state situation would be reached in which the value of ϕ would be described by a distribution function dictated by a balance between the orientation force produced by Γ and the random Brownian motion due to thermal energy.

Effects similar to flow birefringence are obtained when the sample is subject to an expansion such as that produced in a jet or to a series of compressions and expansions originated by a sound wave. This last kind of induced birefringence was discovered in 1938 and it is known as the Lucas effect [294, 295].

The results obtained by flow birefringence depend on the experimental conditions such as kind of flow, gradient Γ, temperature, etc., and on many characteristics of the sample like viscosity, size, shape, and optical anisotropy of the molecules. Thus, this kind of experiment can be used either to study the flow of a known sample or to obtain information about the microscopic characteristics of the sample flowing under a known difference in shear rate. In practice, this second application is more frequent than the first, and, from the first decades of this century, it has been used to study different kinds of samples, from pure liquids [296–298] to colloidal suspensions [299–301]. The first measurements on polymers were performed in 1930 [302] and since then flow birefringence has been applied to characterize many polymers both in solution (especially in the case of biopolymers) and in the melt. Some good reviews on this subject can be found in references [303–309].

A very simple way of producing laminar flow consists in using a capillary tube and some experimental instruments; most of them designed for measurements on molten polymers have been built on this principle [308]. However, laminar flows produced in capillaries have the inconvenience of not being linear because the velocity of the liquid increases from the walls toward the center following a quadratic function of the radius; this behavior complicates the quantitative analysis of the results. A linear gradient can be created on capillaries having rectangular sections if two of the sides of the rectangle are much larger than the other two so that the end effects can be neglected; in fact, the ideal situation would be two infinite plates separated by a small distance. However, this kind of capillary is difficult to prepare under tight geometrical requirements, especially a constant separation between plates.

A simple device commonly used to produce laminar flows with a linear gradient consists of two concentric cylinders with a separation much smaller than the smallest of the two radii (see Figure 10.2(a). The sample is placed in the annulus between the two cylinders so that when one of them rotates, keeping the other one motionless, a laminar flow with a linear gradient increasing from the stationary to moving surfaces is produced. The best results are obtained when the outer cylinder rotates and the inner one stands still. Then the flow is stabilized by centrifugal forces and larger gradients without turbulence can be obtained than with a fixed external cylinder and a rotating internal one. The two cylinders are placed between crossed polarizers and the beam of light travels parallel to the axis of the cylinders, determining the direction of the z axis on the laboratory coordinate system; see Figure 10.2(b).

Figure 10.3 shows the results obtained when a beam of light wide enough as to observe the whole annulus is employed, for instance, using a lens to expand the beam and a second one to focus on the detector, as indicated in Figure 10.2(b).

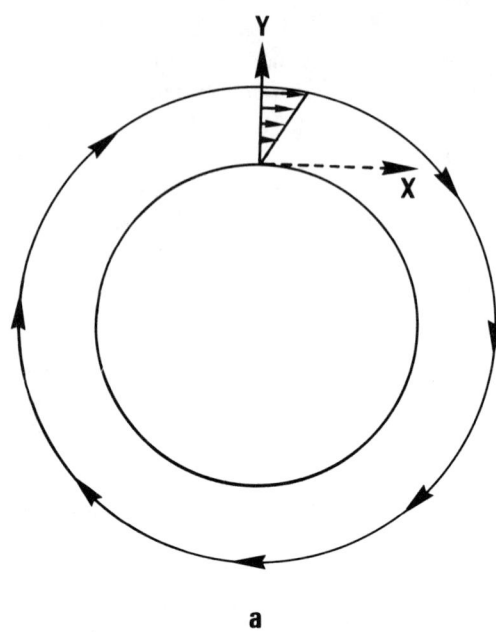

a

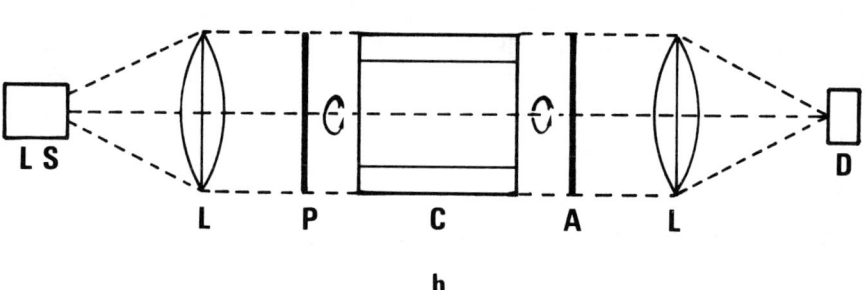

b

Figure 10.2 (a) Schematic of the two-cylinder assembly used to produce laminar flow in a liquid placed in the annulus between the cylinders by rotating the outer one with a constant angular speed. (b) Experimental setup used to measure flow birefringence. The beam of light travels parallel to the axis of the cylinders (i.e., the direction of the z axis): (LS) light source, (L) lens, (P) polarizer, (A) analyzer, (C) sample cell formed by the annulus between the cylinders, and (D) detector.

Assuming that the inner cylinder is either opaque or isotropic, the whole field will be dark when the outer cylinder is at rest because no flow of the liquid is produced then. When the outer cylinder rotates, the liquid becomes birefringent and the whole annulus appears illuminated except in four areas arranged as the arms of a cross (the

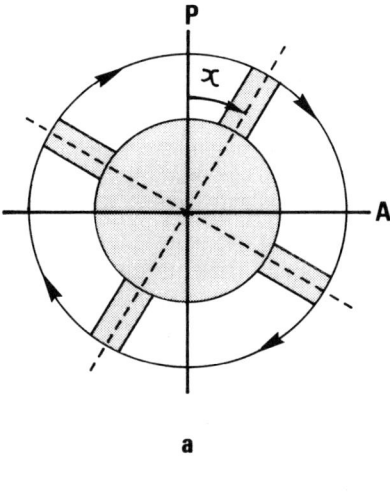

a

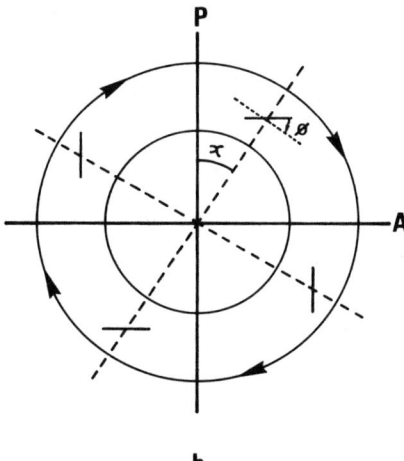

b

Figure 10.3   (a) The cross of isocline obtained when the whole annulus between the two cylinders is observed. P and A represent the planes of the polarizer and analyzer, respectively. (b) Molecular origin of the cross of isocline. The average orientation of the molecules with respect to the flow, $\langle \phi \rangle$, produces four areas in which the molecules are parallel to either polarizer or analyzer.

so-called *cross of isocline)* that is rotated an angle χ with respect to the cross formed by the planes of the polarizer and analyzer; see Figure 10.3(a). χ is called the extinction angle and its value depends on the sample and changes with the shear rate determined by the angular speed of the rotating cylinder.

The molecular origin of the cross of isocline is shown in Figure 10.3(b). The shear rate tends to orientate the molecules in the direction of flow ($\phi = 0$), whereas thermal motions try to randomize this orientation. The balance between these two forces is some kind of distribution of values of ϕ that we will study in what follows, but it is easy to imagine, at a qualitative level, that some kind of average orientation described by an average value $\langle\phi\rangle$, depending on the molecular characteristics of the sample and on the shear rate, will be reached. The light that enters the sample polarized in the yz plane can be resolved into two components, one polarized in the direction of the molecular axis and the second one polarized in a direction perpendicular to it. These two components travel through the sample with different speeds and, therefore, emerge forming an elliptically polarized radiation (i.e., the sample becomes birefringent and appears illuminated when seen through the analyzer).

However, there are four areas in which the molecules are oriented, on average, either parallel or perpendicular to the plane of the polarizer so that the radiation will be polarized in the plane containing the molecular axes or in the plane normal to them. In both cases, the sample will have a single refractive index for this radiation that will emerge polarized in the original plane and, therefore, will be extinguished by the analyzer. It is easy to see in Figure 10.3(b) that these four areas of isotropic sample are arranged like the arms of a cross that is rotated an angle $\chi = \langle\phi\rangle$ with respect to the cross defined by the planes of polarizer and analyzer. Some experimental instruments incorporate a cross wire that can be rotated from the direction of the polarizer–analyzer to the direction of the cross of isocline, thus providing a direct measurement of the extinction angle χ, which gives the average orientation of the molecules.

The qualitative ideas sketched before can be put into a quantitative form by introducing an orientational distribution function $\rho(\phi,t)$ defined in such a way that the product $\rho(\phi,t)\,d(\phi)$ represents the number of molecules per unit volume that, at a given time t, have an orientation between ϕ and $\phi + d\phi$. If the molecules were randomly oriented, $\rho(\phi)$ would be a constant independent of ϕ. On the contrary, if all the molecules were aligned in the direction of flow, $\rho(0) = 1$ and $\rho(\phi) = 0$ for $\phi \neq 0$. In both cases, the distribution would be independent of time.

In actuality, the thermal motions would produce a variation of ρ tending to give the limit of independence with ϕ, whereas the shear rate would modify ρ toward the limit of perfect orientation. A steady state with $\rho(\phi)$ independent of time would be reached when both variations cancel each other out, giving

$$\left[\frac{\delta\rho(\phi,t)}{\delta t}\right]_{\text{thermal}} + \left[\frac{\delta\rho(\phi,t)}{\delta t}\right]_{\text{flow}} = 0 \qquad (10.7)$$

In order to calculate the variation in ρ produced by the flow, let us focus our attention on the molecules whose orientation is comprised between ϕ and $\phi + d\phi$. At any given time, the density number (i.e., the number of molecules per unit volume) in that range of orientations, $\mathcal{N}(\phi,t)$, is given by the definition of the orientational distribution function:

$$\mathcal{N}(\phi,t) = \rho(\phi,t)\,d\phi \qquad (10.8)$$

and, therefore, the variation with time that the flow produces in that population can be obtained by partial derivation as

$$\left[\frac{\delta \mathcal{N}(\phi,t)}{\delta t} \right]_{\text{flow}} = \left[\frac{\delta \rho(\phi,t)\, d\phi}{\delta t} \right]_{\text{flow}} = \left[\frac{\delta \rho(\phi,t)}{\delta t} \right]_{\text{flow}} d\phi \qquad (10.9)$$

During a time dt, the flow motion forces n_1 molecules per unit volume into the range of considered orientations and there are n_2 molecules per unit volume, leaving that orientation for the same kind of motion. Consequently, the net variation of $\mathcal{N}(\phi,t)$ is

$$\left[\frac{\delta \mathcal{N}(\phi,t)}{\delta t} \right]_{\text{flow}} = \frac{n_1 - n_2}{dt} \qquad (10.10)$$

The molecules having an orientation ϕ move with an angular velocity $\omega(\phi)$ and, therefore, during a time dt, their orientation would change in $\omega(\phi)\, dt$. Thus, the number of molecules per unit volume that traverses the orientation ϕ, in the direction of decreasing this angle, during a time dt is composed of those having an orientation comprised between ϕ and $\phi + \omega\, dt$. Therefore,

$$n_1 = \rho(\phi,t)\omega(\phi)\, dt \qquad (10.11)$$

In a similar way, the number of molecules that, in the same time, traverses the orientation $\phi + d\phi$ is

$$n_2 = \rho[(\phi + d\phi),t]\omega(\phi + d\phi)\, dt =$$

$$\left\{ \rho(\phi,t) + \left[\frac{\delta \rho(\phi,t)}{\delta t} \right] d\phi \right\} \left\{ \omega(\phi) + \left[\frac{\delta \omega(\phi)}{\delta \phi} \right] d\phi \right\} dt \qquad (10.12)$$

The angular velocity $\omega(\phi)$ is given by Eq. (10.6), from which one obtains its variation with ϕ as

$$\frac{\delta \omega(\phi)}{\delta \phi} = \Gamma 2 \sin \phi \cos \phi = \Gamma \sin 2\phi \qquad (10.13)$$

Substituting Eqs. (10.6) and (10.13) into (10.12), followed by neglecting the term of $(d\phi)^2$ gives

$$n_2 = \left\{ \rho(\phi,t)\omega(\phi) + \rho(\phi,t)\Gamma \sin 2\phi + \Gamma \sin^2\phi \left[\frac{\delta \rho(\phi,t)}{\delta t} \right] \right\} dt$$

$$(10.14)$$

Substituting Eqs. (10.14) and (10.11) into (10.10) and then into (10.9) gives the final expression of the variation in the orientational distribution function produced by the flow as

$$\left[\frac{\delta \rho(\phi,t)}{\delta t} \right]_{\text{flow}} = -\Gamma \sin^2\phi \left[\frac{\delta \rho(\phi,t)}{\delta \phi} \right] - \Gamma \rho(\phi) \sin 2\phi \qquad (10.15)$$

The thermal motion of the molecules is simply a rotational diffusion process and, therefore, the variation produced on $\rho(\phi,t)$ is governed by an equation similar to Fick's law. Thus, the number of molecules per unit volume that traverse the orientation ϕ in time dt due to thermal motions is

$$\left[d\mathcal{N}(\phi,t) \right]_{\text{thermal}} = -\mathcal{D} \left[\frac{\delta\rho(\phi,t)}{\delta\phi} \right] dt \qquad (10.16)$$

where $\mathcal{D}$ is the rotational diffusion coefficient.

Taking into account that, by definition, $\mathcal{N}(\phi,t)$ is given by Eq. (10.8), Eq. (10.16) can be written as

$$\left[\frac{\delta\rho(\phi,t)}{\delta t} \right]_{\text{thermal}} = -\mathcal{D} \left[\frac{\delta^2\rho(\phi,t)}{\delta\phi^2} \right] \qquad (10.17)$$

Substituting Eqs. (10.15) and (10.17) into (10.7) gives the steady-state condition as

$$\left[\frac{d^2\rho(\phi)}{d\phi^2} \right] + \frac{\Gamma}{\mathcal{D}} \left\{ \sin^2\phi \left[\frac{d\rho(\phi)}{d\phi} \right] + \rho(\phi) \sin 2\phi \right\} = 0 \qquad (10.18)$$

where the time dependence of ρ has been eliminated.

As Eq. (10.18) indicates, the steady-state orientational distribution function $\rho(\phi)$ depends only on the ratio between the shear rate Γ and the diffusion coefficient $\mathcal{D}$. This result is quite intuitive because that ratio represents the relative strengths of orientating forces versus random motions. This equation was first derived by Boeder [310] and has no exact solution; however, a series-expansion solution can be used for small values of the ratio $\Gamma/\mathcal{D}$ representing conditions in which the gradient velocity starts perturbing the random orientation of the molecules. Representing γ by

$$\gamma = \Gamma/\mathcal{D} \qquad (10.19)$$

the series-expansion solution can be written as

$$\rho(\phi) = \xi \left[1 + \gamma \frac{\sin 2\phi}{4} + \gamma^2 \left(\frac{\cos 2\phi}{16} - \frac{\cos 4\phi}{64} \right) + \cdots \right] \qquad (10.20)$$

where ξ is a normalization constant that can be calculated by introducing the conditions that the integral of $\rho(\phi)\, d\phi$ over all possible values of ϕ should be equal to the total concentration.

Figure 10.4 shows the variation of $\rho(\phi)$ with ϕ for several values of γ. The value $\gamma = 0$ represents the situation of no shear rate ($\Gamma = 0$) in which all the molecules are randomly oriented and, therefore, according to Eq. (10.20), $\rho(\phi) = \xi$ = constant, independent of ϕ. For larger values of γ, the function $\rho(\phi)$ shows a maximum that coincides with the extinction angle χ; it is close to 45° for small values of γ and tends to 0 when γ increases. High values of γ indicate that the orientation force Γ dominates the random motion; then all the molecules are ori-

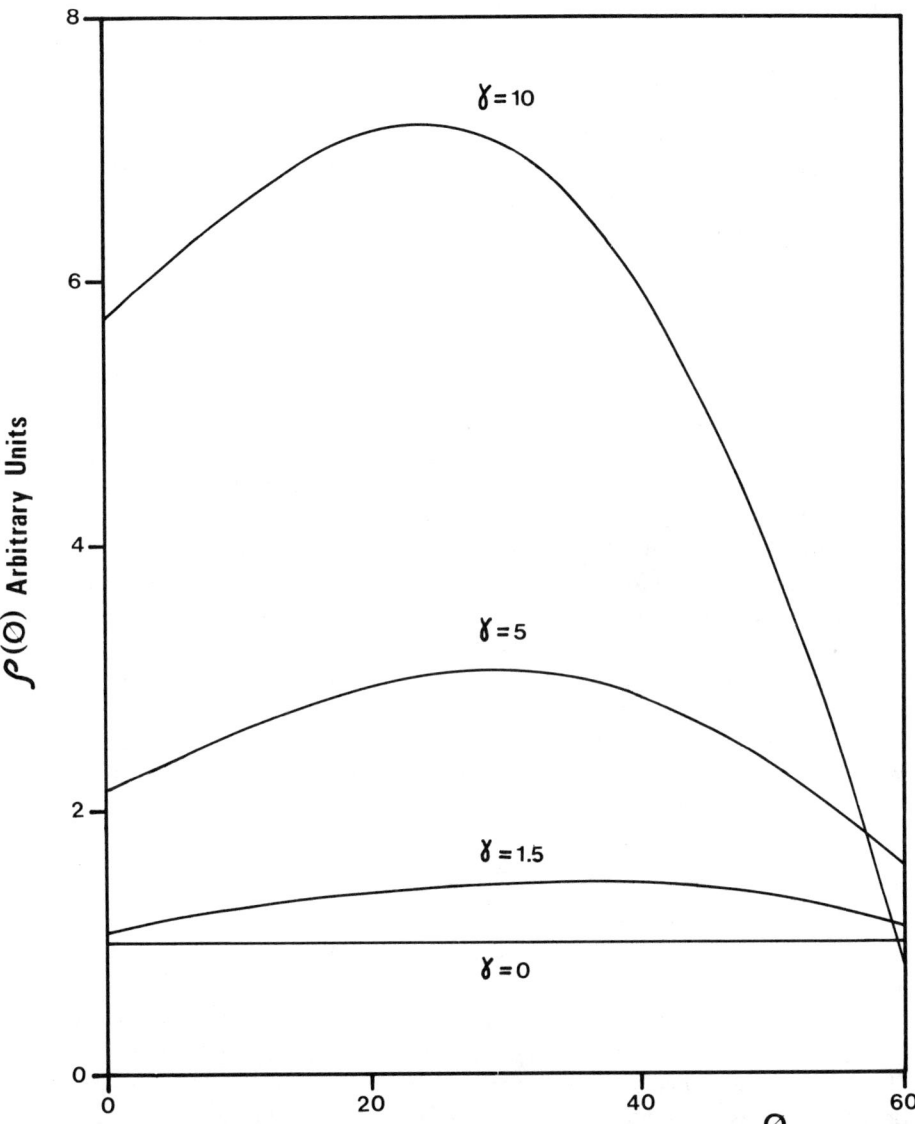

Figure 10.4   Variation of the orientational distribution function $\rho(\phi)$ with the angle ϕ between the molecular axis and the direction of flow at several values of the ratio $\gamma = \Gamma/\mathscr{D}$ between orientation force and random motions.

ented in the direction of flow ($\phi = 0$), as Eq. (10.20) shows with a single nonzero value for $\phi = 0$ at very high values of γ.

However, Eq. (10.18) was obtained with two assumptions that make it useless for macromolecules. Thus, the molecules of the liquid were taken as rigid rods with negligible cross section, which were always confined to the xy plane. Both limitations were removed in the three-dimensional analysis first developed by Peterlin [311] and later improved by other researchers [307, 312–315]. In this theory, all the molecules of the sample are considered equal and are represented by ellipsoids of revolution having a longitudinal axis a and two transverse axes of the same length b. The geometrical shape of the molecules is characterized by the axial ratio p and the anisometry parameter A defined, respectively, as

$$p = a/b \quad \text{and} \quad A = \frac{p^2 - 1}{p^2 + 1} = \frac{a^2 - b^2}{a^2 + b^2} \tag{10.21}$$

Thus, a sphere has $p = 1$ and $A = 0$. Prolate ellipsoids (like a cigar) have $p > 1$ and $A > 0$ with limiting values $p \to \infty$ and $A \to 1$ for an infinitely thin needle. On the contrary, oblate ellipsoids (like flying saucers) are defined by $p < 1$ and $A < 0$ with the limit $p = 0$ and $A = -1$ for a disc.

The main axes of the polarizability tensor of the molecule are assumed to coincide with the revolution axes of the ellipsoid, and the orientation of the molecules is defined by the angle θ between the molecular axis a and the z axis of the coordinate system (defined by the direction of propagation of the beam of light) and the angle ϕ between the projection of a on the xy plane with the direction of the x axis (i.e., the direction of flow).

The shear rate produces movements of the molecules that can be described as variations of the orientation angles θ and ϕ with time. These velocities are [307, 316, 317]

$$\frac{d\theta}{dt} = \frac{1}{4} A\Gamma \sin 2\theta \sin 2\phi \tag{10.22}$$

$$\frac{d\phi}{dt} = \frac{1}{2} \Gamma (1 + A \cos 2\phi) \tag{10.23}$$

According to Eqs. (10.22) and (10.23), the rotation of the molecules depends on the shear rate Γ; the shape, through parameter A; and the orientation. Thus, Eq. (10.22) gives zero angular velocity for spheres ($A = 0$), for zero shear rate ($\Gamma = 0$), or for molecules lying either in the xy plane ($\theta = 90°$ for which a particular case is that in which the molecule is oriented in the direction of the x axis, $\phi = 0$) or in the direction of the z axis ($\theta = 0$). However, the angular velocity of ϕ, given by Eq. (10.23), only vanishes for the trivial case $\Gamma = 0$ or for orientations meeting the condition $\cos(2\phi) = -1/A$.

The orientation distribution function should depend on both θ and ϕ angles, that is, $\rho(\theta,\phi)$. The procedure for the calculation of $\rho(\theta,\phi)$ is similar to the one shown before for the simplest case of just one orientation angle. Thus, the variation with time produced in $\rho(\theta,\phi,t)$ by both shear rate and Brownian motions is evaluated and made to cancel each other out, as in Eq. (10.7), to obtain the steady-state condition in which $\rho(\theta,\phi)$ does not change with time. This condition of independence with time gives an equation similar to Eq. (10.18), which in this case has the form

$$\nabla^2[\rho(\theta,\phi)] - \frac{1}{\mathcal{D}} \text{ div}\left(\frac{d\theta}{dt} , \frac{d\phi}{dt} \right) = 0 \tag{10.24}$$

where ∇ is the Laplacian operator, and div represents the divergence of the vector formed by the two angular velocities.

Substituting Eqs. (10.22) and (10.23) together with the definition of γ, Eq. (10.19), into (10.24) gives

$$\nabla^2 [\rho(\theta,\phi)] - \gamma \text{ div}\left(\frac{A \sin 2\theta \sin 2\phi}{4} , \frac{1 + A \cos 2\phi}{2} \right) = 0 \tag{10.25}$$

Again, this equation has no exact solution; however, it admits a series-expansion solution in powers of A that can be applied to any value of γ, from small values representing situations in which the flow slightly disturbs the random distribution of the molecules to large values of γ for which the orientation forces dominate the Brownian motion. This solution can be written as [307]

$$\rho(\theta,\phi) = \frac{1}{4\pi} (1 + AC_1 + A^2C_2 + \cdots) \tag{10.26}$$

with

$$C_1 = \frac{3\gamma^2 \sin^2\theta}{\gamma^2 + 36} \left(\frac{3 \sin 2\phi}{\gamma} - \frac{\cos 2\phi}{2} \right)$$

$$C_2 = \frac{\gamma^2}{\gamma^2 + 36} \left\{ \frac{15\gamma^2 \sin^4\theta}{16(\gamma^2 + 100)} \left[\frac{(\gamma^2 - 60) \cos 4\phi}{\gamma^2} - \frac{16 \sin 4\phi}{\gamma} \right] \right.$$

$$\left. + \frac{9(35 \cos^4\theta - 30 \cos^2\theta + 3)}{560} - \frac{3(\cos^2\theta - 1)}{14} \right\}$$

For small values of γ, Eq. (10.26) can be transformed into a series expansion of powers of γ as

$$\rho(\theta,\phi) = \frac{1}{4\pi} \left[1 + \gamma \frac{A \sin 2\phi \sin^2\theta}{4} + \gamma^2 \left(\frac{A \cos 2\phi \sin^2\theta}{24} \right. \right.$$

$$\left. \left. + \frac{A^2 \cos 4\phi \sin^4\theta}{64} - \frac{A^2 \sin^4\theta}{64} + \frac{A^2}{120} \right) + \cdots \right] \tag{10.27}$$

In the case of a rigid rod with negligible cross section ($A = 1$), lying in the xy plane ($\theta = 90°$), the distribution function of Eq. (10.27) coincides with that of Boeder, Eq. (10.20).

Once the orientation distribution function $\rho(\theta,\phi)$ has been obtained, it can be applied to calculate the optical properties of the sample. The easiest one to calculate is the extinction angle χ because it is entirely determined by the orientation of the molecules and, therefore, it does not depend on their optical characteristics. Peterlin and Stuart [311, 312] obtained the following relationship:

$$\tan 2\chi = -\frac{\langle \sin 2\phi \, \sin^2\theta \rangle}{\langle \cos 2\phi \, \sin^2\theta \rangle} \tag{10.28}$$

which, using Eq. (10.26) to compute the averages, gives

$$\chi = \frac{\pi}{4} - \gamma \frac{1}{12} + \gamma^3 \left(\frac{1}{1296} + \frac{A^2}{1890} \right) + \cdots \tag{10.29}$$

Thus, the extinction angle χ depends on the shape of the molecules through parameter A and on the ratio γ between the shear rate Γ and the diffusion coefficient $\mathcal{D}$.

Figure 10.5 shows some values of χ as a function of γ computed according to Eq. (10.29). As can be seen in this figure, $\chi \rightarrow 45°$ when $\gamma \rightarrow 0$, whereas $\chi \rightarrow 0$ for high values of γ, although the apparition of turbulence in the flow often precludes experimentally obtaining this limit. Since parameter A is squared in Eq. (10.29), both a needle ($A = 1$) and a disc ($A = -1$) give the same behavior of χ versus γ.

It is interesting to note that for values of γ small enough so as to neglect terms of Eq. (10.29) on powers higher than γ, χ gives a linear variation with γ with a slope of $1/12$, i.e., a linear variation with Γ with a slope of $1/12\mathcal{D}$, independently of the shape of the molecule. This limit of low shear rate is shown as a dashed line in Figure 10.5. Therefore, measurements performed at a low shear rate allow a direct determination of the diffusion coefficient, which can be used to estimate the size and shape of the molecules.

In a similar way, Peterlin and Stuart [311, 312] use Eq. (10.26) to compute the birefringence of a solution containing a concentration c of ellipsoids as

$$\Delta n = n_x - n_y = \left(\frac{2\pi}{n_s} \right) \left(\frac{c}{\rho} \right) f(\gamma,A)(g_a - g_b) \tag{10.30}$$

where ρ is the density of the solute, and n_s is the isotropic value of the refractive index of the solution (i.e., measured when the solution is at rest with $\Gamma = 0$). The ratio c/ρ represents the volume fraction of solute and Eq. (10.30) indicates that the birefringence increases linearly with it. The function $f(\gamma,A)$ is the so-called orientation factor, which indicates the fraction of molecules of solute that are oriented in the direction of flow. It is quite intuitive to figure out that this factor should depend on the ratio $\gamma = \Gamma/\mathcal{D}$ and on the shape of the molecules of solute represented by

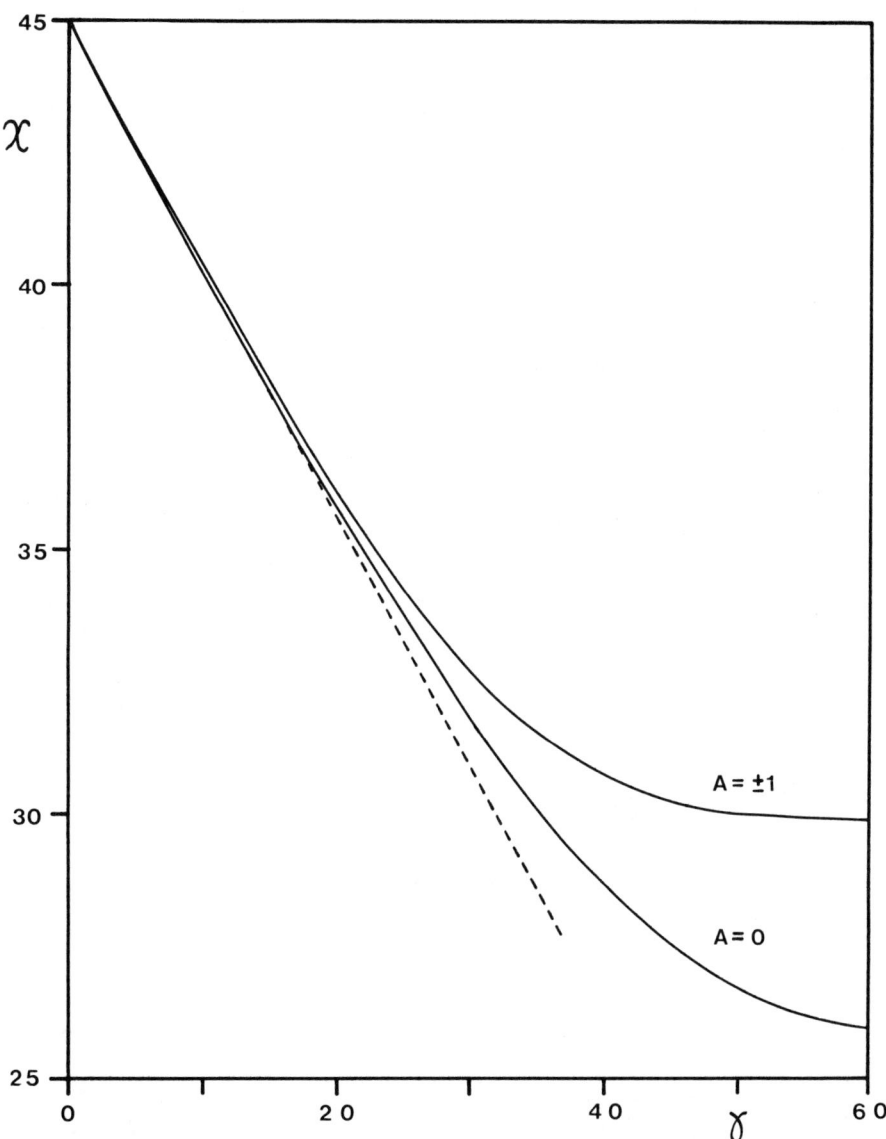

Figure 10.5   The extinction angle χ as a function of the ratio $\gamma = \Gamma/\mathcal{D}$ between the shear rate and the diffusion coefficient for the extreme values of the shape parameter A defined in Eq. (10.21).

their anisometry parameter A; see Eq. (10.21). Finally, the term $g_a - g_b$ represents the optical anisotropy of the molecules of solute.

Therefore, Eq. (10.30) shows that the birefringence of the solution depends on three magnitudes, namely, the concentration expressed as the volume fraction of solute, the anisotropy of the molecules of solute, and the fraction of those molecules that are oriented in the direction of flow.

For small values of γ, the orientation factor $f(\gamma,A)$ can be represented by the following series expansion:

$$f(\gamma,A) = \frac{\gamma A}{15} - \frac{\gamma^3 A}{1080} \left(1 + \frac{6A^2}{35} \right) + \cdots \qquad (10.31)$$

The orientation factor vanishes in the case of a sphere ($A = 0$) because, then, any orientation is equivalent. The factor becomes positive for rods ($A > 0$) and negative for disks ($A < 0$), which indicates that those particles orientate the longitudinal axis parallel and perpendicular, respectively, to the direction of flow. Figure 10.6 shows the variation of $f(\gamma,A)$ with γ for several values of A.

The optical anisotropy term $g_a - g_b$ is written by Peterlin and Stuart as a function of the refractive indices of the molecules of solute along their main axes.

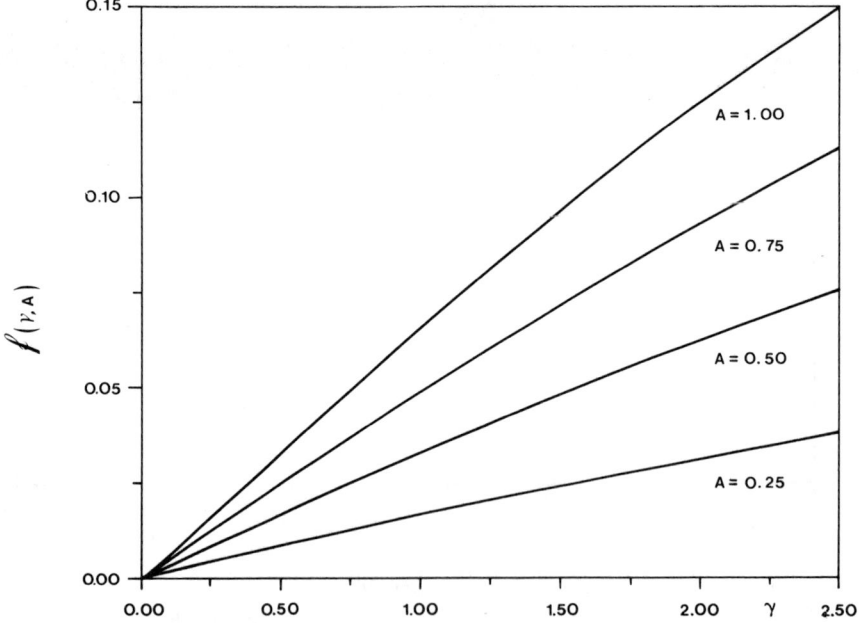

Figure 10.6   Variation of the orientation factor $f(\gamma,A)$ with γ for several values of A. Computation was performed according to Eq. (10.31), which is valid only for small values of γ. In the case of a sphere ($A = 0$), the factor becomes zero for any value of γ.

Thus, with n_s as the isotropic refractive index of the solution and with n_a and n_b as the refractive indices of the molecules of solute along the a and b main axes, the optical anisotropy term can be represented by

$$g_a - g_b =$$

$$\frac{(3n_s)^2}{4\pi} \quad \frac{n_s^2(n_a^2 - n_b^2) + e(n_a^2 - n_s^2)(n_b^2 - n_s^2)}{[(n_a^2 + 2n_s^2) - 2e(n_a^2 - n_s^2)][(n_b^2 + 2n_s^2) - e(n_b^2 - n_s^2)]} \quad (10.32)$$

where e is a shape factor depending exclusively on the geometry of the ellipsoid as characterized by its axial ratio p; see Eq. (10.21). Thus, in the case of prolate ellipsoids (i.e., $p > 1$), representing $r = (p^2 - 1)^{1/2}$, the shape factor is given by

$$e = \frac{1}{4r^2}\left[2p^2 + 4 - \frac{3p}{r} \ln\left(\frac{p + r}{p - r} \right) \right] \quad (10.33)$$

whereas for oblate ellipsoids ($p < 1$), taking $r = (1 - p^2)^{1/2}$, the shape factor is

$$e = \frac{1}{2r^2}\left[\frac{3p}{r} \tan^{-1}\left(\frac{r}{p} \right) - p^2 - 2 \right] \quad (10.34)$$

Figure 10.7 shows the variation of the shape factor e with the axial ratio p of the ellipsoid. As this figure shows, e changes rapidly with p in the vicinity of $p = 1$

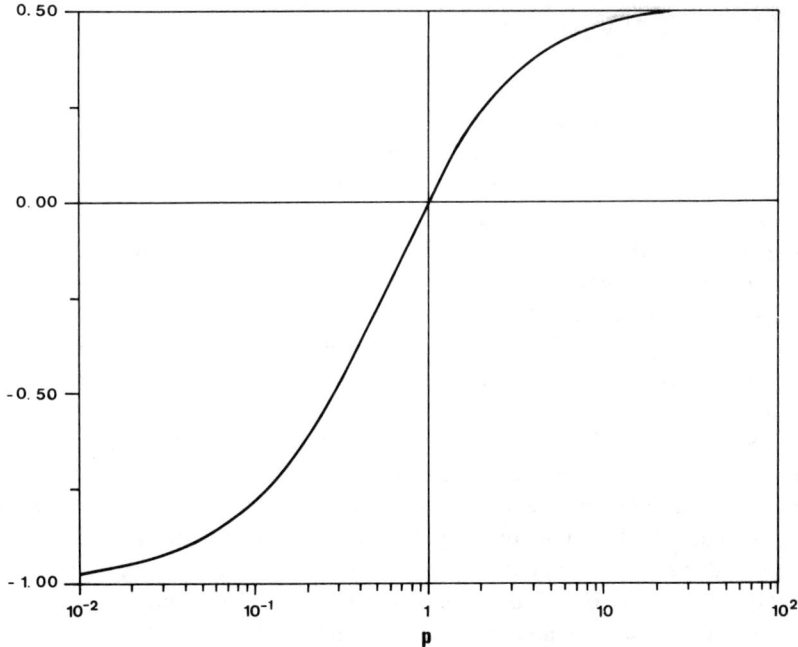

Figure 10.7   Variation of the shape factor e with the axial ratio of the ellipsoid $p = a/b$.

(i.e., geometrical forms close to a sphere), but it soon reaches asymptotic values e = 0.5 for rods ($p > 1$) and $e = -1$ for disks ($p < 1$).

The optical anisotropy term defined by Eq. (10.32) is the sum of two contributions, namely, *intrinsic* and *shape* anisotropies. The *intrinsic anisotropy* depends on the difference between the refractive indices (or optical polarizabilities) of the molecules of solute along their main axes and, therefore, vanishes in the case of isotropic molecules (i.e., $n_a = n_b$). The *shape anisotropy* depends on the shape factor e and the differences between the refractive indices of solute and solution and, therefore, only vanishes for spheres ($e = 0$) or when the solute and solution are isorefractive, that is, $n_a = n_s$ or $n_b = n_s$.

The most interesting conclusion that can be obtained from Eq. (10.32) is that optically isotropic molecules can produce flow birefringence provided that they are not spherical and that their refractive index is different from that of the solution.

As in the case of the extinction angle χ, Eq. (10.29), the values of the birefringence produced by the flow, Eq. (10.30), are usually studied at very low values of shear rate Γ (i.e., with $\gamma < 1$), so that the orientation factor, Eq. (10.31), can be approximated by $f(\gamma, A) \approx \gamma A / 15$. The standard procedure consists in representing values of $\Delta n / c n_s$ versus the shear stress $\Gamma \eta_s$, with η_s representing the viscosity of the solution, and taking the initial slope of this representation, which is the so-called intrinsic birefringence or Maxwell constant, defined as

$$M = [\Delta n]_0 = \lim_{\substack{\Gamma \to 0 \\ c \to 0}} \left(\frac{\Delta n}{c n_s \eta_s \Gamma} \right) = \frac{2\pi}{15} \frac{A}{\rho \mathcal{D} n_s^2 \eta_s} (g_a - g_b) \qquad (10.35)$$

Thus, the simultaneous measurement of the extinction angle χ and the Maxwell constant M gives information on the optical anisotropy factor $(g_a - g_b)$ and the diffusion coefficient $\mathcal{D}$ of the molecules of solute.

When applying Eqs. (10.29) and (10.30) to solutions of macromolecules, two approximations used on its derivation should be taken into account, namely, that no interactions between molecules of solute were allowed and that all the molecules were supposed to be identical.

The first approximation is easy to deal with by measuring at several concentrations and extrapolating to zero concentration before applying Eq. (10.29) to calculate $\mathcal{D}$.

However, the second approximation limits the applicability of this equation to monodisperse samples; otherwise, the values of $\mathcal{D}$ can be severely underestimated. The extinction angle χ and the birefringence Δn produced by a polydisperse sample can be calculated as functions of the values of those magnitudes produced by each of the monodisperse components present on the sample, χ_i and Δn_i, which can be computed according to Eqs. (10.29) and (10.30). The relationships between magnitudes of the whole sample and those of its components are [307]

$$\cot 2\chi = \frac{\sum\limits_{i} \Delta n_i \cos 2\chi_i}{\sum\limits_{i} \Delta n_i \sin 2\chi_i} \tag{10.36}$$

and

$$\Delta n = \left[\left(\sum\limits_{i} \Delta n_i \sin 2\chi_i \right)^2 + \left(\sum\limits_{i} \Delta n_i \cos 2\chi_i)^2 \right]^{1/2} \tag{10.37}$$

where the sums extend to every monodisperse component (i.e., every molecular weight) contained in the polydisperse sample.

When the individual components differ appreciably in Δn_i and χ_i, the values of Δn and χ computed according to Eqs. (10.36) and (10.37) as functions of the shear rate Γ for the polydisperse sample may exhibit a behavior quite different from those of monodisperse fractions [307, 318] computed according to Eqs. (10.29) and (10.30) with the appearance of a minimum and a maximum of Δn and χ at given values of Γ.

Appendix: The Matrix-Multiplication Scheme

Most conformation-dependent properties of polymeric chains are calculated as a sum of contributions from the individual chemical bonds or groups of bonds of which the polymer is constituted. Such properties are known as constitutive properties of the molecule. Thus, assuming that A is a physical property of this type, it can be represented as

$$A = \sum_{i=1}^{n} A_i \qquad (A.1)$$

where A_i is the contribution to A from bond i, and the sum expands over all the bonds in the molecule. Usually, the A property and the A_i contributions are either vectorial or tensorial magnitudes; then the symbol of summation should contain also the transformations of coordinates required to write all those contributions in the same system.

Some typical examples of constitutive properties are the end-to-end distance $\mathbf{r}$, the dipole moment $\boldsymbol{\mu}$, and the anisotropic part of the optical polarizability tensor $\hat{\boldsymbol{\alpha}}$, which can be evaluated as

$$\mathbf{r} = \sum_{i=1}^{n} l_i \qquad \boldsymbol{\mu} = \sum_{i=1}^{n} \boldsymbol{\mu}_i \qquad \hat{\boldsymbol{\alpha}} = \sum_{i=1}^{n} \hat{\boldsymbol{\alpha}}_i \qquad (A.2)$$

The same procedure can be applied to compute any combination of these magnitudes required to calculate properties that could be experimentally measured. For instance,

$$\mu^2 = \boldsymbol{\mu}\boldsymbol{\mu} = \left(\sum_{i=1}^{n} \boldsymbol{\mu}_i \right) \left(\sum_{i=1}^{n} \boldsymbol{\mu}_i \right) \tag{A.3}$$

$$\boldsymbol{\mu}^T \hat{\boldsymbol{\alpha}} \boldsymbol{\mu} = \left(\sum_{i=1}^{n} \boldsymbol{\mu}_i \right)^T \left(\sum_{i=1}^{n} \hat{\boldsymbol{\alpha}}_i \right) \left(\sum_{i=1}^{n} \boldsymbol{\mu}_i \right) \tag{A.4}$$

$$\text{Trace } (\hat{\boldsymbol{\alpha}}\hat{\boldsymbol{\alpha}}) = \text{Trace} \left[\left(\sum_{i=1}^{n} \hat{\boldsymbol{\alpha}}_i \right) \left(\sum_{i=1}^{n} \hat{\boldsymbol{\alpha}}_i \right) \right] \tag{A.5}$$

If the geometrical structure of the polymer is rigid, any of these magnitudes is uniquely specified by the bond contributions and the angles between each pair of bonds. However, if the molecule can assume more than one internal conformational state, the magnitudes should be averaged over all such states. The statistical mechanical average over all the conformations of the molecule is customarily represented by angle brackets, $\langle \ \rangle$. Thus,

$$\langle \mu^2 \rangle = \langle \boldsymbol{\mu}\boldsymbol{\mu} \rangle = \left\langle \left(\sum_{i=1}^{n} \boldsymbol{\mu}_i \right) \left(\sum_{i=1}^{n} \boldsymbol{\mu}_i \right) \right\rangle \tag{A.6}$$

is the mean-squared dipole moment of the polymer.

Of course, the whole procedure outlined before rests upon the assumption of additivity of the contributions A_i from each bond or group of bonds in the polymer. In order to be additive, the contribution a_i from a given bond has to be the same regardless of the environment in which it is found; thus, it should be invariant with the type of molecule and with the conformation that it may adopt.

We have already treated the difficulties concerning the additivity of bond or group contributions, especially in the case of optical anisotropy tensors $\hat{\boldsymbol{\alpha}}$. At this moment, we assume that such contributions can be obtained so that we can proceed to compute the averaged magnitudes for a polymeric chain. For instance, in the case of a vinyl polymer, $(CHR—CH_2—)_x$, such as the one shown in Figure A.1, three kind of contributions are needed: l_i, $\boldsymbol{\mu}_i$, and $\hat{\boldsymbol{\alpha}}_i$ representing the group $CHR—$, and l_{i+1}, $\boldsymbol{\mu}_{i+1}$, $\hat{\boldsymbol{\alpha}}_{i+1}$ for the $CH_2—$.

CALCULATIONS FOR A FIXED CONFORMATION

The magnitude for the whole chain containing n skeletal bonds in a given conformation can be computed using expressions similar to Eqs. (A.2) to (A.5). The transformation of coordinates is customarily handled by assigning a reference frame to each

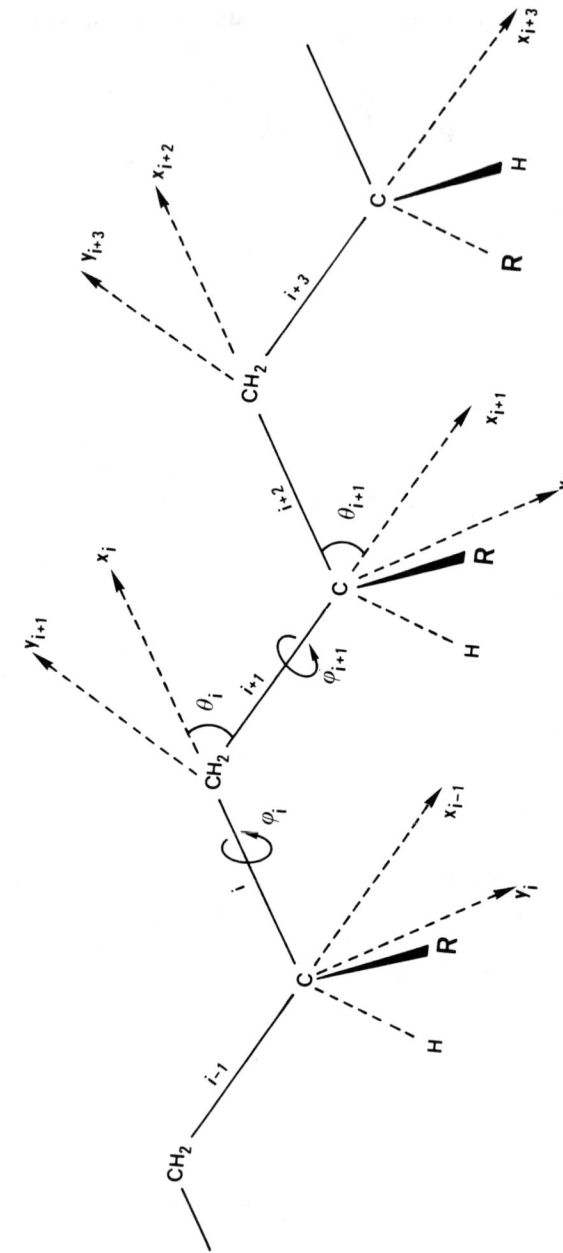

Figure A.1 A vinyl polymer shown in the all-trans ($\phi = 0$) conformation. Coordinate systems affixed to each skeletal bond are indicated.

skeletal bond of the chain (so that every contribution is written in its own coordinate system), and using a matrix operator to transform the frame defined for bond i into coincidence with that of bond $i - 1$, repeating the transformation from $i = n$ to $i = 2$.

For instance, the standard procedure for vinyl polymers [319] consists in the assignment of a d and l character to each skeletal bond. A given bond is taken to be d if the lateral R is found on the right side when looking in the direction of the bond from the CHR group. Conversely, the bond is l when, under the same conditions, the lateral R is found on the left side. Thus, bonds $i + 3$, $i + 2$, and i on Figure A.1 are d, whereas $i + 1$ and $i - 1$ are l.

The internal reference frame for skeletal bond i is defined with x_i having the direction of bond i; y_i in the plane of bonds $i - 1$ and i with a direction such as to have a positive projection over x_{i-1} and z_i completing a right- or left-handed coordinate system for d and l bonds, respectively. Figure A.1 shows the directions of x and y axes; all the z axes are perpendicular to the plane of the drawing and z_{i+3} comes up from that plane and all the others go down.

The matrix that performs the transformation of coordinates from the system $i + 1$ to the system i can be written as

$$\mathbf{T}_i = \begin{bmatrix} \cos\theta_i & \sin\theta_i & 0 \\ \sin\theta_i\cos\phi_i & -\cos\theta_i\cos\phi_i & B\sin\phi_i \\ \sin\theta_i\sin\phi_i & -\cos\theta_i\sin\phi_i & -B\cos\phi_i \end{bmatrix} \tag{A.7}$$

where θ_i is the supplement of the valance angle between bonds i and $i + 1$, and ϕ_i is the rotational angle of bond i with its origin ($\phi = 0$) in the trans conformation. B is -1 for transformations from d to l or l to d bonds and $+1$ for d to d or l to l. Thus, $B = -1$ in all cases except for the transformation from CH_2—CHR to CHR—CH_2 pair of bonds of a racemic dyad such as the pair $i + 3$ and $i + 2$ in Figure A.1. Then, Eq. A.2 can be written in a more explicit form as

$$\begin{aligned} \mathbf{r} &= l_1 + \mathbf{T}_1\{l_2 + \mathbf{T}_2[l_3 + \mathbf{T}_3(l_4 + \cdots)]\} \\ \boldsymbol{\mu} &= \boldsymbol{\mu}_1 + \mathbf{T}_1\{\boldsymbol{\mu}_2 + \mathbf{T}_2[\boldsymbol{\mu}_3 + \mathbf{T}_3(\boldsymbol{\mu}_4 + \cdots)]\} \\ \hat{\boldsymbol{\alpha}} &= \hat{\boldsymbol{\alpha}}_1 + \mathbf{T}_1\{\hat{\boldsymbol{\alpha}}_2 + \mathbf{T}_2[\hat{\boldsymbol{\alpha}}_3 + \mathbf{T}_3(\hat{\boldsymbol{\alpha}}_4 + \cdots)]\}\ldots\mathbf{T}_3^T\mathbf{T}_2^T\mathbf{T}_1^T \end{aligned} \tag{A.8}$$

where the superscript T indicates the transpose of a matrix.

The same procedure can be applied to compute products such as those appearing in Eqs. (A.3) to (A.5). However, substituting Eq. (A.7) into Eq. (A.8) and then into Eqs. (A.3) to (A.5) gives expressions that are too cumbersome to use even for fairly small chains.

A more elegant, and much easier, procedure to calculate these magnitudes is provided by the matrix-multiplication scheme developed by Flory and co-workers [95, 96]. It consists in using generator matrices constructed in such a way that their product generates all the terms required to compute the desired magnitude. The procedure can be used, with appropriate generator matrices, to calculate almost any magnitude of the polymer.

For instance, the generator matrix required to compute the squared dipole moment μ^2 has the form

$$\mathbf{M}_i = \begin{bmatrix} 1 & 2\boldsymbol{\mu}^T\mathbf{T} & \mu^2 \\ \mathbf{O} & \mathbf{T} & \boldsymbol{\mu} \\ \mathbf{O} & \mathbf{O} & 1 \end{bmatrix}_i \tag{A.9}$$

where $\boldsymbol{\mu}$ and $\boldsymbol{\mu}^T$ are the contributions of skeletal bond i to the dipole moment expressed in its own reference frame as a column (3×1) and a row (1×3) matrix, respectively. Subscript i indicates that the contribution of skeletal bond i should be used, and $\mathbf{T}$ represents the matrix for transformation of coordinates from skeletal bond $i + 1$ to i. The zeros appearing in Eq. (A.9) represent null matrices of different sizes as required to complete $\mathbf{M}_i$ as a 5×5 matrix.

With this matrix, the squared dipole moment μ^2 of a polymeric chain containing n skeletal bonds in a given conformation can be computed as

$$\mu^2 = \mathbf{M}_1 \left[\prod_{i=2}^{n-1} \mathbf{M}_i \right] \mathbf{M}_n \tag{A.10}$$

where $\mathbf{M}_1$ represents the first skeletal bond of the chain, and it is the 1×5 matrix appearing as the first row of $\mathbf{M}$ in Eq. (A.9). In a similar way, $\mathbf{M}_n$ represents the last bond of the chain and it is given by the 5×1 last column of the $\mathbf{M}$ matrix.

The construction of the generator matrices is detailed elsewhere [95, 96]; however, it is easy to check that Eq. (A.10) generates the same products as Eq. (A.3) with the advantage of being easily programmable for computer calculations.

The squared end-to-end distance r^2 can be computed with an equation identical to Eq. (A.10) using a generator matrix that is obtained from Eq. (A.9) by substituting dipole moment contributions $\boldsymbol{\mu}^T$, $\boldsymbol{\mu}$, and μ^2 by the respective bond length vectors $\mathbf{l}^T$, $\mathbf{l}$, and l^2.

In a similar way, the generator matrices for the magnitudes appearing in Eqs. (A.4) and (A.5) can be written, respectively, as

$$\mathbf{Q}_i = \begin{bmatrix} 1 & 2\boldsymbol{\mu}^T\mathbf{T} & \hat{\boldsymbol{\alpha}}^R(\mathbf{T}\otimes\mathbf{T}) & (\boldsymbol{\mu}^T\otimes\boldsymbol{\mu}^T)\,(\mathbf{T}\otimes\mathbf{T}) & 2\hat{\boldsymbol{\alpha}}^R(\boldsymbol{\mu}\otimes\mathbf{T}) & \hat{\boldsymbol{\alpha}}^R(\boldsymbol{\mu}\otimes\boldsymbol{\mu}) \\ 0 & \mathbf{T} & 0 & (\mathbf{I}_3\otimes\boldsymbol{\mu}^T)\,(\mathbf{T}\otimes\mathbf{T}) & \hat{\boldsymbol{\alpha}}\mathbf{T} & \hat{\boldsymbol{\alpha}}\boldsymbol{\mu} \\ 0 & 0 & \mathbf{T}\otimes\mathbf{T} & 0 & 2(\boldsymbol{\mu}\otimes\mathbf{T}) & \boldsymbol{\mu}\otimes\boldsymbol{\mu} \\ 0 & 0 & 0 & \mathbf{T}\otimes\mathbf{T} & 0 & \hat{\boldsymbol{\alpha}}^C \\ 0 & 0 & 0 & 0 & \mathbf{T} & \boldsymbol{\mu} \\ 0 & 0 & 0 & 0 & 0 & 1 \end{bmatrix}_i \tag{A.11}$$

and

$$\mathbf{P}_i = \begin{bmatrix} 1 & 2\hat{\boldsymbol{\alpha}}^R\,(\mathbf{T}\otimes\mathbf{T}) & \hat{\boldsymbol{\alpha}}^R\hat{\boldsymbol{\alpha}}^C \\ 0 & \mathbf{T}\otimes\mathbf{T} & \hat{\boldsymbol{\alpha}}^C \\ 0 & 0 & 1 \end{bmatrix}_i \tag{A.12}$$

where the symbol $\otimes$ indicates the direct product of matrices, $\mathbf{I}_3$ is the 3×3 identity matrix, $\hat{\boldsymbol{\alpha}}^C$ and $\hat{\boldsymbol{\alpha}}^R$ are column (9×1) and row (1×9) matrices containing the elements of $\hat{\boldsymbol{\alpha}}_i$ in the "reading order" (i.e., $\hat{\boldsymbol{\alpha}}^R = [\hat{\alpha}_{xx}, \hat{\alpha}_{xy}, \hat{\alpha}_{xz}, \hat{\alpha}_{yx}, \ldots, \hat{a}_{zz}]$). The zeros appearing in Eq. (A.11) and (A.12) represent null matrices of different sizes as required to complete $\mathbf{Q}_i$ and $\mathbf{P}_i$, respectively, as 26×26 and 11×11 matrices.

With these matrices, the magnitudes appearing in Eqs. (A.4) and (A.5) can be computed as

$$\boldsymbol{\mu}^T \hat{\boldsymbol{\alpha}} \boldsymbol{\mu} = \mathbf{Q}_1 \left[\prod_{i=2}^{n-1} \mathbf{Q}_i \right] \mathbf{Q}_n \qquad (A.13)$$

$$\text{Trace} (\hat{\boldsymbol{\alpha}} \hat{\boldsymbol{\alpha}}) = \mathbf{P}_1 \left[\prod_{i=2}^{n-1} \mathbf{P}_i \right] \mathbf{P}_n \qquad (A.14)$$

where $\mathbf{Q}_1$ and $\mathbf{P}_1$ represent the first bond of the chain and are given, respectively, by the 1×26 and 1×11 first rows of the $\mathbf{Q}$ and $\mathbf{P}$ matrices defined by Eqs. (A.11) and (A.12). Subscript n indicates the last bond of the polymeric backbone, and $\mathbf{Q}_n$ and $\mathbf{P}_n$ are, respectively, the 26×1 and 11×1 last columns of the $\mathbf{Q}$ and $\mathbf{P}$ matrices.

The combination $\mathbf{r}^T \hat{\boldsymbol{\alpha}} \mathbf{r}$ appearing in the magnitude Δa characteristic of stress birefringence is formally identical to $\boldsymbol{\mu}^T \hat{\boldsymbol{\alpha}} \boldsymbol{\mu}$ and, therefore, its generator matrix is obtained by replacing dipole moments by bond vectors in Eq. (A.11).

CONFORMATIONAL AVERAGES

The procedure used to compute the statistical average of these magnitudes over all the conformational space depends on the number of conformations allowed to the molecule and the model used to describe its conformational characteristics. Thus, if the number of conformations is small, as happens in exploratory calculations for short sequences of the polymeric chain or in the analysis of model compounds, it is expeditious to use Eqs. (A.10) to (A.14) to compute values of the desired magnitude for every allowed conformation and perform the average of those results with a weighting factor obtained as a Boltzmann exponential of the conformational energy. A similar procedure is employed in Monte Carlo simulations, with the only noteworthy difference that the weighting factors are incorporated in the method used to generate the conformations to be analyzed, so that all the results obtained for different conformations have the same participation in the final average.

In the case of a freely jointed chain, because there is no correlation between the orientation of the different bonds in the chain, the averages represented in Eq. (A.6) can be easily performed in the following way:

$$\langle \mu^2 \rangle = \langle \boldsymbol{\mu}\boldsymbol{\mu} \rangle = \left\langle \left[\sum_{i=1}^{n} \boldsymbol{\mu}_i \right] \left[\sum_{i=1}^{n} \boldsymbol{\mu}_i \right] \right\rangle$$

$$= \left\langle \sum_{i=1}^{n} \sum_{j=1}^{n} [\boldsymbol{\mu}_i\boldsymbol{\mu}_j] \right\rangle \tag{A.15}$$

$$= \left\langle \left(\sum_{i=1}^{n} \boldsymbol{\mu}_i\boldsymbol{\mu}_i \right) \right\rangle + \left\langle \sum_{i \neq j}^{n} \sum^{n} [\boldsymbol{\mu}_i\boldsymbol{\mu}_j] \right\rangle$$

assuming that all the contributions $\boldsymbol{\mu}_i$ are identical. If τ_{ij} is the angle formed by the $\boldsymbol{\mu}_i$ and $\boldsymbol{\mu}_j$ vectors, Eq. (A.15) can be written as

$$\langle \mu^2 \rangle = n\boldsymbol{\mu}_i^2 + \left\langle \boldsymbol{\mu}_i^2 \sum_{i \neq j}^{n} \sum^{n} [\cos \tau_{ij}] \right\rangle \tag{A.16}$$

and, because there is no correlation between the orientations, any value of τ_{ij} from 0 to π has the same probability of occurrence, and the average of its cosine vanishes, giving the well-known values for this model:

$$\langle \mu^2 \rangle = n\mu_i^2 \quad \text{and} \quad \langle r^2 \rangle = nl_i^2 \tag{A.17}$$

When the rotational state of each bond on the polymeric chain is independent of its neighbors, the transformation matrix $\mathbf{T}_i$ defined in Eq. (A.7) can be averaged over all the values of ϕ_i allowed to bond i. Then Eqs. (A.10) to (A.14) give the conformational average for the polymer provided that the average $\langle \mathbf{T} \rangle_i$ is used instead of $\mathbf{T}_i$ in the formulation of the generator matrices. This procedure is often used in calculations performed for polypeptide chains.

However, in most cases, the rotational states of each bond are assumed to be correlated with its first neighbors, neglecting long-range interactions. Then, the *rotational isomeric states model* is used to compute statistical averages. Thus, if each skeletal bond of the chain has ν allowed rotational states represented by α, β, . . . , ν, the statistical weight of each allowed conformation for the pair of bonds $i - 1$ and i is written as a matrix of dimensions $\nu \times \nu$ defined as

$$\mathbf{U}_i = \begin{bmatrix} u_{\alpha\alpha} & u_{\alpha\beta} & \cdots & u_{\alpha\nu} \\ u_{\beta\alpha} & u_{\beta\beta} & \cdots & u_{\beta\nu} \\ \cdot & \cdot & \cdot & \cdot \\ \cdot & \cdot & \cdot & \cdot \\ \cdot & \cdot & \cdot & \cdot \\ u_{\nu\alpha} & u_{\nu\beta} & \cdots & u_{\nu\nu} \end{bmatrix} \tag{A.18}$$

where $u_{\phi\varphi} = \exp(-V_{\phi\varphi}/kT)$ is the Boltzmann factor for the energy of the conformation obtained when bond $i - 1$ is in the rotational state ϕ and i in the φ. With these matrices, the partition function of a chain containing n skeletal bonds can be calculated as

$$Z = \mathbf{U}_1 \left[\prod_{i=2}^{n-1} \mathbf{U}_i \right] \mathbf{U}_n \tag{A.19}$$

where the initial $\mathbf{U}_1$ and the final $\mathbf{U}_n$ are, respectively, a row of one unity followed by $v - 1$ zeros and a column of v unities, that is,

$$\mathbf{U}_i = \text{row } (1\ 0\ 0 \cdots 0) \quad \text{and} \quad \mathbf{U}_n = \text{col } (1\ 1\ 1 \cdots 1) \tag{A.20}$$

The average of any conformation-dependent magnitude can be obtained by serial products of supermatrices formed by combination of the statistical-weight matrices $\mathbf{U}$ and the generator matrices of the desired magnitude. The required supermatrices are obtained as

$$\begin{aligned} \mathcal{M}_i &= (\mathbf{U}_i \otimes \mathbf{E}_m)\|\mathbf{M}_i\| \\ \mathcal{Q}_i &= (\mathbf{U}_i \otimes \mathbf{E}_q)\|\mathbf{Q}_i\| \\ \mathcal{P}_i &= (\mathbf{U}_i \otimes \mathbf{E}_p)\|\mathbf{P}_i\| \end{aligned} \tag{A.21}$$

where $\mathbf{E}_m$, $\mathbf{E}_q$, and $\mathbf{E}_p$ are identity matrices with the same dimensions as $\mathbf{M}$, $\mathbf{Q}$, and $\mathbf{P}$ defined according to Eqs. A.9, A.11, and A.12, respectively, while $\|\mathbf{M}_i\|$, $\|\mathbf{Q}_i\|$, and $\|\mathbf{P}_i\|$ represent diagonal arrays of matrices $\mathbf{M}$, $\mathbf{Q}$, and $\mathbf{P}$ written for each rotational state of bond i. Thus, taking the $\mathbf{P}$ matrix, for example,

$$\|\mathbf{P}_i\| = \begin{bmatrix} \mathbf{P}_i(\alpha) & 0 & \cdots & 0 \\ 0 & \mathbf{P}_i(\beta) & \cdots & 0 \\ \cdot & \cdot & \cdot & \cdot \\ \cdot & \cdot & \cdot & \cdot \\ \cdot & \cdot & \cdot & \cdot \\ 0 & 0 & \cdots & \mathbf{P}_i(v) \end{bmatrix} \tag{A.22}$$

The supermatrices representing the first and last bonds of the chain are constructed, for example, in the case of $\mathbf{Q}$, as

$$\mathcal{Q}_1 = \mathbf{U}_1 \otimes \mathbf{Q}_1 \quad \text{and} \quad \mathcal{Q}_n = \mathbf{U}_n \otimes \mathbf{Q}_n \tag{A.23}$$

where $\mathbf{Q}_1$ and $\mathbf{Q}_n$ are defined in Eq. (A.13), and $\mathbf{U}_1$ and $\mathbf{U}_n$ are given by Eq. (A.20).

Once all these matrices have been formed, the statistical averages over all possible conformations of the polymeric chain can be finally computed as

$$\langle \mu^2 \rangle = Z^{-1} \mathcal{M}_1 \left[\prod_{i=2}^{n-1} \mathcal{M}_i \right] \mathcal{M}_n \tag{A.24}$$

$$\langle \mu^T \hat{\alpha} \mu \rangle = Z^{-1} \mathcal{Q}_1 \left[\prod_{i=2}^{n-1} \mathcal{Q}_i \right] \mathcal{Q}_n \tag{A.25}$$

$$\langle \text{Trace } (\hat{\boldsymbol{\alpha}}\hat{\boldsymbol{\alpha}})\rangle = Z^{-1} \mathcal{P}_1 \left[\prod_{i=2}^{n-1} \mathcal{P}_i \right] \mathcal{P}_n \qquad (\text{A.26})$$

where Z is the partition function evaluated according to Eq. (A.19).

Expressions identical to Eqs. (A.24) and (A.25), with the appropriate generator matrices constructed as indicated before, can be used to compute values of $\langle r^2 \rangle$ and $\langle \mathbf{r}^T \hat{\boldsymbol{\alpha}} \mathbf{r} \rangle$.

Therefore, assuming that the conformational energies and the contributions l_i, $\mathbf{m}_i$, and $\hat{\boldsymbol{\alpha}}_i$ of each group of the polymeric chain are known, it is relatively simple to write a computer program for the generation of all the required matrices and the serial multiplications given in Eqs. (A.24) and (A.26).

References for Part Two

1. J. W. Doane, N. Z. Vaz, B. G. Wu, and S. Zumer, *Appl. Phys. Lett.*, **48**, 269 (1986).
2. J. L. West, *Mol. Cryst. Liq. Cryst.*, **157**, 427 (1988).
3. J. W. Doane, A. Golemme, J. L. West, J. B. Whitehead, Jr., and B. G. Wu, *Mol. Cryst. Liq. Cryst.*, **165**, 511 (1988).
4. J. L. West, J. W. Doane, Z. Domingo, and P. Ukleja, *Polym. Prep.*, **30**(2), 530 (1989).
5. J. Javornicky, *Photoplasticity*, Elsevier, Amsterdam, 1974.
6. A. Kuske and G. Robertson, *Photoelastic Stress Analysis*, Wiley Interscience, London, 1974.
7. J. Kerr, *Phil. Mag.*, **50**, 337 (1875); **50**, 416 (1875); **8**, 85 (1879); **8**, 229 (1879); **9**, 157 (1880); **13**, 153 (1882); **13**, 248 (1882); **37**, 380 (1894); **38**, 144 (1894).
8. J. E. H. Gordon, *Phil. Mag.*, **1**, 203 (1876); *Proc. Roy. Soc. (London)*, **A28**, 346 (1879).
9. J. J. Mackenzie, *Wied. Ann.*, **7**, 356 (1877).
10. G. Quincke, *Wied. Ann.*, **10**, 536 (1880); **19**, 733 (1883).
11. W. C. Röntgen, *Wied. Ann.*, **10**, 77 (1880).
12. M. H. Brongersma, *Wied. Ann.*, **16**, 222 (1882); *Phil. Mag.*, **14**, 127 (1883).
13. T. des Coudres, *Verhandl. Ges. Deutsch. Natuforsch., Arzte*, **65**, 67 (1893).
14. J. Lemoine, *Compt. Rend.*, **122**, 835 (1896).

15. W. Schmidt, *Ann. Physik.*, **7**, 142 (1902).

16. G. W. Elmen, *Phys. Rev.*, **20**, 54 (1905); *Ann. Physik.*, **16**, 350 (1905).

17. H. L. Blackwell, *Proc. Amer. Acad. Arts Sci.*, **41**, 650 (1906).

18. L. B. Morse, *Phys. Rev.*, **23**, 252 (1906).

19. T. H. Havelock, *Proc. Roy. Soc. (London)*, **A80**, 28 (1907).

20. C. F. Hagenow, *Phys. Rev.*, **27**, 196 (1908).

21. H. E. McComb, *Phys. Rev.*, **29**, 525 (1909).

22. R. Lieser, *Abhandl. Bunsen Ges.*, **4**, (1910).

23. L. Chaumont, *Ann. Phys.*, **4**, 61 (1915); **5**, 17 (1916).

24. R. Leiser, *Verhandl. Deutsch. Physik. Ges.*, **13**, 903 (1911); *Physik Z.*, **12**, 955 (1911).

25. G. Szivessy, *Z. Physik*, **2**, 30 (1920); **26**, 323 (1924).

26. J. W. Beams and E. C. Stevenson, *Phys. Rev.* **38**, 133 (1931).

27. C. W. Bruce, *Phys. Rev.*, **44**, 682 (1933).

28. G. G. Quarles, *Phys. REv.*, **46**, 692 (1934).

29. W. M. Breazeale, *Phys. Rev.*, **48**, 237 (1935); **49**, 625 (1936).

30. H. A. Stuart, *Z. Physik.*, **47**, 457 (1928); **55**, 358 (1929); **59**, 13 (1929); **63**, 533 (1930); *Ergeb. Exakt. Naturw.*, **10**, 151 (1931); *Z. Elektrochem.*, **40**, 478 (1934); *Z. Physik. Chem.*, **B27**, 350 (1935); *Naturwissenschaften*, **31**, 123 (1943); *Die Structur des Freien Moleküls*, Springer-Verlag, Berlin, 1952.

31. K. L. Wolf, G. Briegleb, and H. A. Stuart, *Z. Physik. Chem.*, **B6**, 163 (1929).

32. H. A. Stuart and H. Volkmann, *Z. Physik. Chem.*, **B17**, 429 (1932); **80**, 107 (1933); **83**, 444 (1933); *Ann. Physik.*, **18**, 121 (1933); *Physik. Z.*, **35**, 988 (1934).

33. H. A. Stuart and W. Buchheim, *Z. Physik.*, **111**, 36 (1938).

34. E. Kuss and H. A. Stuart, *Physik. Z.*, **42**, 95 (1941).

35. H. A. Stuart and S. Schiezl, *Ann. Physik.*, **2**, 321 (1948).

36. G. Briegleb, *Z. Physik. Chem.*, **B14**, 97 (1931).

37. G. Otterbein, *Physik. Z.*, **34**, 645 (1933); **35**, 249 (1934).

38. G. Sachsse, *Physik. Z.*, **36**, 357 (1935).

39. H. Friedrich, *Physik. Z.*, **38**, 318 (1937).

40. E. Saiz, U. W. Suter, and P. J. Flory, *J. Chem. Soc., Faraday Trans. 2*, **73**, 1538 (1977).

41. J. Errera, J. Th. G. Overbeek, and H. Sack, *J. Chim. Phys.*, **32**, 681 (1935).

42. M. A. Lauffer, *J. Am. Chem. Soc.*, **61**, 2412 (1939).

43. W. Kaye and R. Devaney, *J. Appl. Phys.*, **18**, 912 (1947).

44. N. A. Tolstoi and P. P. Feofilov, *Zh. Eksperim. i Teor. Fiz.*, **19**, 421 (1949); *Dokl. Acad. Nauk. SSSR*, **66**, 617 (1949).

45. H. Benoit, *Comp. Rend.*, **228**, 1716 (1949); *Ann. Physik.*, **6**, 561 (1951); *J. Chim. Phys.*, **47**, 719 (1950); **49**, 517 (1952).

46. C. T. O'Konski and B. H. Zimm, *Science*, **111**, 113 (1950).

47. I. Tinoco, *J. Am. Chem. Soc.*, **77**, 3476 (1955); **77**, 4486 (1955).

48. C. T. O'Konski and J. B. Applequist, *Nature*, **178**, 1464 (1956).

49. A. Norman and J. A. Field, *Arch. Biochem. Biophys.*, **70**, 257 (1957); **71**, 170 (1957).

50. C. T. O'Konski, K. Yoshioka, and W. H. Orttung, *J. Phys. Chem.*, **63**, 1558 (1959).

51. I. Tinoco and K. Yamaoka, *J. Phys. Chem.*, **63**, 423 (1959).

52. S. Krause and C. T. O'Konski, *J. Am. Chem. Soc.*, **81**, 5082 (1959).

53. P. Ingram and R. G. Jerrard, *Nature*, **196**, 57 (1962).

54. R. F. Itzhaki, *Nature*, **194**, 1241 (1962); *Biochem. J.*, **100**, 211 (1966); *Proc. Roy. Soc. B*, **164**, 75 (1966); **164**, 411 (1966).

55. M. J. Shah and C. M. Hart, *IBM J. Res. Devt.*, **7**, 44 (1963).

56. M. J. Shah, D. C. Thompson, and C. M. Hart, *J. Phys. Chem.*, **67**, 1170 (1963).

57. W. H. Orttung and J. A. Meyers, *J. Phys. Chem.*, **67**, 1905 (1963); **67**, 1911 (1963).

58. J. Garcia de la Torre and V. A. Bloomfield, *Quat. Rev. Biophys.*, **14**, 81 (1981).

59. J. Garcia de la Torre, *Dynamic Properties of Macromolecules Assemblies*, S. Harding and A. Rowe (Eds.), The Royal Society of Chemistry, London, 1989.

60. C. G. Le Fevre and R. J. W. Le Fevre, *Rev. Pure Appl. Chem.*, **5**, 261 (1955); **20**, 57 (1970).

61. R. J. W. Le Fevre, *Adv. Phys. Org. Chem.*, **3**, 1 (1965).

62. C. G. Le Fevre and R. J. W. Le Fevre, *Physical Methods of Chemistry*, Part IIIC, Chapter 6, A. Weissberger and B. W. Rossiter (Eds.), Wiley Interscience, New York, 1972.

63. L. V. Cherry, M. E. Hobbs, and H. A. Strobel, *J. Phys. Chem.*, **61**, 465 (1957).

64. H. A. Stuart, *Landolt-Börnstein, Zahlenwerte und Funktionen*, **2**, 8, (1962).

65. H. A. Stuart, *Molekülstruktur*, Springer, Berlin, 1967.

66. H. H. Huang and S. C. Ng, *J. Chem. Soc., B*, 582 (1968).

67. L. H. L. Chia and H. H. Huang, *J. Chem. Soc., B*, 1369 (1968).

68. A. D. Buckingham, *Proc. Phys. Soc.*, **69B**, 344 (1956).

69. M. Paillette, *J. Chim. Phys.*, **65**, 1629 (1968); *Ann. Phys.*, **4**, 671 (1969).

70. J. R. Lalane, F. B. Martin, and P. Bothorel, *J. Colloid Interface Sci.*, **39**, 601 (1972).

71. P. P. Ho and R. R. Alfano, *Phys. Rev.*, *A*, **20**, 2170 (1979).

72. M. Paillette, *C. R. Acad. Sci, Ser. B*, **262**, 264 (1966).

73. P. P. Ho, W. Yu, and R. R. Alfano, *Chem. Phys. Lett.*, **37**, 91 (1976).

74. B. R. Jennings and H. J. Coles, *Nature*, **252**, 33 (1974).

75. H. J. Coles and B. R. Jennings, *Phil. Mag.*, **32**, 1051 (1975); *Biopolymers*, **14**, 2567 (1975).

76. M. Wirth, T. Eriksson, and B. Norden, *J. Phys. Chem.*, **91**, 1957 (1987).

77. P. Langevin, *Radium Lett.*, **7**, 249 (1910); *Compt. Rend.*, **151**, 475 (1910).

78. M. Born, *Ann. Physik.*, **55**, 177 (1918).

79. P. Debye and H. Sack, *Handbuch de Radiologie*, Marx, Leipzig, 1934.

80. R. Gans, *Ann. Physik.*, **64**, 481 (1921); **65**, 97 (1921).

81. R. L. Kronig, *Z. Physik.*, **45**, 458 (1927); **45**, 508 (1927); **47**, 702 (1928).

82. M. Born and P. Jordan, *Elementare Quantenmechanik*, Springer, Berlin, 1930.

83. T. Neugebauer, *Z. Physik.* **73**, 386 (1932); **119**, 114 (1942).

84. I. Tobias and N. L. Balazs, *Chem. Phys.*, **32**, 93 (1978).

85. K. Yoshioka, *J. Chem. Phys.*, **86**, 491 (1987).

86. A. Iniesta and J. Garcia de la Torre, *J. Chem. Phys.*, **90**, 5190 (1989).

87. H. Watanabe and A. Morita, *Adv. Chem. Phys.*, **56**, 255 (1984).

88. W. A. Wegener, R. M. Dowben, and V. J. Koester, *J. Chem. Phys.*, **70**, 622 (1979).

89. W. A. Wegener, *J. Chem. Phys.*, **84**, 5989 (1986); **84**, 6005 (1986).

90. D. B. Roitman and B. H. Zimm, *J. Chem. Phys.*, **81**, 6333 (1984); **81**, 6348 (1984).

91. D. B. Roitman, *J. Chem. Phys.*, **81**, 6356 (1984).

92. F. G. Diaz and J. Garcia de la Torre, *J. Chem. Phys.*, **88**, 7698 (1988).

93. K. Nagai and T. Ishikawa, *J. Chem. Phys.*, **43**, 4508 (1965).

94. M. V. Volkenstein, *Configurational Statistics of Polymeric Chains*, English translation by S. N. Timasheff and M. J. Timasheff, Wiley Interscience, New York, 1963.

95. P. J. Flory, *Statistical Mechanics of Chain Molecules*, Wiley Interscience, New York, 1969.

96. P. J. Flory, *Macromolecules*, **7**, 381 (1974).

97. E. Fredericq and C. Houssier, *Electric Dichroism and Electric Birefringence*, Clarendon Press, Oxford, 1973.

98. C. T. O'Konski (Ed.), *Molecular Electro-Optics, Theory and Methods*, Part 1, Marcel Dekker, New York, 1976; *Applications to Biopolymers*, Part 2, Marcel Dekker, New York, 1978.

99. B. R. Jennings (Ed.), *Electro-Optics and Dielectrics of Macromolecules and Colloids*, Plenum Press, New York, 1979.

100. S. Krause (Ed.), *Molecular Electro-Optics: Electro-Optic Properties of Macromolecules and Colloids in Solution*, Plenum Press, New York, 1981.

101. V. N. Tsvetkov, *Rigid Chain Polymers: Hydrodynamic and Optical Properties in Solution*, translated by E. A. Korolyova, Consultants Bureau, New York, 1989.

102. M. Voigt, *Wiedem. Ann.*, **69**, 297 (1899); *Ann. Physik.*, **4**, 197 (1901).

103. R. J. W. Le Fevre and D. A. A. S. N. Rao, *J. Chem. Soc.*, 708 (1956).

104. A. D. Buckingham and J. A. Pople, *Proc. Phys. Soc. (London)*, **A68**, 905 (1955).

105. L. L. Boyle, A. D. Buckingham, R. L. Disch, and D. A. Dunmur, *J. Chem. Phys.*, **45**, 1318 (1966).

106. A. D. Buckingham and B. J. Orr, *Quart. Rev. (Chem. Soc., London)*, **21**, 195 (1967).

107. A. D. Buckingham and D. A. Dunmur, *Trans. Faraday Soc.*, **64**, 1776 (1968).

108. A. D. Buckingham and B. J. Orr, *Proc. Roy. Soc. (London)*, **A305**, 259 (1968).

109. C. V. Raman and K. S. Krishnan, *Proc. Roy. Soc. (London)*, **A117**(1), 589 (1928); *Phil. Mag.*, **5**, 498 (1928).

110. H. Mueller, *Phys. Rev.*, **50**, 547 (1936).

111. C. G. Le Fevre and R. J. W. Le Fevre, *J. Chem. Soc.*, 4041, (1953).

112. A. D. Buckingham, *Proc. Phys. Soc. (London)*, **68A**, 910 (1965).

113. D. A. Dows, *J. Chem. Phys.*, **41**, 2656 (1964).

114. K. Sauer and M. Calvin, *J. Mol. Biol.*, **4**, 451 (1962).

115. R. T. Ingwall, E. A. Czurylo, and P. J. Flory, *Biopolymers*, **12**, 1137 (1973).

116. J. O. Ellis and J. P. Llewellyn, *J. Phys., E, Sci. Inst.*, **10**, 1249 (1977).

117. C. Houssier and C. T. O'Konski, Chapter 15 of reference [100].

118. F. Menduti and E. Saiz, *An. Quim.*, **80**, 483 (1984).

119. F. Menduti, *Rev. Sci. Instrum.*, **59**, 728 (1988).

120. E. D. Baily, *Rev. Sci. Instrum.*, **46**, 1105 (1975).

121. H. J. Coles, *Polymer*, **18**, 554 (1977).

122. J. Crossley, B. K. Morgan, and M. Rujimethabhas, *Rev. Sci. Instrum.*, **50**, 1400 (1979).

123. F. Menduti and E. Saiz, *Polymer Bull.*, **11**, 533 (1984); *Makromol. Chem.*, **187**, 2483 (1986).

124. M. J. Aroney, M. Battaglia, R. Ferfoglia, D. Millar, and R. K. Pierens, *J. Chem. Soc., Faraday Trans.*, 2, **72**, 194 (1976).

125. C. T. O'Konski and S. Krause, Part 1, Chapter 3, of reference [98].

126. M. J. Shah, *J. Phys. Chem.*, **67**, 2215 (1963).

127. C. V. Raman and S. C. Sirkar, *Nature*, **121**, 794 (1928).

128. C. T. O'Konski and A. J. Haltner, *J. Am. Chem. Soc.*, **79**, 5634 (1957).

129. S. Ogawa and S. Oka, *J. Phys. Soc. Jpn.*, **15**, 658 (1960).

130. G. B. Thurston and D. I. Bowling, *J. Colloid Interface Sci.*, **30**, 34 (1969).

131. H. A. Stuart and A. Peterlin, *J. Polym. Sci.*, **5**, 551 (1950).

132. See Chapter 7 in reference [94].

133. P. J. Flory and R. L. Jernigan, *J. Chem. Phys.*, **48**, 3823 (1968).

134. E. Saiz, J. P. Hummel, P. J. Flory, and M. Plavsic, *J. Phys. Chem.*, **85**, 3211 (1981).

135. M. M. Rodrigo, M. P. Tarazona, and E. Saiz, *J. Phys. Chem.*, **90**, 2236 (1986); **90**, 5565 (1986).

136. P. J. Flory, E. Saiz, B. Erman, P. A. Irvine, and J. P. Hummel, *J. Phys. Chem.*, **85**, 3215 (1981).

137. E. H. L. Meyer and G. Otterbein, *Phys. Z.*, **32**, 290 (1931).

138. K. G. Denbigh, *Trans. Faraday Soc.*, **36**, 936 (1940).

139. A. López-Piñeiro, M. P. Tarazona, and E. Saiz, *J. Chim. Phys.*, **80**, 529 (1983); *Polym. Bull.*, **10**, 373 (1983).

140. A. E. Tonelli, *Macromolecules*, **10**, 153 (1977).

141. T. K. Ha, Part 1, Chapter 14, of reference [98]; J. M. Andre, C. Barbier, V. Bodart, and J. Delhalle, in *Nonlinear Optical Properties of Organic Molecules and Crystals*, Chapter III-5, D. S. Chemula and J. Zyss, (Eds.), Academic Press, Orlando, Fla., 1987.

142. J. Waite and M. G. Papadopoulos, *J. Phys. Chem.*, **94**, 1755 (1990).

143. U. W. Suter and P. J. Flory, *J. Chem. Soc., Faraday Trans. 2*, **73**, 1521 (1977).

144. G. D. Patterson and P. J. Flory, *J. Chem. Soc., Faraday Trans. 2*, **68**, 1098 (1972); **68**, 1111 (1972).

145. C. W. Carlson and P. J. Flory, *J. Chem. Soc., Faraday Trans. 2*, **73**, 1505 (1977).

146. R. S. Armstrong, M. J. Aroney, B. S. Higgs, and K. R. Skamp, *J. Chem. Soc., Faraday Trans. 2*, **77**, 55 (1981).

147. M. A. Llorente, I. Fernández de Piérola, and E. Saiz, *Macromolecules*, **18**, 2663 (1985).

148. I. R. Gentle, M. R. Hesling, and G. L. D. Ritchie, *J. Phys. Chem.*, **94**, 1844 (1990).

149. G. Khanarian and L. Kent, *J. Chem. Soc., Faraday Trans. 2*, **77**, 495 (1981).

150. S. P. Liebman and J. W. Moskowitz, *J. Chem. Phys.* **54**, 3622 (1971).

151. M. J. Aroney, *Angew. Chem., Int. Ed. Engl.*, **16**, 663 (1977).

152. R. J. W. Le Fevre, D. V. Radford, G. L. D. Ritchie, and R. J. Stiles, *J. Chem. Soc., B*, 148 (1968).

153. P. A. Hopkins, R. J. W. Le Fevre, L. Radom, and G. L. D. Ritchie, *J. Chem. Soc., B*, 574 (1971).

154. S. B. Bulgarevich, V. S. Bolotnikov, V. N. Sheinker, O. A. Osipov, and A. D. Garnovskii, *J. Org. Chem., USSR*, **11**, 2244 (1975).

155. M. S. Beevers, *Mol. Crys., Liq. Cryst.*, **31**, 333 (1975).

156. W. Pyzuk and H. Majgier-Baranowska, *Chem. Phys. Lett.*, **63**, 184 (1979).

157. B. R. Jennings, *Adv. Polym. Sci.*, **22**, 61 (1977).

158. I. R. Gentle and G. L. D. Ritchie, *J. Phys. Chem.*, **93**, 7740 (1989).

159. C. G. Le Fevre and R. J. W. Le Fevre, *J. Chem. Soc.*, 1577 (1954).

160. R. J. W. Le Fevre and B. Purnachandra Rao, *J. Chem. Soc.*, 1465 (1958).

161. M. J. Aroney, G. Cleaver, R. K. Pierens, and R. J. W. Le Fevre, *J. Chem. Soc. Perkin Trans. 2*, 3 (1974).

162. M. P. Bogaard, A. D. Buckingham, R. K. Pierens, and A. H. White, *J. Chem. Soc., Faraday Trans. 1*, **74**, 3008 (1978).

163. J. W. Lewis and W. H. Orttung, *J. Phys. Chem.*, **82**, 698 (1978).

164. J. M. Eckert and R. J. W. Le Fevre, *J. Chem. Soc.*, 1081 (1962).

165. R. J. W. Le Fevre and A. Sundaram, *J. Chem. Soc.*, 3904 (1962).

166. G. Rowe, M. P. Tarazona, and E. Saiz, *An. Quim.*, **77**, 363 (1981).

167. W. L. Mattice and E. Saiz, *J. Am. Chem. Soc.*, **100**, 6308 (1978).

168. M. S. Beevers, *J. Chem. Soc., Faraday Trans. 2*, **75**, 679 (1979).

169. S. Filipczuk and G. Khanarian, *J. Chem. Soc., Faraday Trans. 2*, **77**, 477 (1981).

170. K. E. Calderbank, R. J. W. Le Fevre, and R. K. Pierens, *J. Chem. Soc., B*, 968 (1969).

171. M. J. Aroney, K. E. Calderbank, R. J. W. Le Fevre, and R. K. Pierens, *J. Chem. Soc., B*, 159 (1969); 1120 (1970).

172. P. A. Irvine, B. Erman, and P. J. Flory, *J. Phys. Chem.*, **87**, 2929 (1983).

173. G. Floudas, A. Patkowski, G. Fytas, and M. Ballauff, *J. Phys. Chem.*, **94**, 3215 (1990).

174. F. Mendicuti, M. M. Rodrigo, M. P. Tarazona, and E. Saiz, *Macromolecules*, **23**, 1139 (1990).

175. D. A. Dunmur, M. R. Manterfield, W. H. Miller, and J. K. Dunleavy, *Mol. Cryst. Liq. Cryst.*, **45**, 127 (1978).

176. H. J. Coles and B. R. Jennings, *Mol. Phys.*, **36**, 1661 (1978).

177. P. P. Kolinsky and B. R. Jennings, *Mol. Phys.*, **40**, 979 (1980).

178. D. A. Dunmur and A. E. Tomes, *Mol. Cryst. Liq. Cryst.* **97**, 241 (1983).

179. P. Navard and P. J. Flory, *J. Chem. Soc., Faraday Trans. 1*, **82**, 3367 (1986).

180. P. J. Flory and P. Navard, *J. Chem. Soc., Faraday Trans. 1*, **82**, 3381 (1986).

181. M. J. Aroney, R. J. W. Le Fevre, and G. M. Parkins, *J. Chem. Soc.*, 2890 (1960).

182. R. J. W. Le Fevre and K. M. S. Sundaram, *J. Chem. Soc.*, 1880 (1963).

183. R. J. W. Le Fevre and K. M. S. Sundaram, *J. Chem. Soc.*, 3188 (1963).

184. R. T. Ingwall and P. J. Flory, *Biopolymers*, **11**, 1527 (1972); **12**, 1123 (1973).

185. M. J. Aroney, R. J. W. Le Fevre, and A. N. Singh, *J. Chem. Soc.*, *B*, 3179 (1965).

186. C. Dolcet, M. P. Tarazona, and E. Saiz, *An. Quim.*, **80**, 476 (1984).

187. C. Dolcet and M. P. Tarazona, *Int. J. Peptide Protein Res.*, **30**, 548 (1987); *Biophys. Chem.*, **29**, 253 (1988); **32**, 247 (1988).

188. C. T. O'Konski and N. C. Stellwagen, *Biophys. J.*, **5**, 607 (1965).

189. D. W. Ding, R. Rill, and K. E. Van Holde, *Biopolymers*, **11**, 2109 (1972).

190. E. Neumann and A. Katchalsky, *Proc. Natl. Acad. Sci. USA*, **69**, 993 (1972).

191. A. Revzin and E. Neumann, *Biophys. Chem.*, **2**, 144 (1974).

192. G. Schwarz and U. Schrader, *Biopolymers*, **14**, 1181 (1975).

193. D. Porschke, *Biopolymers*, **15**, 1917 (1976).

194. M. Pollak and H. A. Glick, *Biopolymers*, **16**, 1007 (1977).

195. H. Watanabe and K. Yoshioka, Chapter 37 of reference [99].

196. K. Kikuchi and K. Yoshioka, *Biopolymers*, **12**, 2667 (1973); **15**, 1669 (1976).

197. K. Yoshioka, K. Kikuchi, and M. Fujimori, *Biophys. Chem.*, **11**, 369 (1980).

198. J. V. Champion, G. H. Meeten, and C. D. Whitle, *Trans. Faraday Soc.*, **66**, 2671 (1970).

199. J. V. Champion, G. H. Meeten, and G. W. Southwell, *J. Chem. Soc., Faraday Trans. 2*, **71**, 225 (1975).

200. J. V. Champion, G. H. Meeten, and G. W. Southwell, *Polymer*, **17**, 651 (1976).

201. K. M. Kelly, G. D. Patterson, and A. E. Tonelli, *Macromolecules*, **10**, 859 (1977).

202. F. Mendicuti, M. P. Tarazona, and E. Saiz, *Polymer Bull.*, **13**, 263 (1985).

203. E. Riande, E. Saiz, and J. E. Mark, *Macromolecules*, **13**, 448 (1980).

204. B. Erman, D. C. Marvin, P. A. Irvine, and P. J. Flory, *Macromolecules*, **15**, 664 (1982).

205. B. Erman, D. Wu, P. A. Irvine, D. C. Marvin, and P. J. Flory, *Macromolecules*, **15**, 670 (1982).

206. S. J. Mumby and M. S. Beevers, *J. Chem. Phys.*, **83**, 5260 (1985).

207. M. Faraday, *Phil. Trans. Roy. Soc.*, 1 (1846); *Phil. Mag.*, **28**, 294 (1846); **29**, 153 (1846).

208. E. Verdet, *Ann. Chim. Phys.*, **41**, 570 (1854); **52**, 151 (1858).

209. Q. Majorana, *Compt. Rend.*, **135**, 235 (1902).

210. W. Voigt, *Magneto- und Electro-Optic*, Teubner, Leipzig, 1908.

211. A. Cotton and M. Mouton, *Compt. Rend.*, **141**, 317, 349 (1905); **145**, 229, 870 (1907).

212. R. J. W. Le Fevre, P. H. Williams, and J. M. Eckert, *Austral. J. Chem.*, **18**, 1133 (1965).

213. C. L. Cheng, D. S. N. Murthy, and G. L. D. Ritchie, *Mol. Phys.*, **22**, 1137 (1971).

214. H. B. Serreze and R. B. Goldner, *Rev. Sci. Instrum.*, **45**, 1613 (1974).

215. G. Maret, M. V. Schickfus, A. Mayer, and K. Dransfeld, *Phys. Rev. Lett.*, **35**, 397 (1975).

216. J. V. Champion, D. Downer, G. H. Meeten, and L. F. Gate, *J. Phys. E*, **10**, 1137 (1977).

217. M. R. Battaglia and G. L. D. Ritchie, *Mol. Phys.*, **32**, 1481 (1976); *J. Chem. Soc., Faraday Trans. 2*, **73**, 209 (1977); *J. Chem. Soc., Perkin Trans. 2*, **1977**, 897, 901.

218. A. D. Buckingham and J. A. Pople, *Proc. Phys. Soc.*, **B69**, 1133 (1956).

219. A. D. Buckingham, W. H. Prichard, and D. H. Whiffen, *Trans. Faraday Soc.*, **63**, 1057 (1967).

220. R. J. W. Le Fevre and D. S. N. Murthy, *Austral. J. Chem.*, **19**, 179 (1966); **22**, 1415 (1969).

221. G. Weill, Chapter 21 of reference [99].

222. R. J. W. Le Fevre, D. N. S. Murthy, and P. J. Stiles, *Austral. J. Chem.*, **22**, 1421 (1969).

223. G. L. D. Ritchie and J. Vrbancich, *Austral. J. Chem.*, **35**, 869 (1982).

224. A. A. Boethner-By and J. A. Pople, *Ann. Rev. Phys. Chem.*, **16**, 43 (1965).

225. R. Ditchfield, *MTP Int. Re. Sci. Phys. Chem. Ser. One*, **2**, 91 (1972).

226. B. R. Appleman and B. P. Dailey, *Adv. Magn. Reson.*, **7**, 231 (1974).

227. W. H. Flygare, *Chem. Rev.*, **74**, 653 (1974).

228. D. H. Sutter and W. H. Flygare, *Topics in Current Chemistry*, **63**, 89 (1976).

229. R. C. Benson and W. H. Flygare, *J. Chem. Phys.*, **53**, 4470 (1970).

230. J. H. S. Wang and W. H. Flygare, *J. Chem. Phys.*, **53**, 4479 (1970).

231. W. H. Flygare and R. C. Benson, *Mol. Phys.*, **20**, 225 (1971).

232. T. G. Schmalz, C. L. Norris, and W. H. Flygare, *J. Am. Chem. Soc.*, **95**, 7961 (1973).

233. J. Hoarau, N. Lumbroso, and A. Pacault, *Compt. Rend.*, **242**, 1702 (1956).

234. M. A. Lasheen, *Phil. Trans. A*, **256**, 375 (1964).

235. M. P. Bogaard, A. D. Buckingham, M. G. Corfield, D. A. Dunmur, and A. H. White, *Chem. Phys. Let.*, **12**, 558 (1972).

236. K. E. Calderbank, R. L. Calvert, P. B. Lukins, and G. L. D. Ritchie, *Austral. J. Chem.*, **34**, 1835 (1981).

237. M. P. Brereton, M. K. Cooper, G. R. Dennis, and G. L. D. Ritchie, *Austral. J. Chem.*, **34**, 2253 (1981).

238. P. B. Lukins, A. D. Buckingham, and G. L. D. Ritchie, *J. Phys. Chem.*, **88**, 2414 (1984).

239. P. B. Lukins and G. L. D. Ritchie, *J. Phys. Chem.*, **89**, 1312 (1985).

240. C. Rosenblatt and A. C. Griffin, *Macromolecules*, **22**, 4102 (1989).

241. M. Arpin, C. Strazielle, G. Weill, and H. Benoit, *Polymer*, **18**, 262 (1977).

242. L. R. G. Treloar, *The Physics of Rubber Elasticity*, 3rd ed., Clarendon Press, Oxford, 1975.

243. W. Kuhn and F. Grün, *Kolloid Z.*, **101**, 248 (1942).

244. J. E. Mark and B. Erman, *Rubberlike Elasticity. A Molecular Primer*, John Wiley & Sons, Inc., New York, 1989.

245. Yu Ya Gotlib, M. V. Volkenstein, and E. K. Byutner, *Dokl. Akad. Nauk. SSSR*, **99**, 935 (1954).

246. Yu Ya Gotlib, *Zh. Tekhn. Fiz.*, **27**, 707 (1957).

247. K. Nagai, *J. Chem. Phys.*, **40**, 2818 (1964); **47**, 2052 (1967).

248. P. J. Flory, R. L. Jernigan, and A. E. Tonelli, *J. Chem. Phys.*, **48**, 3822 (1968).

249. B. Erman and P. J. Flory, *Macromolecules*, **16**, 1601 (1983); **16**, 1607 (1983).

250. R. S. Stein, *Rubb. Chem. Technol.*, **49**, 458 (1976).

251. J. E. Mark and M. A. Llorente, *Polym. J. (Tokyo)*, **13**, 543 (1981).

252. J. E. Mark and M. A. Llorente, *J. Polym. Sci., Polym. Phys. Ed.*, **19**, 1107 (1981).

253. E. Riande, J. Guzmán, E. Saiz, and M. P. Tarazona, *J. Polym. Sci., Polym. Phys. Ed.*, **23**, 1031 (1985).

254. A. L. Andrady, M. A. Llorente, E. Saiz, *J. Polym. Sci., Polym. Phys. Ed.*, **25**, 1935 (1987).

255. R. S. Stein, S. Krimm, and A. V. Tobolsky, *Text. Res. J.*, **19**, 1, (1949).

256. R. S. Stein and A. V. Tobolsky, *J. Polym. Sci.*, **11**, 285 (1953).

257. R. S. Stein, F. H. Holmes, and A. V. Tobolsky, *J. Polym. Sci.*, **14**, 443 (1954).

258. M. Fukuda, G. L. Wilkes, and R. S. Stein, *J. Polym. Sci.*, A-2, **9**, 1417 (1971).

259. R. S. Stein and S. D. Hong, *J. Macromol. Sci. Phys.*, **B12**, 125 (1976).

260. V. N. Tsvetkov, O. V. Kallistov, Y. V. Korneyeva, and I. K. Nekrasov, *Vysokomol. Soedin*, **5**, 1538 (1963).

261. A. Y. Grishchenko, M. G. Vitovskaya, V. N. Tsvetkov, Y. P. Vorob'eva, N. N. Saprykina, and L. I. Mezentseva, *Vysokomol. Soedin*, **A9**, 1280 (1967).

262. V. N. Tsvetkov and A. Y. Grishchenko, *Polym. Sci. USSR*, **7**, 902 (1965); *J. Polym. Sci., Part C.*, **16**, 3195 (1968).

263. D. W. Saunders, D. R. Lightfoot, and D. A. Parsons, *J. Polym. Sci.*, A-2, **6**, 1183 (1968).

264. N. J. Mills and D. W. Saunders, *J. Macromol. Sci. B*, **2**, 369 (1968).

265. N. J. Mills, *Polymer*, **12**, 658 (1971).

266. A. N. Gent and V. V. Vickroy, Jr., *J. Polym. Sci.*, A-2, **5**, 47 (1967).

267. A. N. Gent, *Macromolecules*, **2**, 262 (1969).

268. A. N. Gent and T. H. Kuan, *J. Polym. Sci.*, A-2, **9**, 927 (1971).

269. K. Nagai, *J. Chem. Phys.*, **47**, 4690 (1967).

270. T. Ishikawa and K. Nagai, *J. Polym. Sci.*, A-2, **7**, 1123 (1969); *Polym. J.*, **1**, 116 (1970).

271. T. Ishikawa, *Polymer. J.*, **5**, 227 (1973).

272. Y. Abe, A. E. Tonelli, and P. J. Flory, *Macromolecules*, **3**, 294 (1970).

273. M. H. Liberman, Y. Abe, and P. J. Flory, *Macromolecules*, **5**, 550 (1972).

274. M. H. Liberman, L. C. De Bolt, and P. J. Flory, *J. Polym. Sci., Polym. Phys. Ed.*, **12**, 187 (1974).

275. M. Ilavsky, J. Hasa, and K. Dusek, *J. Polym. Sci., Polym. Symp.*, **53**, 239 (1975).

276. M. A. Llorente, J. E. Mark, and E. Saiz, *J. Polym. Sci., Polym. Phys. Ed.*, **21**, 1173 (1983).

277. E. Saiz, E. Riande, and J. E. Mark, *Macromolecules*, **17**, 899 (1984).

278. E. Riande, E. Saiz, and J. E. Mark, *J. Polym. Sci., Polym. Phys. Ed.*, **22**, 863 (1984).

279. E. Riande, J. Guzmán, M. P. Tarazona, and E. Saiz, *J. Polym. Sci., Polym. Phys. Ed.*, **22**, 917 (1984); E. Riande and J. Guzmán, *J. Polym. Sci., Polym. Phys. Ed.*, **23**, 1235 (1985); E. Riande and J. Guzmán, *J. Polym. Sci., Polym. Phys. Ed.*, **24**, 2805 (1986); E. Riande, J. Guzmán, and J. G. de la Campa, *Macromolecules*, **21**, 2128 (1988).

280. E. Saiz, M. P. Tarazona, E. Riande, and J. Guzmán, *J. Polym. Sci., Polym. Phys. Ed.*, **22**, 2165 (1984).

281. E. Saiz and J. E. Mark, *Makromol. Chem.*, **188**, 2185 (1987).

282. M. Ilavsky, E. Saiz, and E. Riande, *J. Polym. Sci., Polym. Phys. Ed.*, **27**, 743 (1989).

283. L. C. De Bolt and E. Riande, *Makromol. Chem.*, **187**, 2497 (1986).

284. G. Floudas, G. Fytas, B. Momper, and E. Saiz, *Macromolecules*, **23**, 498 (1990).

285. J. C. Maxwell, *Proc. Royal Soc. (London)*, **A22**, 46 (1873). *Collected Papers*, Cambridge University Press, 1890, Vol. 2.

286. A. von Kundt, *Ann. Physik.*, **13**, 110 (1881).

287. G. de Metz, *Ann. Physik.*, **35**, 497 (1888).

288. K. Umlauf, *Ann. Physik.*, **45**, 304 (1892).

289. J. E. Almy, *Phil. Mag.*, **44**(5), 499 (1897).

290. B. V. Hill, *Phil. Mag.*, **48**(5), 485 (1899).

291. C. Zakrzewski, *Ann. Akad. Krakau*, **50** (1904).

292. C. Zakrzewski and C. Kraft, *Ann. Akad. Krakau*, **506** (1905).

293. V. Bernatzki, *Physik. Z.*, **6**, 730 (1905).

294. R. Lucas, *Compt. Rend.*, **206**, 827 (1938); *J. Phys. Radium*, **10**(7), 151 (1939); *Rev. Acoust.*, **8**, 121 (1939).

295. M. Kawamura, *Kagaku (Nature)*, **7**, 6, 54, 139 (1938).

296. D. Vorlander and R. Walter, *Z. Physik. Chem.*, **118**, 1 (1925).

297. Ch. Sadron, *J. Phys. Radium*, **7**(7), 263 (1936).

298. W. Buchheim, H. A. Stuart, and H. Mentz, *Z. Physik.*, **112**, 407 (1939).

299. D. Diesselhorst and H. Freundlich, *Physik., Z.*, **16**, 419 (1915).

300. H. Freundlich, D. Diesselhorst, and K. Leonhardt, *Elster-Geitel Festschr.*, Braunschweig, 1915.

301. H. Freundlich, F. Stapelfeldt, and H. Zocher, *Z. Physik. Chem.*, **114**, 161, 190 (1925).

302. R. Signer, *Z. Physik. Chem. (A)*, **150**, 247, 257 (1930).

303. R. Cerf and H. A. Scheraga, *Chem. Rev.*, **51**, 185 (1952).

304. A. Peterlin, in *Rheology: Theory and Applications,* Chapter 15, F. R. Eirich (Ed.), Academic Press, New York, 1956.

305. H. G. Jerrard, *Chem. Rev.,* **59,** 345 (1959).

306. H. Janeschitz-Kriegl, *Adv. Polym. Sci.,* **6,** 170 (1969).

307. A. Peterlin and P. Munk, in *Physical Methods of Chemistry,* Part IIIC, Chapter 4, A. Weissberger and B. W. Rossiter (Eds.), Wiley Interscience, New York, 1972.

308. H. Janeschitz-Kriegl, *Polymer Melt Rheology and Flow Birefringence,* Springer-Verlag, Berlin, 1983.

309. See Chapters 5 and 6 of reference [101].

310. P. Boeder, *Z. Physik.,* **75,** 258 (1932).

311. A. Peterlin, *Z. Physik,* **111,** 232 (1938).

312. A. Peterlin and H. A. Stuart, *Z. Physik.,* **112,** 1, 129 (1939).

313. O. Snellman and Y. Björnstahl, *Kolloid-Beih,* **52,** 403 (1941).

314. H. A. Scheraga, J. T. Edsall, and J. O. Gadd, Jr., *J. Chem. Phys.,* **19,** 1101 (1951).

315. H. A. Scheraga, *J. Chem. Phys.,* **23,** 1526 (1955).

316. G. B. Jeffery, *Proc. Roy. Soc. (London),* **A102,** 161 (1922).

317. H. J. Workman and C. A. Hollingsworth, *J. Colloid Interface Sci.,* **29,** 664 (1969).

318. R. Signer and H. Liechti, *J. Chim. Phys.,* **44,** 58 (1947).

319. P. J. Flory, P. Sundararajan, and L. C. De Bolt, *J. Am. Chem. Soc.,* **96,** 5015 (1974).

Index